Animal Husbandry, Dairying and Fisheries

Animal Husbandry, Dairying and Fisheries

Jhansi Rani Mallam

RANDOM

RANDOM PUBLICATIONS

NEW DELHI (INDIA)

Animal Husbandry, Dairying and Fisheries

ISBN 978-93-5111-787-2

Published in 2016 in India by

RANDOM PUBLICATIONS

4376-A/4B, Gali Murari Lal, Ansari Road
New Delhi-110 002
Phone : +9111-43580356, 011-23289044, 011-43142548
e-mail: sales@randompublications.com,
info@randompublications.com, randomexports@gmail.com

Reprinted 2025

Type Setting by : Friends Media, Delhi-110089
Digitally Printed at : Replika Press Pvt. Ltd.

Preface

Today, herd managers often oversee thousands of animals and many staff. Farms, stations and ranches may employ breeders, her health specialists, feeders, and milkers to help care for the animals. Animal Husbandry is intended primarily as a text for the basic course in animal industries and farm animals, as the case may be. It should prove useful as a reference book for those involved in livestock production.

Animal husbandry is the management and care of farm animals by humans for profit, in which genetic qualities and behavior, considered to be advantageous to humans, are further developed. The term can refer to the practice of selectively breeding and raising livestock to promote desirable traits in animals for utility, sport, pleasure, or research.

Dairy farming is a class of agriculture for long-term production of milk, which is processed (either on the farm or at a dairy plant, either of which may be called a dairy) for eventual sale of a dairy product. The study attempted in this book pin points the affects of dairying on rural development through the socio-economic changes among the dairy households. The study evaluates the nature, magnitude and dimension of dairy development.

Dairy farming has been part of agriculture for thousands of years. Historically it has been one part of small, diverse farms. In the last century or so larger farms doing only dairy production have emerged. Large scale dairy farming is only viable where either a large amount of milk is required for production of more durable dairy products such as cheese, butter, etc. or there is a substantial market of people with cash to buy milk, but no cows of their own.

A fishery may involve the capture of wild fish or raising fish through fish farming or aquaculture. Directly or indirectly, the livelihood of over 500 million people in developing countries depends on fisheries and aquaculture. Overfishing, including the taking of fish beyond sustainable levels, is reducing fish stocks and employment in many world regions.

Fisheries form an important sector of the country's economy in terms of food supply, employment, income and foreign exchange earnings. Fishes are one of the important sources of cheap protein food for the people and millions of fishermen and several industries depend on this source.

The book will be useful not only to the students of animal husbandry but also to general reader.

– Author

Contents

1

Animal Care and Husbandry

The principal scientist or animal management supervisor should make all animal care personnel aware of their responsibilities during both normal work hours and emergencies. A programme of special husbandry procedures in case of an emergency should be developed. It is the reserach facility management's responsibility to ensure that personnel caring for agricultural animals used for research or teaching are appropriately qualified or trained. This responsibility may be delegated to an IACUC. Qualification by experience and training must be documented.

The animal facility manager must ensure that all animal care personnel are aware of their responsibilities during and outside normal work hours. Protocols for emergency care must be developed and made available to all personnel.

OBSERVATION

Animals in intensive accommodations should be observed and cared for daily by trained and experienced caretakers. Illumination must be adequate to facilitate inspection. In some circumstances, more frequent observation or care may be needed. Under extensive conditions, such as range or pasture, observations should be frequent enough to detect illness or injury in a timely fashion, recognize the need for emergency action, and ensure adequate availability of feed and water.

A disaster plan must be developed for observing animals and providing care during emergency weather or health situations. Regardless of accommodations, animal observations should be documented and husbandry or health concerns reported to the animal facility manager or attending veterinarian as appropriate.

EMERGENCY, WEEKEND, AND HOLIDAY CARE

There must be a means for rapid communication in case of an emergency. In emergencies, facility security and fire personnel must be able to contact staff members responsible for the care of agricultural animals. Names and contact information for those individuals should be posted prominently in the animal facility and provided to the security department or telephone center. If posting names and contact information poses privacy or security issues, a contact number for a security or command center should be used instead.

The institution must ensure that emergency services can be contacted at any time by staff members. The institution must assure continuity of daily animal care, to encompass weekends, holidays, unexpected absences of assigned personnel, and emergency situations. Staff assigned to weekends and holidays must be qualified to perform assigned duties. Cross-training of staff and establishment of standard operating procedures is encouraged to assure consistent, high-quality care.

Emergency veterinary care must be readily available after daily work hours, on weekends, and on holidays. In the event that weather conditions or natural disasters make feeding temporarily impossible, every attempt should be made to provide animals with a continuous supply of water. Absence of feed for up to 48 h should not seriously endanger the health of normal, well-nourished juvenile or adult cattle, sheep, goats, horses, poultry, or swine. Feed should be provided within 24 h to very young animals that are not nursing their dams.

EMERGENCY PLANS

A site-specific emergency plan must be developed to care for agricultural animals that are used for research and teaching. The goal for a plan should be to provide proper management and care for the animals regardless of the conditions. However, some conditions may be so unusual and extreme that it will not be possible to provide immediate care for the animals and to simultaneously ensure employee safety.

Thus, emergency plans should define proper animal management and care and parameters to ensure employee safety. Emergency plans should name employees or positions that are considered essential for providing proper animal management and care. Those employees should a priori understand that responding to emergencies is a condition of employment and that they will be held accountable should they fail to care properly for the animals.

Plans should focus on emergencies that are most likely to occur in the specific geographic area or the research or teaching facility. Emergency plans should include animal evacuation plans specific to the research or teaching facility and actions to be taken if transportation is interrupted.

ANIMAL IDENTIFICATION AND RECORDS

Animals should be permanently identified by a method that can be easily read. Identification of individual animals is desirable, but, in some circumstances, it is acceptable to identify animals by group, cage, or pen. Individual birds may be wing-banded or leg-banded. Ear-notching, ear tattooing, electronic transponders, and branding may be used for individual identification of other species, and each has its advantages and disadvantages. Ear notches and tattoos are permanent and effective, but notching constitutes elective surgery and tattoos generally cannot be read without restraining animals.

Electronic transponders require special sensor units or stations, but should be considered when possible. Cattle and horses are most consistently identified using freeze-branding on the hip, shoulder, rear leg, or side. In addition, when freeze branding is used on more than one breed of horse, branding is performed under the mane. Some states require that cattle be permanently identified by branding with a hot iron; however, this procedure is more stressful than freeze-branding.

Ear and neck chain tags, although readable at some distance, can become lost and are therefore not necessarily permanent. In addition, neck chains and straps should be avoided in situations in which the animal could become entangled in a fence, rock outcropping, or other feature of the environment. Any associated pain and distress should be considered when determining the method of identification. In some cases, it may be necessary to identify animals in multiple ways. Individual records are needed for most animals.

These records should include information about the animal, its source and location, its productivity, its reproductive performance, protocols the animal is assigned to, and its ultimate disposition. Records for individual animals or groups should also include dates of vaccination, parasite control measures used, blood testing dates and results, and notations as to whether castration, spaying, or other elective procedures have been performed.

Applicable veterinary data to be recorded include dates of examination/treatment, clinical information/diagnosis, names of medications and amounts and routes of administration, descriptions of surgical procedures, and resolution of surgical procedures or illnesses. Principal scientists or animal facility managers may wish to record nutritional information. Research protocols often dictate that additional information be recorded.

VERMIN CONTROL

Programmes should be instituted to control infestation of animal facilities by vermin. The most effective control in facilities prevents entry of vermin into the facility by screening openings and ceilings; sealing cracks; eliminating vermin breeding, roosting, and refuge sites; and limiting access of vermin to feed supplies and water sources. Building openings should be screened with 1.3-cm mesh, and ceilings with ridge vents should be screened with 1.9-cm mesh to minimize rodent and bird entry. Smaller mesh sizes are recommended where they will not interfere with airflow.

Mesh may need to be installed along foundations below ground level, especially with wood foundations. Pesticides should be used only as approved.

Particular caution should be exercised with respect to residues in feedstuffs, which could injure animals and eventually pass into the meat, milk, or eggs. Pesticides should be used in or around animal facilities only when necessary, only with the approval of the scientist whose animals will be exposed to them, and with special care.

A pesticide applicator or a commercial service may be used. In some regions, wildlife and stray cats and dogs may spread zoonotic diseases, including rabies, to agricultural animals. In highrisk locations, institutions should implement an educational programme that includes training scientific and animal care personnel to recognize the signs of rabies in both wildlife and agricultural species and to handle and report potentially rabid animals.

Inoculation may be advisable for humans who may come into contact with animals in regions where rabies is endemic. Many agricultural institutions keep cats for pest-control purposes. Although the use of free-roaming cats is a traditional form of pest control for agricultural facilities, cats may limit the ability for baiting and may present hygiene or accident risks or serve as disease vectors. However, when cats are present, proper veterinary care and oversight should be provided to these animals. Veterinary care should include vaccinations, parasite control, and neutering.

MILK AND MEAT YIELDING ANIMALS

Depending upon the availability and regional considerations different animals are reared for the purposes of yielding milk and meat in India. India is the world's largest producer of milk. The majority of the milk consumed is also in liquid form in India. Over 53% of milk produced in India is from the water buffalo and a majority of milk processing plants in the country depend upon buffalo milk. The *National Dairy Development Board (NDDB)* is the main agency behind the cooperative movement in India. India is now seeking joint ventures and financial participation from the private sector including foreign investment for production of milk and milk products in India.

CATTLE

Cattle mainly include cow, bull, oxen, goat, sheep etc. The females of the species provide milk, which in turn contribute animals protein to the diet of people. While the female species of these cattle are used for milk, the male species play an important role in the agricultural economy by providing labour, meat and hide. Milk itself is taken in many forms like ghee, curd, butter and cheese etc. The excreta of these animals (dung) is used as manure, in biogas and as fuel. There are several important breeds of cattle in India and abroad.

MILK YIELDING ANIMALS WHAT IS A BREED

A breed is a group of one species of animals, which have the same descent and are similar in body shape, size and structure.

Categories of Important Breeds

There is following three categories:

1. Indian breeds
2. Exotic Breeds
3. Improved breeds

Indian Breeds

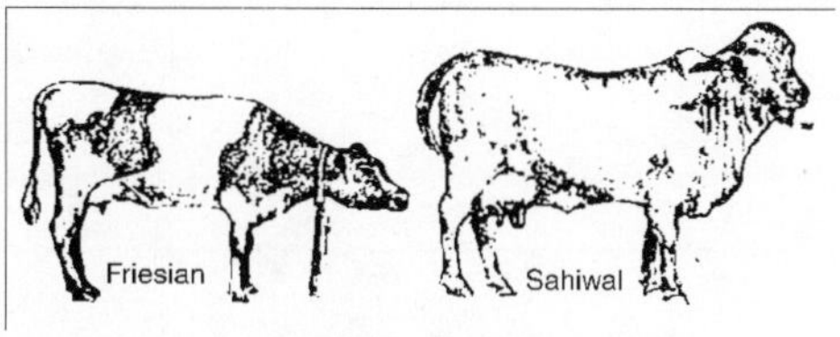

Fig. Indian Breeds of High Milk Yielding

Gir, Sahiwal, Red Sindhi, Thararkar, Kankrej etc. are some high yielding varieties of Indian cattle.

Exotic Breeds (Imported Breeds)

Hilstein, Friesian, Jersey, Swiss etc. are some of the high yielding varieties that have been imported from abroad and reared widely in India.

Improved breeds of Indian Cattle

Certain improved breeds have been developed by making a cross between two desired breeds. A cross between Sahiwal and Friesian varieties has been named as Friewal, Karan Swiss is another improved breed for milk production in large quantities. Table shows some indian breeds, their milk yield and distribution.

Table. Some Indian Breeds and their Milk Yield

Breed	Milk Yield (litres) per Lactation Period	Distribution
Gir	1200-2200	Gujarat, Rajasthan, Maharastra
Red Sindhi world including	700-2200	Andhra Pradesh, all part of India and Pakistan
Sahiwal Pradesh	1100-3100	Haryana, Punjab. Uttar
Kankrej	1400	Gujarat
Tharparkar	700-2200	Rajasthan
Mewati	1100	Rajasthan
Ongole	700	Andhra Pradesh
Hariana	500	Gujarat, Rajasthan
Hallikar	227-1134 litres	South India
Kangayam	665 litres	Tamil Nadu
Murrah Pradesh	20-22 litre/day	Punajb, Haryana, Uttar

Lactation Period is the period of milk production between birth of a young one and the next pregnancy and it usually lasts about 300 days.

Table. Breeds of Buffaloes

Breed	Distribution
Murrah	Haryana and Punjab
Bhadawari	Uttar Pradesh and Madhya Pradesh
Jaffarabadi	Gujarat
Surti	Gujarat
Mehsana	Gujarat (cross breed between Surti and Murrah)
Nagpuri	Maharastra
Nill Ravi	Punjab
Porlakmedi	Orissa

CATTLE FEED

The main feed of cows and buffaloes are grass but this does not provide them all the nourishment. They require balanced diet in the form of roughage which is fibrous food containing large amount of fibres such as hay fodder, leguminous plants-soyabeans, peas and cereals like maize, jowar etc. The diet of cattle mainly consists of roughage (dry or green fodder or fibrous food) and concentrates like grains, oil cakes and seeds, mineral salts and vitamins.

DAIRY PRODUCTS

Milk as drawn from the animals is known as full cream milk. When the cream is separated and the remaining milk is called toned milk. This milk contains no fat and is known as skimmed milk.

On the basis of fat contents the various milk product are as follows:

- *Cream*: It is prepared by churning milk, the fat comes on the top which is separated by draining out the liquid. It is known as cream with 10-70% fat contents.
- *Curd*: Milk is converted to curd due to bacterial activities.
- *Butter Milk*: It is the left over liquid after removal of butter.
- *Ghee*: After heating butter, the water evaporates and fat contents are almost 100%.
- *Condensed Milk*: Milk is concentrated by removing water contents with or without adding sugar. It has 31% milk solids with 9% fats.
- *Powdered Milk*: It is the powdered form of milk.
- *Cheese*: It is coagulated milk protein-casein with fat and water.
- *Khoya*: A desicated milk product prepared by evaporating water contents and reducing the bulk to about 70-75%.
- *Cattle Dung*: Cattle dung is mainly used to make dung cakes for burning as fuels. It is used mainly in villages of India. The farmers also use cattle dung to produce bio gas and the leftover residue as manure.

Biogas Plant (Gobar Gas Plant)

Bio gas plant is a chamber where animal excreta (Cow dung, buffalo dung etc) and some anaerobic bacteria are fed into airtight biogas chamber. Decomposition of excreta produces methane gas used as a smoke free gas for cooking. This gas can also be utilized for lighting. The left over solid residue serves as a good manure.

VETERINARY CARE

ATTENDING VETERINARIAN

The agricultural animal health care programme is the responsibility of the attending veterinarian. The Institute for Laboratory Animal Research (ILAR), National Research Council *Guide for the Care and Use of Laboratory Animals* defines the attending veterinarian as "a veterinarian who has direct or delegated authority" and who "should give research personnel advice that ensures that humane needs are met and are compatible with scientific requirements."

Animal Welfare Act regulations and the Public Health Service policy require that the attending veterinarian have the authority to oversee the adequacy of other aspects of animal care and use, including animal husbandry and nutrition, sanitation practices, zoonoses control, and hazard containment. Research and teaching institutions must provide investigators and instructors with access to a veterinarian who has experience in the care of agricultural animals. The veterinarian can be full-time or part-time and must have authority to ensure that the provisions of the programme are met. The attending veterinarian must be provided access to all research and teaching animals and to any related documents including health care records.

The attending veterinarian also must be involved in the development and oversight of the veterinary care programme, as well as in other aspects of animal care and use such as protocol review, establishment of anesthetic and analgesic guidelines, study removal criteria, training of animal users, and responsible conduct of research activities. Veterinary involvement in these activities helps to ensure animal health and well-being. The attending veterinarian is not required to be the sole provider of veterinary care and can delegate authority to another qualified veterinarian. However, the attending veterinarian must communicate with, and oversee veterinary care provided by, other veterinarians. When necessary, the attending veterinarian should utilize the expertise of other professionals when making determinations about agricultural animal care. Trained nonveterinary staff may administer treatments according to standard operating procedures approved by the attending veterinarian.

PREVENTIVE MEDICINE

Adequate agricultural animal health care in research and teaching involves a written and implemented programme for disease prevention, surveillance, diagnosis, treatment, and endpoint resolution. The objectives of such a programme are to ensure animal health and wellbeing, minimize pain and distress, maintain animal production, prevent zoonoses, provide assistance to investigators on study-related animal health issues, and avoid contaminants or residues in animal products.

The programme should include training for animal users regarding animal behaviour, humane restraint, anesthesia, analgesia, surgical and postsurgical care, and euthanasia. A mechanism for direct, frequent, and regular communication must be established among personnel who are responsible for daily animal care and observation, animal users, and the attending veterinarian. This will help ensure that timely and accurate animal health information is effectively communicated.

SICK, INJURED, AND DEAD ANIMALS

Animal care personnel must be trained to recognize signs of illness and injury. In general, sick and injured animals should be segregated from the main group to protect them and the other animals, observed at least once daily, and provided with veterinary care as appropriate. When animals are separated, a mechanism should be in place to communicate to staff the status of the animals and to ensure proper daily, weekend, holiday, and emergency care.

In some circumstances, segregation is not feasible or may disrupt the social hierarchy, cause additional stress to the animal, or adversely affect research. The advantages of segregation should be weighed against its disadvantages, especially for mild illnesses or injuries that can be easily managed. Care should be taken to minimize spread of pathogens from ill animals to healthy animals by observing appropriate biocontainment measures. Incurably ill animals or ill or injured animals with unrelievable pain or distress should be humanely killed as soon as possible.

Unexpected deaths should be reported to the attending veterinarian. Dead animals are potential sources of infection and should be disposed of promptly by a commercial rendering service or other appropriate means following applicable state and local ordinances and regulations.

Postmortem examination of fresh or well-preserved animals may provide important animal health information and aid in preventing further losses. When warranted, waste and bedding that have been removed from a site occupied by an animal that has died should be moved to an area that is inaccessible to other animals and the site appropriately disinfected.

MEDICAL RECORDS

An important component of an agricultural animal health programme is maintaining records that can be used to monitor animal health events, both physical and behavioural health events, as well as outcomes and levels of production. Medical records should comply with the American College of

Laboratory Animal Medicine (ACLAM) statement on medical records. Group health records may be appropriate for animals that are kept as cohorts, particularly when animals undergo periodic evaluation by means of examining several representative individuals of the group. The institution, under the guidance of the attending veterinarian, should determine the method(s) by which medical records are maintained. Oversight of medical records is the responsibility of the attending veterinarian and the IACUC.

When institutional representatives determine that a medical record should be created, the record typically contains the following information:

- Identification of the animal(s) or group(s);
- Clinical information, such as the animal's behaviour, results of physical examinations, and observed abnormalities, illnesses, and/or injuries;
- Immunizations and other prophylactic treatments and procedures;
- Documentation and interpretation of diagnostic tests;
- Documentation of research interventions;
- Treatments prescribed and administered;
- Clinical response and follow up information;
- Descriptions of surgical procedures, anesthesia, analgesia, and perioperative care;
- Methods used to control pain and distress;
- Documentation of resolution;
- Documentation of euthanasia or other disposition; and
- Necropsy findings if necropsy is indicated.

The record system must be structured so that information is easily collected, gathered, analysed, summarized, and available to the veterinarian, the principal scientist, and the IACUC.

The ACLAM statement on *Medical Records for Animals used in Research Teaching and Testing* suggests that:

- Notations in the medical record should be made by individuals who have administered treatments, or made direct observations or evaluations of the animal(s) or their diagnostic results, or their designee. Individuals typically responsible for making notations in the record include veterinary staff, animal husbandry staff, and research staff. All entries in the record should be dated, indicate the originator of the entry and be legible to someone other than the writer.

SOCIAL ENVIRONMENT OF HUSBANDRY

Dairy cattle are social animals that exist within a herd structure and follow a leader. Cows exhibit wide differences in temperament, and their behaviour is determined by inheritance, physiology, prior experience, and training. Cattle under duress may bellow, butt, or kick; however, cows are normally quiet and thrive on gentle treatment by handlers. Cows learn to discriminate among people and react positively to pleasant handling. Aversive handling leads to more incidents during handling and transport for calves than positive handling. Similarly, heifers and cows exposed to aversive handling took longer to traverse and more force to move than those handled more gently. Although the presence of an aversive handler reduced kicking during udder preparation, residual milk was 70% greater than for the control milkings.

Cows have higher milk yields if handlers touch, talk to, and interact with them frequently. Cows should have visual contact with one another and with animal care personnel. Handling procedures are more stressful for isolated cattle; therefore, attempts should be made to have several cows together during medical treatment, artificial insemination, or when cows are being moved from one group to another.

This was verified by increased heart rate, hypothalamic-pituitary-adrenocortical axis activity, and vocalizations. Pain sensitivity is reduced during isolation, suggesting a stress-induced analgesia. Care should be taken to minimize the negative impact of moving cows to new groups by avoiding frequent regrouping and by always moving more than one animal at a time to a new group. The use of a trainer cow can have a positive impact on adjustment to feedlot environments where many heifers are raised.

However, dairy calves had few indicators that repeated regrouping and relocations stressed calves. Aggression was rare and the calves seemed to habituate to the repeated mixing. Calves from larger groups after weaning had fewer incidences of displacement of other calves from the feed barrier, were more active, and had more positive interactions with familiar calves. Calves, like cows, prefer familiar calves to unfamiliar calves during stressful situations, and a familiar companion calf improved cows' reaction to separation.

Social status can affect health issues such as lameness. Low-ranking cows spent more time standing and standing half in cubicles than did middleand high-ranking cows. Standing half in cubicles correlated positively with the number of soft tissue lesions related to lameness. Dairy cattle have traditionally been kept in groups of 40 to 100 cows, although specific research protocols may require smaller or larger group sizes.

Variation in group size—small, medium and large—does not cause a problem *per se*. Expansion to a larger herd size, however, can affect management decisions because overcrowding with an insufficient number of headlocks or inadequate manger space per cow, irregular or infrequent feeding, and excessive walking distance to and from the milking parlor have a greater impact on behaviour and well-being than does group size. Cattle of all ages are gregarious.

Socially isolated cattle show clear signs of stress: increased heart rate, vocalization, defecation/urination, and cortisol levels. In addition, there are benefits of housing cattle together. For example, pairs of calves are more likely to play than isolated calves, a behaviour thought to be associated with positive welfare. Young calves should be kept in groups from 2 to 7 animals in order for animals to benefit from social contact, but larger groupings are associated with health problems and morbidity.

Management of resources is an important part of reducing aggression and other problems, such as cross sucking, in groups of animals. Adult dairy cattle should have 1 freestall/cow to reduce competition. Similarly, dairy cattle with more space at the feedbunk engage in fewer aggressive interactions and the reduction of competitive behaviour associated with more feeder space is particularly marked in post-and-rail feeder design.

Cross sucking in calves is an undesirable behaviour performed in groups. Calves are typically fed 10% body weight during the milk-fed period and there is clear evidence that this feeding level is insufficient. A combination of slower milk flow, hay feeding, and access to a nonnutritive artificial teat are also recommended to reduce cross sucking. Providing additional objects for oral manipulation, such as tires, has also been shown to reduce other problems such as stereotypic tongue rolling in calves.

Restraint and Handling

Vaccination schedules that are appropriate for the location and dynamics of the individual herd should be established with the advice of the attending veterinarian. Certain dairy cattle behaviours put at risk the health and well-being of herdmates as well as the humans handling the cattle. These behaviours can be reduced or modified by implementing principles of low-stress handling and restraint that include appropriate movement of people, well-designed facilities, optimal lighting, nonslip flooring, and smooth, quiet restraint devices.

Stanchions, head gates, and squeeze chutes can be modified to function optimally, but acclimation and positive reinforcement by individuals trained in low-stress handling can minimize the need for additional restraint by halters, rope, tail hold, and nose tongs. Hobbles and casting ropes should be used selectively and only when necessary. Chemical sedation is always preferable to excessive use of force or application of electrical prods. Information about calving management is given by Albright and Grandin.

First-calf heifers should be bred to calving-ease bulls and be of appropriate stature and body condition to minimize the chances of dystocia or the need for calving assistance. Optimal calving conditions in a clean, quiet environment with employees appropriately trained to follow calving protocols will result in more live calves and fewer calving injuries and illness.

Calving injuries should be assessed immediately so that appropriate footing is provided and proper treatment is implemented. Cows that are unable to stand should be moved to a soft-bedded pack and examined by a veterinarian within 2 to 4 h of calving. Calves require special handling and care from the time they are born. Colostrum should be fed or ingested within the first 5 h after birth always before milk is fed. Between 1.89 L and 3.79 L of colostrum are necessary to impart adequate immunity to the calf.

In the absence of colostrum, a colostrum replacement product that delivers at least 125 g of immunoglobulin should be given by bottle, bucket, or tube feeder. Colostrum is rich in nutrients and provides the calf with vital immunoglobulins and other important immune factors. Clean navels can be dipped in a dilute chlorhexidine solution as soon as possible after birth. Good nutrition as supplied by a combination of milk starter grain, and fresh water along with proper handling and close monitoring starts a calf on its way towards a healthy life.

STANDARD AGRICULTURAL PRACTICES

All animals should be individually identified. Heifer calves should have supernumerary teats removed at an early age. Removal may be performed in the first 3 mo of life with a scalpel or sharp scissors. Older calves and heifers close to calving that have supernumerary teats should be examined by a qualified person. The removal of extra teats at this advanced age is necessary if they will later disrupt the milking process or be at risk of becoming infected.

If so, they can be removed with proper restraint and use of appropriate anesthesia by a qualified and trained person. Milking procedures should follow National Mastitis Council guidelines. Routine breeding programmes should include housing and handling facilities that allow for effective implementation of artificial insemination programmes. Castration may be performed on male calves.

Foot Care

Lameness in dairy cattle is a major source of economic loss to the farmer and a serious cause of pain and discomfort to the cow. It is perhaps the most important condition affecting the welfare of cows on dairy farms. Lame cows suffer lowered milk production and reduced fertility, and are culled at 2 to 4 times the rate of healthy control cows.

The pain associated with lameness results in changes in the animal's gait that include:

- Arching of the back;
- Shortening of the stride length on the affected limb;
- Sinking of the dew claws on the unaffected contralateral limb;
- Head bob in a vertical plane;
- Reduction in walking speed, and frequent stops; and
- Swinging the affected limb in or out depending on the location of the painful lesion.

These alterations can be used to provide a locomotion score for each animal, and the most commonly used system in North America utilizes a 5-point system of scoring where 1 is nonlame and 5 is severely lame. Herd workers should be taught how to score locomotion so that they can identify cows with scores >2 for treatment by an attending veterinarian or hoof-trimmer.

Around 85% of lameness in dairy cattle is associated with lesions in the rear feet, particularly the outer claw, because of the overgrowth of horn resulting from the redistribution of weight as the cow walks on hard concrete surfaces, with a large udder occupying the space between her rear legs. This overgrowth of the outer claw may be removed and the weight transferred equally between the inner and outer claw by regular hoof-trimming.

Trimming to restore a normal toe length along the dorsal hoof wall of around 75 mm for mature Holstein cattle, combined with balancing weight between the inner and outer claw, lasts around 4 mo on average. Therefore, it is recommended that cattle be trimmed at 6-mo intervals, typically at the time of dry off and in mid-lactation around 90 to 150 d in milk. Some cows with pre-existing hoof disease may require attention more frequently.

Hoof lesions causing lameness may be broadly classified into 2 groups:

1. Infectious and
2. Claw horn.

Infectious lesions include digital dermatitis, interdigital phlegmon, and heel horn erosion. These lesions are associated with poor feet and leg hygiene and are a particular problem in free-stall environments, where the cow is exposed to alleyways contaminated with wet manure when she is not occupying a stall. Putative agents such as several species of *Treponema* and *Fusobacterium necrophorum* are involved in the pathogenesis of these conditions, but hydropic maceration of the skin of the interdigital space appears to be a prerequisite for the development of disease.

Infectious causes of lameness are controlled by improving leg hygiene by removing manure from the walkways and by the use of a topical antibacterial administered either directly to the lesion by a hand-held spray or via a footbath. The frequency of foot bathing is dependent on the degree of manure contamination of the cows, and a variety of chemicals are available for use, such as copper sulfate, zinc sulfate, and formalin.

Use of any of these chemicals should be done under veterinary direction. Claw horn lesions include sole hemorrhage, sole ulcer, toe and heel ulcer, and white line disease. These are clinical signs on the surface of the claw that represent the result of several possible causative pathways. Sinking of the third phalanx within the claw horn capsule, due to a breakdown in the connective tissue of the suspensory apparatus, may be caused by hormonal changes at calving time and nutritional events such as subacute ruminal acidosis.

Sinking of the third phalanx compresses the corium below, interrupting the flow of blood and nutrients to the cells responsible for horn growth. As a result, a defect develops that becomes apparent several months later as the sole horn continues to grow. Excessive removal of sole horn, either through poor hoof trimming or due to excessive wear from walking long distances on rough concrete will also contribute to lesion development.

Flooring surfaces should be nonslip, avoid excessive trauma to the claw surface and be dry. Concrete should be grooved to improve traction; a pattern that utilizes parallel grooves 3/4 inch wide and deep, spaced 3 inches on center appears to provide a good compromise between sufficient traction to reduce injury while limiting the amount of wear.

For transfer lanes between milking centers and the living accommodation, a 1-m wide strip of rubber flooring has been used successfully to reduce trauma and wear, and rubber flooring has been used in parlor holding areas to provide cushion for cows that have to stand for long periods of time. The severity of the claw horn lesions that develop is influenced by the time spent standing each day, which results in increased loading of the claw and increased compression of the tissues below the third phalanx.

Time spent standing may be increased by:

- Poor stall designs that fail to provide surface cushion, room to lunge, and sufficient resting area;
- *Overstocking*: Providing fewer usable stalls than there are cows in a pen;
- Excessively prolonged milking times;
- Time spent locked up away from the stalls for management tasks; and
- *Heat Stress:* Cows may stand more in an attempt to cool off.

In addition, lame cows struggle to use stalls with hard surfaces because the act of rising and lying down becomes more challenging due to foot pain. These cows stand more in the stall and fail to gain adequate rest for lesion healing. For this reason, deep sand-bedded stalls provide the gold standard in cow comfort. If sand stalls are unavailable, lame cows should be treated and returned to a bedded pack area for rest and recuperation until normal ambulation returns.

Failure to identify a claw horn lesion early in its course may result in deep digital sepsis. This is a complication caused by infection of the deeper structures of the claw, including the distal interphalangeal joint and tendon sheaths. Such animals are usually severely lame and require euthanasia or extensive surgery. Seeking veterinary assistance is recommended for individual cows that show signs of lameness or if a significant lameness issue exists for the herd.

Disbudding and Dehorning

A review of horn anatomy and growth and dehorning and disbudding of cattle was provided by the AVMA. The AVMA also provides guidance on use of sedation, anesthesia, and analgesia and alternatives to horn removal. The AVMA policy on dehorning/disbudding should be followed. Calves should be observed closely for 1 to 2 h following dehorning. No food or water should be offered until the sedation is completely worn off or reversed. Persistence of a depressed attitude, head-pressing, or an abnormal head tilt for more than 2 h should result in a complete examination.

Tail Docking

The bovine tail has several physiological and behavioural functions including dissipation of heat, and facilitation of visual communication among cattle and with human caretakers; the tail often serves as a primary mechanism of fly control. Removal of the lower portion of a cow's tail is commonly referred to as "tail docking" and the use of tail docking as a routine dairy farm management tool apparently originated in New Zealand.

New Zealand farmers responding to a 1999 survey believed removal of tails resulted in faster milking, reduced risks to the operator, and reduced rates of mastitis. Similar unsubstantiated claims have been made for the US dairy industry. Several European countries, some Australian states, and California have prohibited tail docking. Both the Canadian and American veterinary medical associations have policy statements that oppose the practice of tail docking for routine management of dairy cattle.

The policy statement of the *American Association of Bovine Practitioners* (AABP) indicates that scientific evidence to support tail docking is lacking and recommends that "if it is deemed necessary for proper care and management of production animals in certain conditions, veterinarians should counsel clients on proper procedures, benefits, and risks.

Scientific studies have been performed to evaluate both the potentially negative and positive aspects of tail docking. Important welfare issues that have been evaluated have included pain caused by tail docking, changes in fly avoidance behaviour, immune responses, and changes in levels of circulating plasma cortisol. Experiments that have been performed on both calves and preparturient heifers have consistently concluded that the process of tail docking does not induce significant acute or chronic changes in plasma cortisol or other selected physiological measures. Modest changes in general behaviour of calves that have been docked using rubber rings or cautery irons have been reported but these changes have not been associated with significant differences in normal feeding, ruminating, or grooming behaviours. Likewise, few significant differences in general behaviour of docked preparturient heifers have been noted.

However, greater changes have been observed in tail surface temperatures of docked heifers compared with heifers with intact tails, indicating that heifers may experience chronic pain similar to the phantom pain reported by human amputees. Research has demonstrated that tail-docked heifers flick their tails more often and are forced to use alternative behaviours such as rear leg stomps, feed tossing, and head turning to try to rid themselves of flies.

More flies settle on tail-docked cows than on intact cows, and the proportion of flies settling on the rear of the cow increases as tail length decreases. In another study, there were no significant differences in the numbers of stable flies found on the front legs of cows but docked cows had nearly twice as many flies on their rear legs compared with those with intact tails.

Fly avoidance behaviours were increased in the docked animals, whereas tail swinging was increased in the control animals. Foot stamping was identified only in docked animals and, overall, fly numbers and fly avoidance behaviours were increased in docked animals. Researchers have been unable to identify improvements in udder health or udder cleanliness for animals in commercial herds that have docked tails. In one study, the effect of tail docking on cow cleanliness and *somatic cell counts* (SCC) was evaluated over an 8-wk period for lactating cows that were housed in a free-stall facility.

Standardized cleanliness scores obtained from the rump, midline of the back, or rear udder were not significantly different between docked and intact animals nor was there any significant difference in SCC or the number of teats containing obvious debris. In another study, SCC, occurrence of *intramammary infections* (IMI), and udder and leg hygiene scores were evaluated over an 8-mo period for lactating dairy cows that had been blocked by farm and randomly allocated to tail-docked or control groups.

No significant differences were found in SCC or udder and leg hygiene scores. The prevalence of contagious, environmental, and minor pathogens was not significantly different between cows with docked or intact tails. Although current studies do not indicate that the process of tail docking modifies physiological indicators of stress, several studies have documented changes in fly avoidance behaviour and recent research has suggested that docked tails have enhanced sensitivity to heat.

No benefits to cattle welfare have been associated with tail docking. The routine use of tail docking in research or teaching herds should be discouraged, and alternatives to tail docking are recommended when appropriate. Any use of tail docking, other than for medical reasons, should be reviewed and approved by the IACUC.

SPECIAL CONSIDERATIONS OF MILKING MACHINE AND UDDER SANITATION

The milking facility should have a programme for regular maintenance of milking machines and follow the recommended mastitis control programme of the National Mastitis Council. Appropriate equipment and competent personnel should be available for milking. Personnel responsible for milking should receive ongoing training about proper milking procedures as the frequency of training has been associated with adequacy of milking performance. Animal care facilities should be designed and operated to standards meeting or exceeding those of grade A dairies as defined in the Pasteurized Milk Ordinance. Areas where milking takes place must be designed and constructed in accordance with the 3-A Sanitary Standards Inc. Accepted Practices. Cows should be maintained in housing areas that provide for adequate hygiene to ensure that udders are visibly clean. Cows should be milked on a regular schedule that is appropriate for the goals of the herd or specific research project.

Written operating procedures should be established to control potential contamination of milk with antibiotics or other pharmaceutical agents. Antimicrobial treatments should be administered based on approved defined protocols. All extra-label treatments must be administered under the supervision of a veterinarian that has an appropriate veterinary-client-parent relationship.

Milking machine and udder sanitation are vital to an effective preventive programme against mastitis and follow guidelines as established by the NMC. Care should be used to minimize the excessive use of water before and during udder preparation. Emphasis should be placed on ensuring that cows enter the milking parlor with clean, dry teats. Udders, especially teat ends, should be clean and dry when teat cups are applied for milking.

The removal of foremilk before teat disinfection is encouraged as a means to detect mild cases of clinical mastitis. Teat sanitation, predipping, and wiping immediately before machine attachment reduce udder infection caused by environmental pathogens. Postmilking disinfection of teats is an essential management practice that greatly reduces the incidence of mastitis. Milkers handling cows should pay meticulous attention to their own personal hygiene and wash their hands thoroughly before milking and frequently during milking.

The use of clean nitrile or latex gloves during milking is highly encouraged to prevent contamination of the udder. Cows with subclinical cases of contagious mastitis should be milked last to reduce the spread of mastitis throughout the herd. Udder hair removal is recommended as a means to improve milking hygiene and udder health. Cleaning of milk handling equipment is accomplished by a combination of chemical, thermal and physical processes and cleaning regimens should be designed to meet appropriate regulatory standards.

Recommended cleaning and sanitizing practices are a balance between the cleaning temperatures, cleaning chemical concentration, contact time and mechanical action. Effective cleaning programmes for milking machines include use of hot water; use of disinfectant solutions and other chemical agents effective for removing mineral, milk fat, and protein deposits from equipment between milkings; disinfection of teat cups between cows; and flushing of teat cups with warm water, cold water, boiling water, or chemical disinfectant solution. The most common routine in the United States is a combination of prerinse, alkaline detergent, acid rinse and premilking sanitize. Very small herds may utilize manual cleaning and disinfecting that involves hand-cleaning of some or all of the milk harvesting and storage equipment. Small to medium herds commonly use automatic washing equipment.

This equipment will automatically mix the chemicals with the appropriate water volume and temperature and circulate these solutions through the milking machine. On large farms, an attendant may be present to mix chemical solutions and operate valves for circulation. The effectiveness of milking system cleaning can be evaluated by examination of standard plate counts and laboratory pasteurized counts performed on bulk tank milk samples.

Stray Voltage

The term stray voltage describes a special case of voltage that develops on grounded metal objects on farms. If this voltage reaches sufficient levels, animals coming into contact with grounded devices may receive a mild electrical shock that can cause a behavioural response. At voltage levels that are just perceptible to the animal, behaviours indicative of perception such as flinches may result, with little change in normal routines.

Studies by numerous independent research groups in several countries are in agreement that the most sensitive cows begin to react to 50 or 60 Hz electrical current of 2 mA applied from muzzle to hooves or from hoof to hoof. This corresponds to a contact voltage level of about 1 V. As the voltage and current is increased, a greater percentage of cows will react with behavioural responses becoming more pronounced. Numerous studies have documented avoidance behaviours at levels above the first reaction threshold.

The median avoidance threshold for 50 or 60 Hz current flowing through a cow is about 8 mA. Even when the threshold is exceeded not all cows would be expected to show a behavioural response but as the voltage increases, signs in a herd would be expected to be more widespread and uniform. The scientific evidence strongly suggests there is no relationship between behavioural responses to stray voltage and physiological or hormonal responses.

There is no apparent relationship among behavioural modifications, milk production, and animal health. The only studies that have documented adverse effects of voltage and current on cows had both sufficient current applied to cause aversion and forced exposures. It is typical for voltage levels to vary considerably at different locations on a farm. Decreased water and feed intake or undesired behaviours will result only if current levels are sufficient to produce aversion at locations that are critical to daily animal activity.

These locations include feeders, waterers, and milking areas. Controlled research has shown that if an aversive voltage was administered to a water bowl once per second, water intake was reduced. However, when the same voltage was applied once every 10 min and once per day, no reduction in water intake was observed. If an aversive current occurs only a few times per day, it is not likely to have an adverse effect on cow behaviour.

The more often an aversive voltage occurs in areas critical to cows' normal feeding, drinking, or resting, the more likely it is to affect cows. No one sign is pathognomonic; a variety of signs has been reported in cows exposed to different levels of voltage. Documented signs are behavioural changes and decreased drinks of water per day and length of time per drink. The amount of water consumed may not be affected even when behavioural modification occurs.

Intermittent periods of poor performance, poor milk letdown, and incomplete or uneven milk-out, abnormal behaviour during milking, increased milking time, refusal of feed or water, increased SCC in milk, and increased mastitis are signs often attributed by farmers to stray voltage; however, none of these signs were evident in numerous controlled studies. These signs are often caused by other factors such as abusive cow handling, faulty milking machine, poor milking techniques and hygiene, and nutritional deficiencies.

Therefore, animal behaviour or other symptoms cannot be used to diagnose stray voltage problems. The only way to determine if stray voltage is a potential cause of abnormal behaviours or poor performance is to perform electrical testing. A thorough investigation of the entire production unit should be conducted to determine other sources of problems. Electrical systems should comply with wiring codes and standards at all times to protect both animals and people.

Whenever suggestive signs cannot be attributed to other causes, measurements should be taken to determine if a voltage potential exists, and the results recorded for future comparisons. A diagnostic confirmation of stray voltage must include a competent electrical measurement indicating at least 2 to 4 V between 2 points that a cow might contact, with some cows should exhibiting avoidance behaviours at this location. Voltage levels may need to be monitored at different times of the day and on different days because the threshold level may be exceeded intermittently.

All voltage readings should be made with a 500 to 1,000 Ω resistor across the 2 measuring leads to the cow contact points in addition to open circuit measurements. Readings without the use of a shunt resistor are meaningless. Although the resistance of cow and human tissues is similar, the contact resistance is generally lower for cows than for humans, particularly if cows are in a wet environment. The resistance of a cow's body plus the contact resistance with the floor is commonly estimated as 500 Ω. This is a reasonable value for a cow standing on a wet floor.

Cows standing on a dry surface will typically produce 1,000 Ω resistance or higher. Cows standing or lying on dry bedding will have a resistance many times higher than this. The resistance of a human can be as low as 1,000 Ω for wet hand-foot contact to $>$10,000 Ω for dry handfoot contact. The contact voltage to produce sensation can therefore be higher for humans than for cows, depending on the conditions of the contact points. If more than 1 V is detected at the cow contact points, it is advisable to have a qualified electrician or the local power supplier evaluate the situation.

Bulls

The feeding and watering of growing and mature bulls should meet requirements of the National Research Council. Bulls should be housed in clean, well-lit, and ventilated buildings or outside in facilities that protect them from inclement conditions and allow them to remain clean and dry. Young bulls kept in small and uniform groups should be observed carefully as they mature to make certain that one or more individuals are not injured.

A panel can be installed in the center of group-housing pens to allow subordinate bulls to escape aggressive behaviour of dominant pen mates. Aggressive behaviour increases with age, and group housing should be discontinued by around 3 yr of age. Smaller or subordinate bulls should be removed from the group, and a bull removed from a group for over a few hours should never be returned to the group. Visual and vocal social interactions with other bulls may be stressful. The safety of humans and animals is the chief concern underlying bull management practices. By virtue of their size and disposition, bulls may be considered as one of the most dangerous domestic animals. Management procedures should be designed to protect human safety and to provide for bull welfare. Electroejaculation of bulls is sometimes necessary and should be performed by a qualified person using equipment that functions properly and is in good repair.

A programme of annual self-regulation should be followed for:

- Semen identification and sire health auditing service and
- Minimum requirements for health of bulls producing semen for artificial insemination.

HOUSING FOR FINISHING ANIMALS

The provision of accommodation systems for finishing beef animals is essential for efficient herd management. The type of housing provided will depend on a range of factors including geographic location, availability of straw, size of the unit and on the traditional methods of fattening in the particular region. Depending on the finishing system animals may remain in the house until slaughter or return to pasture following the winter period.

In the latter case these animals may be finished outside or spend a subsequent period in the accommodation prior to being slaughtered. Loose housing systems predominate for beef cattle. When designing accommodation facilities for fattening cattle consideration should be given to labour availability, feeding system, type of diet, group size, drinking system, and facilities for handling and storage of the manures produced. The requirement for housing during the finishing period may be due to land conditions that do not facilitate outdoor feeding due to soil type and climatic factors. In certain situations housing is provided to facilitate the structured feeding of the animals under controlled management conditions. Also, housing facilities should provide a suitable working environment for the farmer and any employees involved in taking care of the animals.

HOUSE TYPES

Traditionally, beef cattle were housed in straw bedded facilities. The unavailability of straw in some areas (combined with its increased cost), the need to reduce labour requirements and the necessity to ensure manure was efficiently managed to avoid pollution risks, has encouraged the development of housing systems utilising liquid manure storage.

Variations in the design have evolved including houses with partially slatted floors, houses with sloped floors inclining towards narrow slurry channels and houses with slats over shallow tanks from which the slurry flows by gravity to an adjacent storage facility. Cubicle houses can also be used for finishing animals where no straw is available but such facilities are not recommended for male animals because, they foul the lying area with urine.

Where straw is avail-able for bedding typical systems include fully bedded pens, facilities with bedded lying areas and solid unbedded or slatted feeding stands and facilities with sloped floors, which depend on the animals to move the fouled bedding down the slope for collection. Tethering of animals in stalls is still used as a management system in some beef units but this method of housing is not recommended.

Bedded House

Such houses consist of bedded pen(s) with the total living area covered in bedding material, which normally is straw. The facilities are roofed but are sometimes referred to as "straw bedded yards". Cattle should be housed in groups of not more than 20 to aid management. This type of house requires 4 to 6 kg of straw per animal per day which equates to approximately 1 (metric) ton straw per animal for a winter housing period.

The completely bedded system does not prepare the hooves of the animals for subsequently walking on harder surfaces such as concrete. Sometimes the hooves may become overgrown and misshapen, which may lead to lameness problems. Although the systems are relatively cheap to construct the high straw requirement with associated labour costs must also be taken into account when making comparisons with other systems. The manure/bedding mixture is allowed to build up over the housing period and is normally removed once at the end of the season. Depending on the length of the housing season the accumulated material may rise up to greater than 1m above floor level. It is necessary to take this into account when installing gates, partitions and feeding barriers and troughs. A typical layout of such a facility is presented in *Figure*.

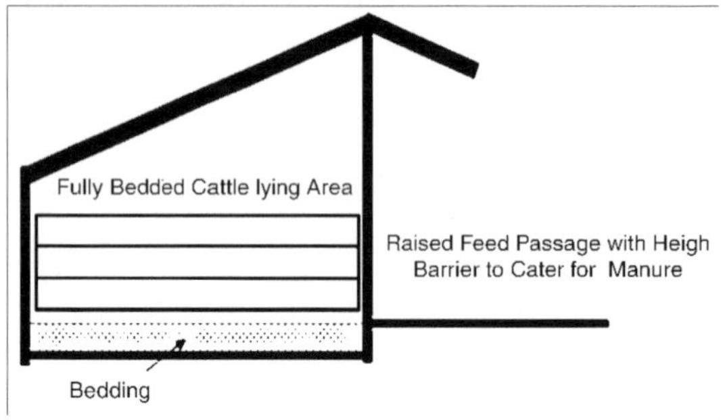

Fig. Straw Bedded Facility Cross-section Showing Raised Passageway

Bedded House with Concrete or Slatted Feeding Stand

In this type of facility the animals come to feed on an area of solid concrete or an area covered with slats. In the case of solid concrete the area is cleaned by an electric, hydraulic or tractor powered scraper. Where a tractor is used additional gates are used to close off the feeding stand from the bedded area during cleaning.

The design has the advantages that lower quantities of straw are required and that the geometry of the feeding stand does not change as the manure builds up over the housing period. Straw usage is in the order of 2 to 3 kg per animal per day. However, both a liquid and solid manure is also produced with the system. If slats are used in the feed stand it is important to minimise the quantity of straw entering the tanks to avoid problems with slurry agitation. A cross sectional view is shown in *Figure*.

Bedded House with Sloped Concrete Floor

This housing system involves the frequent removal of manure but daily straw requirements can be as low as 1 to 3 kg per animal per day. The floor is laid with a slope of 5 to 10%.

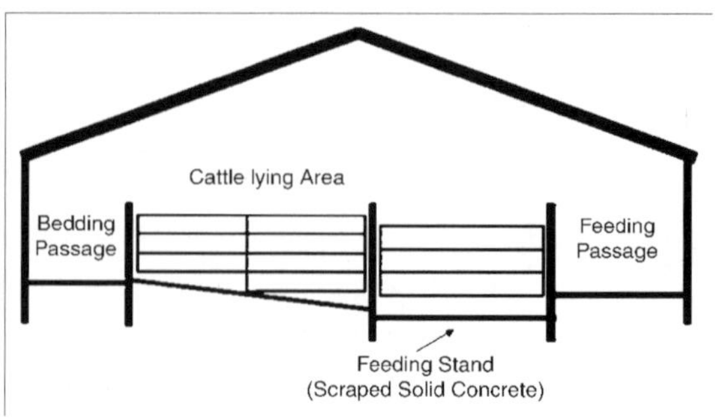

Fig. Bedded House with Sloped Concrete Floor

The system operates on the principal that the movement of the animals will transfer the manure down the slope where it is removed by scraping. A cross section of a typical layout is shown in *Figure*.

Cubicle House

A cubicle house provides an animal with an individual safe lying area. The system is widely used in the dairy industry for cows and the house type provides a clean lying area without the requirement for bedding material. Sometimes a little straw is used on the beds, which are normally constructed with concrete. Artificial mats are widely used in the dairy industry to improve the lying environment in cubicle beds.

Major limitations with cubicles for beef finishing animals are:

- The fact that as animal size is changing it is difficult to optimise the dimensions of the cubicle
- Male cattle foul the cubicle bed with urine
- Animals may only be housed for a relatively short period *e.g.* one winter season and often it is difficult in these situations to train animals to use the facilities

Slatted Floor House

The development of housing systems for beef animals utilising liquid manure storage has been promoted in order to overcome the unavailability of straw in many areas (combined with its increased cost), the need to reduce labour requirements and the necessity to ensure manure is efficiently managed to avoid pollution risks. The majority of such systems use concrete slatted floors with the liquid manure or slurry falling through the floor perforations into a below ground concrete tank.

The depth of the tank is such so as to provide adequate waste storage capacity for the housing period and is now typically about 2,5 m. A central covered feeding passage is typically used with confinement pens on each side. Silage can either be fed independently from concentrates or in combination. Feed face length is typically 0.3 m per adult animal for silage only feeding; 0.45 m where complete diets are fed and 0.6 m where concentrates are separately fed from silage.

The ratio of pen depth to width can be manipulated at the design stage to ensure adequate lying area and feeding space is provided. The superstructure is typically constructed from steel stanchions, steel trusses and timber purlins. Portal frame configurations are also used. Roof sheeting and cladding is typically galva-nised steel or fibre cement sheeting. Ventilation openings are provided at the sides of the building and at the ridges.

The size of the opening is dictated by the exposure of the site. More recently spaced roof sheeting has successfully been used and it has the advantage in that it provides a more uniform removal of stale air from over the animals throughout the buildings. To improve the interface of the floor with the animal the ribs of the slats can be covered with rubberised materials to provide cushioning for the animals. A cross sectional view of a typical unit is shown in *Figure*.

Sloped Floor System without Bedding

This design is based on solid floors, which slope towards narrow manure collection channels covered with slats. The optimum length of the sloping floor section to avoid manual cleaning is approximately 1.5 m. The slope should be 1:12 (8%). The manure channels should be 300 mm wide and at least 300 mm deep and up to 450 mm deep for long channel runs.

A flushing system or mechanical scraper can be used to effectively remove the manure from the channel to an outside storage facility. While the system offers the advantages of the slatted system (no straw costs and effective collection and management of liquid manure) the animal's hides will be dirtier, resting behaviour of the animals disturbed and the risk for pressure injuries on the animals will increase with this design of facility.

2

Keeping Animals

HOUSING BUFFALOW

Housing for water buffaloes should give protection against thermal stress - particularly direct exposure to sun, heavy rains and cold weather. It must allow good ventilation. Housing may therefore be different in different areas of the world, due to differences in climate.

Below are some considerations and solutions when planning housing in hot and cold climates respectively. Common for all housing is that enough space should be allowed for each buffalo.

The outdoor yard should preferably be covered with grass or maybe concrete, in order to prevent it from becoming an unhygienic mud hole in rainy periods.

Hot climate

Buffaloes may appear to be misplaced in a hot and humid environment. They have a dark skin and few sweat glands and are more or less dependent on water for their cooling. This is not entirely true, buffaloes protected from direct sunlight do very well even during hot and humid days, partly because their ability to loose heat through the respiratory tract.

High milk production requires a high feed intake which leads to higher metabolic heat production. High yielding buffaloes thus have a disadvantage over lower yielding animals, and need more cooling facilities. The following points are guidelines to have in mind when giving advice on management.

- The feeding, watering and milking place should always give shade and protection from heavy rains, either by trees or by a roof.
- Cool water either from a clean river or served in an earthen pit, helps the animals to maintain temperature. Water trough should always be placed in the shade.
- A paddock with trees gives a very cheap and effective protection from sun. However, the trees may need to be protected from the buffaloes also.

- A shelter of a simple construction with only a roof. In hot humid climates it is better not to have walls. Walls may lead to inadequate ventilation and thereby favouring bacteria and mold growth, thus making the stable unhygienic. To protect the inside from sunshine (or heavy rain), curtains made from straw, textile or other suitable material, can be used.
- Providing the animals with a wallow. However, the wallow should be one with clean water and not far from the farm. Spending time walking in the sun to and from the wallowing costs more than it saves.
- Showering of the buffaloes with cool water for 3 minutes twice a day has proven to be an efficient way for them to get rid of excess heat.

Cold Climate

- A shelter should protect the animals from rain, snow and strong wind. It may be a simple construction with a roof and three walls. This system will allow the buffaloes to go outside to graze when the weather allows it. There should be a feeding area inside the shelter in case of several days with bad weather. A separate heated milking area is advisable in this case.
- A dry and clean bedding is important in cold weathers to maintain animal health.
- In case of extremely cold climate, (Caucasia and Balkan) with several months with a temperature below 0?C, a heated barn may be necessary.

Pens for Calves

Calves should be kept in individual pens for the first month. The pens should be easy to keep clean, with shelter from direct sunlight, rain, snow and draught. By keeping the calves individually it is easier to check that they eat and grow properly and to detect illnesses. Also, naval suckling is avoided and spread of diseases is more difficult.

The calves should have access to fresh and clean water at all times. Preferably, the buckets for milk and water should be outside the pen, in a steady holder within easy reach for the calf. Hereby, the calves can not splash it on the bedding. A humid bedding will facilitate growth of germs and parasites. The pen should contain a holder for hay and concentrate. These holders should be placed above the ground so that the calf cannot step or defecate in them.

Chicken coop

A chicken coop is a building where chickens are kept on a farm or homestead. The inside of the chicken coop normally consists of several nesting boxes just large enough for the chickens to sit in while laying their eggs, as well as perches for the chickens to use while sleeping. Often the floor of the chicken coop is covered with a material, such as straw or wood chips, to contain the chicken waste, and to allow easier cleanup of the chicken coop. Sometimes the chicken coop contains feeding and watering devices for the chickens, but these may also be kept outside of the chicken coop. Most have some kind of ventilation to help air out any odors that may occur.

Fig. Chicken coop

Since chickens are often prone to illness from drafts and poor weather, chicken coops are best designed to be warm at cold times, cool at hot times, and as draft free as possible. Many people, especially in rural areas, keep a small flock of chickens to provide themselves with eggs and meat. These small coops are often surrounded by a fence, usually made with wire fencing to allow the chickens an area to roam, peck, and hunt insects, to keep them contained. Contrary to popular use of the term, fencing known as "chicken wire" is often too weak to pose a barrier to predators and is therefore seldom used to confine the birds.

Free range

Free range poultry farming consists of poultry permitted to roam freely instead of being contained in any manner. The principle is to allow the animals as much freedom as possible, to live out their instinctual behaviours in a reasonably natural way, regardless of whether or not they are eventually killed for meat. In the UK, the Department for Environment, Food and Rural Affairs says that a free range chicken must have daytime access to open-air runs during at least half of their life. Unlike in the United States, this definition also applies to eggs. The European Union regulates marketing standards for egg farming which specifies a minimum condition for Free Range Eggs states that "hens have continuous daytime access to open-air runs, except in the case of temporary restrictions imposed by veterinary authorities".

NEST BOX

A nest box also called birdhouse is a man-made box provided for animals to nest in. Nest boxes are most frequently utilized for wild and domesticated birds, but certain species of mammal may also use them. Boxes are usually wooden, though some for birds are made from mixtures of wood and concrete, called *woodcrete*. Metal nest boxes are also marketed, but these are generally unsuitable for outdoor use, as they can overheat easily in sunshine. The majority are cuboid, and many have a sloping roof. Many have a hinged top to provide access for cleaning, bird ringing or, when used for domesticated species, to give the breeder access to the young. Boxes may either have an entrance hole or be open fronted. Some nest boxes can be highly decorated and complex. Designs mimicking human house design or other structures are common in some gardens.

The size of a nest box should be suited for the guest inside it. Quite large boxes are used for ducks and owls. Nest boxes for wild birds should be cleaned annually to remove old nests and kill parasites. The nest boxes of domestic birds and other animals should be removed and cleaned after the breeding season is over and no more young are expected. Budgerigars and other birds raised domestically will not usually mate unless a nest box is present. Bats houses are becoming popular sources of natural mosquito and insect control. A single bat can eat 500 to 1,000 mosquitoes a night as well as numerous garden pests. Directions for making the open bottom bat houses for small and large colonies, as well as locations to purchase them are available on the internet. Bat houses are an ecologically friendly way of controlling mosquitoes as pesticides mostly kill the mosquito's predators rather than mosquitoes. Nest boxes are marketed not only for birds but also for bats, some other mammals and bees. Depending on the animal, these boxes are used for roosting, breeding, or both. Most small mammals will use a nest box for raising young if one is provided.

Kennel

A kennel is the name given to any structure or shelter for dogs. A *kennel* is a doghouse, run, or other small structure in which a dog is kept.

Fig. Kennel for Dog

American English often uses the word "doghouse" in this context; in British English the term "kennel" is preferred, but the phrase "in the doghouse" is understood but does not mean "in the kennel", rather it is a colloquial phrase meaning (of a person) "in trouble" or "being shunned".

Used in the plural, *the kennels*, the term means any building, collection of buildings or a property in which dogs are housed, maintained, and bred.

Stable

A stable is a building in which livestock, especially horses, are kept. It most commonly means a building that is divided into separate stalls for individual animals. There are many different types of stables in use today such as the American barn which is a large barn with a door each end and individual stalls inside or free standing stables with the classic top and bottom opening doors.

Historically, the headquarters of a unit of cavalry rather than just their horses' accommodation, would be called a stable.Used metaphorically from this origin, a stable is a collection of people (e.g. professional wrestlers) working under a single manager or animals kept by one owner.

Yarding

While often confused with free range farming, yarding is actually a separate method by which a hutch and fenced off area outside are combined when farming poultry. The distinction is that free-range poultry are either totally unfenced, or the fence is so distant that it has little influence on their freedom of movement. This is common technique used by small farms in the Northeastern portion of the US. Daily releases out of hutches or coops allows for instinctuial nature for the chickens with protections from predators. The hens usually lay eggs either on the ground of the coop or in baskets if provided by the farmer.

This technique can be complicated if used with roosters though, mostly because of difficulty getting them into the coop and to clean the coop while it is inside. This territorial nature is apparent while outside in which they have a brood of hens and sometimes even informal land claims. This can endanger people unaware of the existence of the territories who are attacked by the larger birds.

Housing goats

Housing for dairy goats does not have to be elaborate, but it must satisfy the health and comfort of the animals. There are five requirements of good housing for goats:

- The building should be adequately ventilated but not drafty;
- The walls and ceiling should be free from condensation;
- The bedded area should be relatively dry and clean;
- The hay, grain and water receptacles must be well built and located so that feed is not wasted or contaminated; and

- The facilities should provide easy access to the animals and require a minimum amount of labour.

The comfort zone for dairy goats is between 55 and 70 degrees Fahrenheit. Non-sweating animals are much less sensitive to declining temperatures than to rising temperatures. Milk production, feed consumption and comfort are not affected by temperatures between 0 and 55 degrees Fahrenheit, but temperatures over 80 degrees Fahrenheit seriously reduce feed intake and milk output. Therefore, the object is not how to keep the goats warm in winter, but how to keep them cool in summer.

The movement of air, either by mechanical or natural means, to remove heat, moisture and odors is a necessary part of a housing plan. Most pneumonia problems with dairy goats can be traced to inadequate ventilation. Wet walls and ceilings are the result of improper ventilation, poor insulation or a combination of the two. The rate of air movement is influenced by the amount of animal heat produced and the temperature you wish to maintain in the building. Additional heat and insulation may be required to keep the stable air fresh and to prevent water pipes from freezing in the winter. An air inlet system also must be provided for good air distribution. Ventilation entails more than installing a fan to move some air.

Proper ventilation during the summer may require moving 150 to 200 cubic feet of air per minute per animal. Winter weather may reduce the amount of air to be exhausted to as little as 20 cubic feet per goat. Contact your local MU Extension specialist for assistance in planning the right ventilation system.

With proper ventilation, you are protecting the animals, the building and the quality of milk produced. Windows are essential in a closed barn. They permit sunlight for warmth and drying and provide a source of vitamin D for the animals. Well-lighted barns usually are kept cleaner. In summer, open windows are important for air movement.

Housing for Young Stock and the Milking Herd

There are two housing systems often used for goats:

- Loose housing, where the animals run loose in a pen or shed
- Stall barns, where each animal is confined in a small box stall or tie stall

Loose housing is an old system with many desirable features. With plenty of bedding, the manure pack for a goat herd can be kept fairly clean and dry. Heat produced within the pack makes a warm bed. There is ample exercise area and an opportunity for the goats to move around.

This type of housing, however, presents a few problems. A built-up manure pack, even though dry on the surface, releases a lot of moisture into the barn air. In addition, hay racks should be located off the bedded area to reduce parasite infestations. Watering devices should be placed in an area where spilled water will not mess up the bedded pack. Many goat owners use a shed or barn that is open on the south or southeast side, thus eliminating the need for mechanical ventilation.

These are cold housing facilities that do not require additional heat or insulation and use normal air movement for ventilation. Goats move freely in or out of the housing area and into the paddock or feeding area. Hay feeders, watering devices, mineral feeders and grain bunks are located on concrete pads some distance from the built-up manure pack.

Housing for the kids and other young stock may be included in the plan for the milking herd. However, young goats must be kept separate from the milking herd. Frequently, the kids are kept in 4-foot-square box stalls with at least one side slatted to permit air movement. Do not make a box stall from plywood with all four walls solid; air movement will not reach the kids.

Usually a gated front contains enough openings to allow air drainage. Provide a heat lamp for newborn kids and for kids that are ill. Older kids may be kept in a corner of the stable area inside a pen made from movable hurdles. Buck housing must be separate and downwind from the milking herd. It does not have to be elaborate and often is no more than a 5- or 6-foot shed with an open side facing the south to give the bucks free access to an exercise lot.

The milking area is part of the housing plan, but it should be separated from the stable area. It should have a concrete floor to make cleaning easier, and the milking platform should be 15 to 18 inches higher than the floor to permit easier milking. The platform should allow 18 inches in width and 3-1/2 feet in length for each animal to be tied. The manger should be 6 inches deep and 1 foot wide. If several goats are to be fastened and milked at once, a lever that will open and close the stanchion head locks is helpful.

Some goat owners who keep their milking herds in stall barns milk their animals on a milking stand located in one corner of the stable. Dusty air and flies in the stable, however, may require you to place the milking stand in a separate screened-in facility to guarantee clean milk. A 5-by-8-foot room with a concrete floor and drain is adequate for milking a small herd, but be sure there is plenty of light either from natural sources or electric bulbs.

If the milk is sold to the public or to a processor, there will be state inspection of the operation just as is required for production of Grade A milk from cows. Check with your local health authorities and the state inspector before starting construction.If the milk is to be used by the family, the degree of sanitation and cleanliness maintained and the size and type of building constructed can be based on the owner's good sense, pride, conscience and finances. Equip the milk house with a double sink for washing utensils, a hot water heater, refrigeration, a small table or work space, and a rack for drying and storing utensils. At least 50 percent of the milk house should be open floor space with ample room for equipment.

Regardless of the herd size, refrigeration must be available to quickly cool the milk. A household refrigerator may be used, but cold water is more efficient in cooling milk than cold air. The refrigerator or cooler should be large enough to hold a pan of water. The pan should hold enough water to match the amount of milk in the milk bucket. If the herd is big enough and milk production is 5, 10 or more gallons per day, it will be necessary to purchase a water-immersion cooler or a bulk tank for cooling purposes.

Cooling is critical to milk flavor and quality. All milk contains bacteria, some of which comes from the air and the utensils. If milk remains warm for a short period, the bacteria begin to multiply and the quality of milk deteriorates. Therefore, cool milk immediately after milking to a temperature under 40 degrees Fahrenheit and hold it at that temperature until processed and/or consumed. Plans for building a milking barn and milking stand and for construction of the milk house area are available from your local MU Extension dairy specialist.

Fences serve two purposes that make them an important part of a housing plan — they keep goats in and dogs out. They also keep foraging goats away from your trees and shrubs as well as your neighbour's flower patch. Set 7-foot posts, either wooden or steel, on 12-foot centres.

Then use either a 4-foot-high woven wire fence topped with an electric wire 12 inches above the woven wire or use a completely electric fence. If you use a completely electric fence, place the first wire 12 inches from the ground, then four more wires on 6-inch spacings to a total height of 36 inches. Use insulators on the post to attach the wires. The electric fence works well when weeds, grass and brush are not allowed to touch the wires and short it out or when the ground isn't bone dry during periods of drought. Your local farm supply store can recommend the proper equipment.

In summary, keep the plans as simple and economical as possible. Consider the health of your animals as well as the conveniences a housing facility might offer. Contact your local MU Extension specialist for additional ideas on housing facilities for your dairy goat enterprise.

BEEKEEPING

FIXED FRAME HIVES

There are considerable regional variations in the type of hive in which bees are kept. A hive is a set of rectangular wooden boxes filled with moveable wood or plastic frames, each of which holds a sheet of wax or plastic foundation. The bees build cells upon the sheets of foundation to create complete honeycombs. Foundation comes in two cell-sizes: worker foundation - which enables the bees to create small, hexagonal worker cells and 'drone foundation' which allows the bees to build much larger cells - drone cells, for the production of male bees.

The bottom box, or brood chamber, contains the queen and most of the bees; the upper boxes, or supers, contain just honey. Only the young nurse bees can produce wax flakes which they secrete from between their abdominal plates; they build honeycomb using the artificial wax foundation as a starting point, after which they may raise brood or deposit honey and pollen in the cells of the comb. These frames can be freely manipulated and honey supers with frames full of honey can be taken and extracted for their honey crop.

MODERN BEEKEEPING MOVABLE FRAME HIVES

In the USA, the Langstroth hive is commonly used. The Langstroth was the first successful top-opened hive with movable frames, and other designs of hive have been based on it. Langstroth hive was however a descendant of Jan Dzierzon's Polish hive designs. In the United Kingdom, the most common type of hive is the British National Hive, but it is not unusual to see some other sorts of hive.

Straw skeps, bee gums, and unframed box hives are now unlawful in most US states, as the comb and brood cannot be inspected for diseases. However, straw skeps are still used for collecting swarms by hobbyists in the UK, before moving them into standard hives.

Top Bar Hives

A few hobby beekeepers are adopting various top bar hives of the type commonly found in Africa. These have no frames and the honey filled comb is not returned to the hive after extraction, as it is in the Langstroth hive. Because of this, the production of honey in a top bar hive is only about 20 per cent that of a Langstroth hive, but the initial costs and equipment requirements are far lower.

Top-bar hives also offer some advantages in interacting with the bees and the amount of weight that must be lifted is greatly reduced. Top Bar Hives are being widely used in developing countries in Africa and Asia as a result of 'Bees For Development' programme.

Protective Clothing

Beekeepers often wear protective clothing to protect themselves from stings. While knowledge of the bees is the first line of defence, most beekeepers also wear some protective clothing. Novice beekeepers usually wear gloves and a hooded suit or hat and veil. Experienced beekeepers sometimes elect not to use gloves because they inhibit delicate manipulations. The face and neck are the most important areas to protect, so most beekeepers will at least wear a veil.

Defensive bees are attracted to the breath, and a sting on the face can lead to much more pain and swelling than a sting elsewhere, while a sting on a bare hand can usually be quickly removed by fingernail scrape to reduce the amount of venom injected.

The protective clothing is generally light coloured and of a smooth material. This provides the maximum differentiation from the colony's natural predators (bears, skunks, etc.) which tend to be dark-coloured and furry.

AQUACULTURE

Aquaculture is the cultivation of aquatic organisms. Unlike fishing, aquaculture, also known as aquafarming, implies the cultivation of aquatic populations under controlled conditions. Mariculture refers to aquaculture practiced in marine environments. Particular kinds of aquaculture include algaculture (the production of kelp/seaweed and other algae); fish farming; shrimp farming, shellfish farming, and the growing of cultured pearls.

Fig. Aquaculture

Aquaculture has been used in China since circa 2500 BC. When the waters lowered after river floods, some fishes, namely carp, were held in artificial lakes. Their brood were later fed using nymphs and silkworm feces, while the fish themselves were eaten as a source of protein. By a fortunate genetic mutation, this early domestication of carp lead to the development of goldfish in the Tang Dynasty.

The Hawaiian people practiced aquaculture by constructing fish ponds. A remarkable example from ancient Hawaii is the construction of a fish pond, dating from at least 1,000 years ago, at Alekoko. According to legend, it was constructed by the mythical Menehune. The Japanese practiced cultivation of seaweed by providing bamboo poles and, later, nets and oyster shells to serve as anchoring surfaces for spores. The Romans often bred fish in ponds.

The practice of aquaculture gained prevalence in Europe during the Middle Ages, since fish were scarce and thus expensive. However, improvements in transportation during the 19th century made fish easily available and inexpensive, even in inland areas, causing a decline in the practice. The first North American fish hatchery was constructed on Dildo Island, Newfoundland

Canada in 1889, it was the largest and most advanced in the world. Americans were rarely involved in aquaculture until the late 20th century, but California residents harvested wild kelp and made legal efforts to manage the supply starting circa 1900, later even producing it as a wartime resource.

In contrast to agriculture, the rise of aquaculture is a contemporary phenomenon. According to professor Carlos M. Duarte About 430 (97 per cent) of the aquatic species presently in culture have been domesticated since the start of the 20th century, and an estimated 106 aquatic species have been domesticated over the past decade. The domestication of an aquatic species typically involves about a decade of scientific research.

Current success in the domestication of aquatic species results from the 20th century rise of knowledge on the basic biology of aquatic species and the lessons learned from past success and failure. The stagnation in the world's fisheries and overexploitation of 20 to 30 per cent of marine fish species have provided additional impetus to domesticate marine species, just as overexploitation of land animals provided the impetus for the early domestication of land species

In the 1960s, the price of fish began to climb, as wild fish capture rates peaked and the human population continued to rise. Today, commercial aquaculture exists on an unprecedented, huge scale. In the 1980s, open-netcage salmon farming also expanded; this particular type of aquaculture technology remains a minor part of the production of farmed finfish worldwide, but possible negative impacts on wild stocks, which have come into question since the late 1990s, have caused it to become a major cause of controversy.

TYPES OF AQUACULTURE

Algaculture

Algaculture is a form of aquaculture involving the farming of species of algae. The majority of algae that are intentionally cultivated fall into the category of microalgae, also referred to as phytoplankton, microphytes, or planktonic algae. Macroalgae, commonly know as seaweed, also have many commercial and industrial uses, but due to their size and the specific requirements of the environment in which they need to grow, they do not lend themselves as readily to cultivation on a large scale as microalgae and are most often harvested wild from the ocean.

Fish Farming

Fish farming is the principal form of aquaculture, while other methods may fall under mariculture. It involves raising fish commercially in tanks or enclosures, usually for food. A facility that releases juvenile fish into the wild for recreational fishing or to supplement a species' natural numbers is generally referred to as a fish hatchery. Fish species raised by fish farms include salmon, catfish, tilapia, cod, carp, trout and others. Increasing demands on wild fisheries by commercial fishing operations have caused widespread overfishing. Fish farming offers an alternative solution to the increasing market demand for fish and fish protein.

Freshwater Prawn Farming

A freshwater prawn farm is an aquaculture business designed to raise and produce freshwater prawn or shrimp for human consumption. Freshwater prawn farming shares many characteristics with, and many of the same problems as, marine shrimp farming.

Unique problems are introduced by the developmental life cycle of the main species. The global annual production of freshwater prawns (excluding crayfish and crabs) in 2003 was about 280,000 tons, of which China produced some 180,000 tons, followed by India and Thailand with some 35,000 tons each. Additionally, China produced about 370,000 tons of Chinese river crab (*Eriocheir sinensis*).

Integrated Multi-trophic Aquaculture

Integrated Multi-Trophic Aquaculture (IMTA) is a practice in which the by-products (wastes) from one species are recycled to become inputs (fertilizers, food) for another. Fed aquaculture is combined with inorganic extractive (e.g. seaweed) and organic extractive (e.g. shellfish) aquaculture to create balanced systems for environmental sustainability (biomitigation), economic stability (product diversification and risk reduction) and social acceptability (better management practices).

"Multi-Trophic" refers to the incorporation of species from different trophic or nutritional levels in the same system. This is one potential distinction from the age-old practice of aquatic polyculture, which could simply be the co-culture of different fish species from the same trophic level. In this case, these organisms may all share the same biological and chemical processes, with few synergistic benefits, which could potentially lead to significant shifts in the ecosystem.

Some traditional polyculture systems may, in fact, incorporate a greater diversity of species, occupying several niches, as extensive cultures (low intensity, low management) within the same pond. The "Integrated" in IMTA refers to the more intensive cultivation of the different species in proximity of each other, connected by nutrient and energy transfer through water, but not necessarily right at the same location.

Ideally, the biological and chemical processes in an IMTA system should balance. This is achieved through the appropriate selection and proportions of different species providing different ecosystem functions.

The co-cultured species should be more than just biofilters; they should also be harvestable crops of commercial value. A working IMTA system should result in greater production for the overall system, based on mutual benefits to the co-cultured species and improved ecosystem health, even if the individual production of some of the species is lower compared to what could be reached in monoculture practices over a short term period.

Sometimes the more general term "Integrated Aquaculture" is used to describe the integration of monocultures through water transfer between organisms. For all intents and purposes however, the terms "IMTA" and "integrated aquaculture" differ primarily in their degree of descriptiveness. These terms are sometimes interchanged. Aquaponics, fractionated aquaculture, IAAS (integrated agriculture-aquaculture systems), IPUAS (integrated peri-urban-aquaculture systems), and IFAS (integrated fisheries-aquaculture systems) may also be considered variations of the IMTA concept.

Mariculture

Mariculture is a specialized branch of aquaculture involving the cultivation of marine organisms for food and other products in the open ocean, an enclosed section of the ocean, or in tanks, ponds or raceways which are filled with seawater. An example of the latter is the farming of marine fish, prawns, or oysters in saltwater ponds. Non-food products produced by mariculture include: fish meal, nutrient agar, jewelries (e.g. cultured pearls), and cosmetics.

Shrimp Farming

A shrimp farm is an aquaculture business for the cultivation of marine shrimp for human consumption. Commercial shrimp farming began in the 1970s, and production grew steeply, particularly to match the market demands of the U.S., Japan and Western Europe. The total global production of farmed shrimp reached more than 1.6 million tonnes in 2003, representing a value of nearly 9,000 million U.S. dollars. About 75 per cent of farmed shrimp is produced in Asia, in particular in China and Thailand. The other 25 per cent is produced mainly in Latin America, where Brazil is the largest producer. The largest exporting nation is Thailand.

Shrimp farming has changed from traditional, small-scale businesses in Southeast Asia into a global industry. Technological advances have led to growing shrimp at ever higher densities, and broodstock is shipped world-wide. Virtually all farmed shrimp are penaeids (i.e., shrimp of the family *Penaeidae*), and just two species of shrimp—the *Penaeus vannamei* (Pacific white shrimp) and the *Penaeus monodon* (giant tiger prawn)—account for roughly 80 per cent of all farmed shrimp.

These industrial monocultures are very susceptible to diseases, which have caused several regional wipe-outs of farm shrimp populations. Increasing ecological problems, repeated disease outbreaks, and pressure and criticism from both NGOs and consumer countries led to changes in the industry in the late 1990s and generally stronger regulation by governments. In 1999, a programme aimed at developing and promoting more sustainable farming practices was initiated, including governmental bodies, industry representatives, and environmental organizations.

SAFETY AND ANIMAL HANDLING

Stockmen working in animal housing and associated handling facilities are exposed to a number of potential risks and health hazards. Many stockmen work alone, handling big, sometimes dangerous animals as well as large groups of animals.

Several work tasks involve daily lifting, carrying of heavy loads and awkward working postures, which may cause overload injuries on the worker's musculo-skeletal system. Other potential risk factors contributing to accidents as well as to physical health problems and physiological stress for the worker is the fact that they operate in an environment where they are exposed to airborne particles, gases, noise.

The work environment is partly the same environment, which is provided for the animals, and it should be noted that there is a strong relationship between the animal welfare issues and the handling of the work environment problems in cattle houses.

HEALTH AND INJURIES AMONG LIVESTOCK FARMERS

Accident Agents

In an accident situation there is a complex interaction between man and injury agents. In animal husbandry the human is subjected to a high degree of exposure to a large number of factors that singly or in combination can be accident provoking, such as animals, equipment and building structures. Risky situations are often due to wet and slippery floors as well as systems for material and animal handling.

Prevention of Accidents

With reference to accident statistics some accident preventing facilities and measures can be listed:

- Good cattle handling and handling facilities
- Pens from which the herdsman can easily escape if necessary in case of for instance frightened cattle and aggressive bulls
- Personnel passages through yard fences, properly placed
- Wide feeding passage or possibility to shut out the cattle from the feeding area

- Non slip flooring
- Satisfactory lighting considering quality and quantity
- Well designed ladders and stairways
- Protective railings around the fodder hay-chute in two storey houses
- Handles and bolts that do not stand out in alleys and pathways
- Protectors for the knuckles on barrows
- Emergency stops on machinery

Table. Recommended Illumination Level for Different Work Areas in Buildings Used for Beef Production. Lux Values Refer to the Work Plane. In Addition to the Quantity, the Quality of Light must be Considered

Building or Work Area/Task	Intensity of Illumination, Lux
Animal housing area	150
Animal handling area	250
Veterinary treatment	1,000
Feed processing, feed alley	250
Silo, feed storage	100
Ladders and stairs	250

Ergonomics

Any normally fit adult person, regardless of sex, physical strength and body dimensions should be able to work comfortably in livestock buildings. Despite an increasing degree of mechanisation and automation many jobs in animal husbandry are still associated with lifting heavy loads, moving and carrying equipment, feed and other materials with manual methods. An important step towards improved human welfare within cattle housing is to eliminate unnecessary loads on the human musculoskeletal system.

Health Aspects Related to Air Quality

It has been known for decades that farm workers suffer health risks due to inhalation of gases and airborne particles commonly referred to as dust. The particles may be liquid droplets or solid. Several respiratory disorders have been reported, which are associated with dust and work in confined livestock buildings such as hypersensitivity, pneumonia, acute inflammation, chronic bronchitis, occupational asthma and toxin fever. The respiratory problems connected to gases range from mild irritations of the respiratory tract to lethal effects. For the most frequent components in the air environment in livestock buildings there are statutory threshold limit values, which differ from country to country. It should be noted that threshold limit values in connection with animal welfare in many countries often are lower than the work environment legislation.

Dust and Dust Reducing

Dust in animal houses is generated inside the building and arises primarily from the animals and bedding and feed material. The airborne particles are often carriers of biologically active material like endotoxins, pathogens and allergens that can highly be disease provoking. Of special interest is the concentration (*i.e.* number of particles per air volume) of respirable particles with a diameter less than 5 μm that can be inhaled and deposited in the lung.

The smaller particle the higher possibility that the particle reaches the lower respiratory tract and the deeper the particle is deposited the higher is the risk for illness. Even if "common dust" can be seen by human eyes (larger diameter than 100 ìm), the very small respirable particles cannot be, so we are indeed dealing with a danger our senses are not able to detect.

There are basically four approaches that can be used for airborne dust reducing:

1. Prevent particle formation
2. Prevent particle release
3. Remove suspended particles from enclosed workspaces
4. Isolate workers from dust clouds in workspace.

Of course the best approach in this matter is to have the cattle and the working sites outdoors or in open barns. In the cattle barn the most effective way to obtain a low dust concentration level is to prevent formation by always using hygienically perfect forage and bedding material. Providing an effective ventilation system in a barn is a good practice to remove suspended respirable particles as well as gases.

Gases and Reduction of Gas Levels

In animal buildings over 150 different gases and volatile compounds have been identified. Most of the volatile compounds originate from biodegradation of animal excreta or are produced by the animals themselves such as carbon dioxide (by the respiration of the animal). The gases that are found in the highest elevated concen-trations are ammonia, carbon dioxide, hydrogen sulphide and methane.

It must be emphasised that it is dangerous to enter any manure tank without either using a self-contained air supply or combining several measures: testing the air, constant and adequate ventilation of fresh air and using an harness and lifeline on the person entering the tank. As well, human and animals positioned over a slatted floor with manure storage underneath are at risk during agitation, mixing or pumping of manure due to the gases, especially hydrogen sulphide. A good strategy is to evacuate dangerous zones during these manure-handling actions.

Ammonia

Ammonia is the most common polluting gas in the atmosphere of the cattle barn. Animal manure is the main source of ammonia. The gas together with hydrogen sulphide a major component in what is termed odour, noticeable at ammonia concentrations of 5 ppm or more. Higher concentrations than that cause irritation to the respiratory organs and can aggravate the negative health effects of high dust concentrations.

Many factors affect the emission and concentration of ammonia such as air flow rate, manure and air temperature, manure surface area, density of animals, the degree of mixing of urine and faeces, time intervals between manuring, the pH-value, carbon/nitrogen ratio, mois-ture content and type of bedding. Consequently, many measures can be taken such as: proper ventilation and sufficient number of air changes, low air temperature in the barn and good urine drainage characteristics of the floor.

Hydrogen Sulphide

Hydrogen sulphide is the most toxic gas in animal confinement houses and 200 ppm and more are lethal for humans. The presence of the gas is of concern in buildings with liquid manure systems particularly. During slurry clearing operations concentrations in the range of 1,500 to 2,000 ppm have been reported. Since hydrogen sulphide is lethal the aim should be to keep it below detectable concentration at all working sites.

Special care has to be taken when pumping or agitating the manure. During emptying a slurry cellar the workers and the animal should be evacuated from the barn. In mechanical ventilated, closed animal houses continuously evacuation of the gases in an indoor slurry pit through a perforated duct under the floor and along the pit can reduce the gas concentrations above the slatted floor. By removing the slurry daily from the cattle barn to an outdoor storage it is possible to hold the concentration at an acceptable level.

Methane

Methane is generated when manure is stored under anaerobic conditions. This occurs in all non-aerated manure storages. Primarily, methane is dangerous because of its flammability. If methane is mixed in a proper proportion with oxygen a spark will be able to set off an explosion.

Carbon Dioxide

Carbon dioxide is present in all air, indoor as well as outdoors. Only extremely high concentrations (70,000 ppm or 7% and over) could have severe health effects. Mostly, incidents and accidents involving carbon dioxide are connected with asphyxiation due to oxygen deficiency rather than the direct effect of carbon dioxide itself. For instance, it is dangerous to enter a haylage silo during storing. In animal buildings the gas is more an indicator of ventilation efficiency than a health risk.

Noise

Noise can be defined as "unwanted sound, rapid, annoying pressure vibration in the surrounding air". Ventilation fans, grinders, vacuum pumps and mechanical feeding and manure systems generate noise. High sound levels are a real stress factor. Exposure to noise will be able to cause temporary deafness and permanent hearing loss after a variable period if the level is over 85 dB(A). The unit, dB(A), is exponential and the human sense of hearing will perceive the sound level as halving if a reduction of 10 dB(A) occurs.

It is fairly easy to reduce the sound level of ventilation fans and other mechanical noise generators. Maximum noise level recommendations and regulations differ from country to country. Furthermore, it is important to reduce noise when handling cattle in order to improve animal movement and to make cattle handling safer. Cattle are more sensitive to high frequency noise than human. Animals will be calmer and easier to handle, if noise level is reduced. Clanging and banging metal parts should be silenced with rubber pads. Equipment operated with hydraulics should be engineered for quietness.

CATTLE HANDLING AND CATTLE HANDLING FACILITIES

Handling facilities are an essential part of a safe, easy and rapid handling of cattle. Appropriate handling and handling facilities remove much of the stress and frustration of the workmen, which inevitably occurs with excited, stubborn or aggressive animals. Properly constructed facilities confine cattle safely and efficiently with minimal animal stress and risk of injury to both cattle and workers.

Animal Behaviour and Improving Animal Handling

Understanding cattle behaviour can help farm workers to avoid dangerous situations and minimise accidents to handlers, as well as to design the handling facilities appropriately. Animals have natural boundary called the Flight Zone.

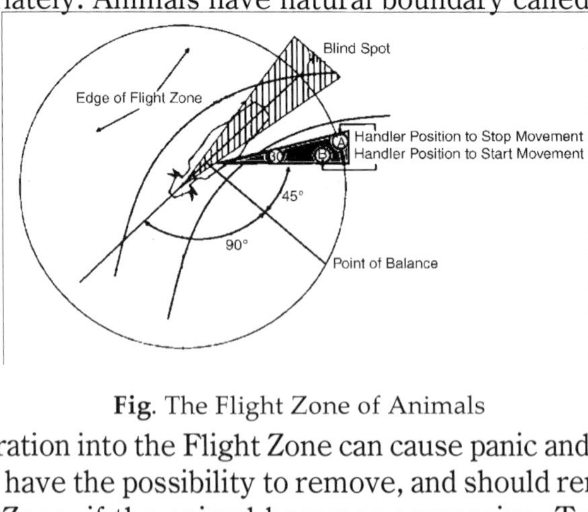

Fig. The Flight Zone of Animals

Deep penetration into the Flight Zone can cause panic and escape attempts. Handlers should have the possibility to remove, and should remove themselves from the Flight Zone, if the animal becomes aggressive. To move an animal forward: stand in the shaded area position B, behind the Point of Balance at the shoulder. Keep out of the Blind Spot at the rear of the animal. To stop movement: back off to position A. To make an animal back up: stand in front of the Point of Balance. To make an animal turn left or right: approach the animal head on. Factors reducing an animal's Flight Zone size: frequent contact with people; history of gentle handling; calm environment. Factors enlarging the Flight zone: infrequent contact with people; history of rough or abusive handling; excitement.

To improve animal handling and animal movement, many technical measures can be taken such as:

- Appropriate handling facility design.
- Animals tend to move from a dark to a more brightly lighted area. The light should illuminate the chute up ahead. Eliminate shadows and patches of light and dark, which may confuse animals. An approach is to illuminate the entire working area. Lamps should not shine into the eyes of approaching animals because glaring and blinding light impedes movement. Illumination should be uniform and diffuse.
- Prevent distractions, such as a chain hanging down in an entrance. Avoid sparkling reflection in a puddle, a moving reflection on a sheet of metal or bars of shadow across an otherwise sunlit alleyway. Dark colours can create shadow effects. Bright colours such as white and light yellow have been proven satisfactory.
- Reduce noise
- Animals might refuse to move, if they can see people ahead. Install shields to prevent animals from seeing farther ahead. Gates can be rigged with motor controls so a handler standing behind the cattle can open them.
- Solid sides that prevent the cattle from seeing outside the fences should be provided in the races and crowding pens. The crowding gate on the crowding pen should be solid as well, preventing animals from attempting to turn back to were they come from.
- Cattle are sensitive to changes in type and texture of floors and fences. Changes in type of flooring can cause balking. Use the same type of flooring throughout a facility, if possible. Use non-slip flooring. Drains should be located outside main drive alleys, chutes and crowd pens.

3

Farm and Farming Systems

FARMING SYSTEMS RESEARCH

Because FSD has its roots in farming systems research, the following comments on approaches within FSR are applicable to the technology research activities of FSD. The term FSD often is used loosely. An activity is considered to be a legitimate part of the technology research function of FSD if the following situations hold:

- The whole farm is viewed as a system.
- Research is conducted with a recognition and emphasis on the choice of priorities that reflect the whole farm.
- Research on a farm sub-system is legitimate FSD, provided the connections with other sub-systems are recognized and taken into account.
- Evaluation of research results explicitly takes into account linkages between sub-systems.
- As long as the concept of the whole farm and its environment is preserved, not all the factors determining the farming system need to be considered as variables — some may be treated as parameters or constants.

A distinctive feature of FSR, or rather FSD, is that a systems perspective is constantly borne in mind. Within this framework, one can look at either a small or a large number of variables in the system. In general, because of the complexities of simultaneously handling a large number of variables, most FSR programmes have tended to limit the number of variables they study and to regard the other factors that influence the farming system as parameters or constants.

Three general approaches to FSR or FSD with a technology generation focus, have been defined relating to the number of variables being investigated. These approaches can be described as FSD 'with a pre-determined focus', 'in the small', and 'in the large', In a sense, these types represent an evolution in FSD as techniques have been developed to handle progressively more complex situations— defined as involving a higher ratio of variables to parameters, The progression, which is illustrated, where it has been expanded to four phases, can be articulated as follows:

- In its early days, FSR (*i.e.*, FSD) type activities concentrated on how farm yields of particular crops could be raised. This form of on-farm research stood in sharp contrast to earlier multi-locational on-farm trials, in which only the effects of the physical (*i.e.*, natural or technical) environment were tested, Even though this FSD approach did involve the inclusion of socio-economic elements and, hence, had a farming systems perspective, it was done with a 'predetermined focus' on the productivity of a particular commodity. Thus, this approach involves looking at one facet of an enterprise or one specific enterprise and identifying improvements within that focus that are compatible with the whole farming system. With the creation of distinct FSD teams — in contrast to an on-farm testing component located within each station based commodity team — FSD activities generally have evolved beyond this level. However, it is important to recognise that this type of activity still is a legitimate part of any FSD in the sense that it can provide a very useful input into the activities of commodity research teams.
- In the pursuit of greater farmer participation, research shifted to addressing farmers' articulated problems directly. Still, farmers' problems were very much con fined to crop production, although links with other components of the whole farm system were studied. An alternative name for this 'farming systems in the small' focus could be 'farming systems with a whole farm problem focus', Many FSD efforts currently tend to be at this point, However, currently, on-farm research projects that seek to improve whole farm performance through manipulation of the linkages between all enterprises are :
  - — Looking further ahead and to enable addressing issues of sustainable agriculture effectively, concerns for maximizing the economic and biological performance of enterprises will be modified by concerns relating to ensuring sustainability, including the rehabilitation and regeneration of natural resource systems. This natural resource system dimension to sustainable agriculture can be thought of as a type of 'farming systems in the large, or 'farming systems with a natural resource systems focus',
  - — Similarly, attention to ecological sustainability will need expanding to include the many off-farm and non-farm activities that make up sustainable livelihoods. This also can be thought of as a type of 'farming systems in the large, or 'farming systems with a livelihood systems focus'.

These notions, particularly the last two, have been largely unexplored until recently, when techniques have evolved to handle complex situations in which the ratio of variables to parameters is high. Formal modelling techniques by themselves generally have limited value in addressing such situations because of the complexity of the relationships and the degree of understanding that is required initially to develop a realistic model. However, the major breakthrough that has occurred is the ability to pursue a less formal or more informal modelling approach through application of farmer participatory type techniques that, in essence, involve using farmers' minds as computers!

Introduction to FSD

Details concerning the characteristics and methodology of the FSD approach, as developed by the Farm Management and Production Economics Service of FAO, are available elsewhere. Thus, the following discussion represents only a brief summary. The primary objective of FSD is to improve the well-being of individual farming families by increasing the overall productivity of the farming system in the context of both the private and societal goals, given the constraints and potentials imposed by the factors that determine the existing farming system. It is based on the development principles of improving productivity, increasing profitability, ensuring sustainability, and guaranteeing an equitable distribution of the results of production,

The farm household is the principal system and focus of FSD and consists of three basic subsystems, which are closely interlinked and interactive: the household, the farm, and off-farm activities. The two major categories of activities or thrusts of FSD, both of which involve intensive interaction with farmers? are:

- Farming Systems Analysis. This involves studying, together with the farmers the natural (*i.e.*, technical) and socio-economic (*i.e.*, human) environments in which farm households operate. The aim is to identify the constraints limiting farm productivity and production and hindering improvement in the welfare of the farm households themselves. Potential solutions to these problems are identified, and the results of this analysis, formulated as suggestions for further action then are passed on to the relevant 'actors'. These could include researchers, extension and support service staff, and/or policy makers.
- Farming Systems Planning, Monitoring, and Evaluation. These involve testing, monitoring, and evaluating improvements on-farm, with the direct involvement of farmers. Examples of such activities include those regarding proposed technological improvements, proposed revisions in farm plans, and improvements in support services and farm-level impact of proposed policy changes. Those improvements thought to be potentially useful then are disseminated to other farmers

via the extension service. Prior to this stage, it may be necessary to negotiate adjustments in the policy/support services that will facilitate the dissemination and adoption of the improvement. After dissemination, it is of primary importance to monitor and evaluate the adoption rate of the proposed improvements by the farming community. This provides an indication of the actual value of the improvement being disseminated.

- Many individuals in the last 15 to 20 years have written extensively to clarity the concepts of the farming systems, particularly with respect to the development of relevant improved technologies. There are many ways of presenting a diagram involving the application of farming systems methodology to research (*i.e.*, technology development). One such way of emphasizing the technical research aspects of FSD. There are four fundamental stages in the FSD approach.
- The Descriptive or Diagnostic Stage, in which the actual farming system is examined in the context of the 'total, environment — to identify constraints farmers face and to determine the potential flexibility in the farming system in terms of timing, unused resources, etc. An effort also is made to understand the goals and motivation of farmers that may affect their efforts to improve the farming system, The following sequence of activities is usually undertaken:
  - After deciding on the location of the work, start-up activities include reviewing secondary sources of information, making the necessary contacts, assembling the professional team, and making logistical arrangements,
  - Farming families are classified tentatively into homogeneous groups. The farming families within each group or domain usually practice the same farming system(s), face the same constraints, and have the same potential solution(s) to their problems. This tentative classification can be modified as additional information becomes available.
  - An informal exploratory diagnosis or survey is undertaken to obtain a qualitative understanding of the determinants of the existing farming systems (*i.e.*, both biophysical and socioeconomic). At the same time, efforts are made with farmers to ascertain the constraints, flexibility, and potential opportunities in the farming systems they are using currently.
  - Sometimes a verification survey involving a structured formal survey, suitable for statistical analysis, is undertaken to quantitatively confirm insights obtained in the exploratory diagnosis. It also can be used to provide a database for farm and development planning, policy analysis, monitoring, and evaluation.

- The Design Stage, in which a range of strategies is identified that are thought to be relevant in dealing with the constraints determined in the descriptive or diagnostic stage.
  The usual sequence of activities involves the following:
  - Together with the farmers, the constraints are ranked according to their severity, and potential solutions are identified after determining what flexibility exists in the farming systems currently practiced. Information for designing such strategies (*i.e.*, particularly relating to technological issues) comes from experiment station work, researcher managed and researcher implemented (RMRI) type work on farmers' fields, and from other farmers.
  - An evaluation is made of the proposed solutions before putting them into practice. These solutions can be technological or institutional in nature. A number of analytical techniques can be used to evaluate the potential technical feasibility? economic viability, social acceptability, and ecological sustainability of the proposed solutions before they are put into practice. One or more of the proposed solutions are selected for actual evaluation on-farm.
- The Testing and Implementation Stage, in which one or more promising strategies, arising from the design stage, are examined and evaluated under term conditions to determine their suitability for producing desirable and acceptable changes in the existing farming system.
  Activities at this stage involve the following:
  - On-farm testing or evaluation with farmers determining how well the potential improvements fit into the system, whether or not they are acceptable to farming households, and what modification may be needed to make them acceptable. In terms of technological issues, this stage conventionally has consisted of two steps:
  - Researcher managed but farmer implemented (RMFI) tests to establish whether previously determined technical relationships are altered by farmers' management of non-treatment variables.
  - Farmer managed and implemented (FMFI) tests, when the team is confident that technical relationships will hold but needs to evaluate the proposed technologies under local socio-economic circumstances. In addition to the on-farm evaluation of the proposed technologies, evaluations also can be made of the proposed farm plans and of proposed changes in support systems and/or policies.

- Positive results of such evaluations provide justification for the FSD team to advocate further action. The tested technologies can be disseminated through the extension service to other similar farming households, The same applies to the farm plans that have been tested successfully. Favourable test results for proposed changes in support services and/or policies provide valuable farm-level information on necessary programme or policy adjustments.

- The Dissemination and Impact Evaluation Stage, in which the strategies that were identified and screened during the design and testing stages are extended to farmers. In terms of activities at this stage, impact/adoption studies can be very important. These potentially can be very useful, not only in giving some idea of the impact of, for example, agricultural research, but also in giving some idea of future priorities for agricultural researchers and indicating what adjustments are required in the policy/support systems to ensure better rates of adoption. Thus, through monitoring and evaluating the impact and rate of adoption of the changes that have been implemented and proposed/tested earlier, it is potentially possible to provide some indication of further desirable activities by both researchers and planners.

There are often no clear boundaries between the various stages, Design activity, for example, may begin before the descriptive and diagnostic stages and may continue into the testing stage, as promising alternatives emerge from RMFI trials.

Similarly, testing by farmers may mark the beginning of dissemination activities. Also, going through all stages may not always be necessary. FSD team confidence in transferability during design/planning activities can sometimes mean going straight to FMFI work or even to recommendation/ dissemination activities.

Thus, the process of FSD is recognized as being dynamic and iterative, with linkages in both directions between farmers, researchers, extension staff and policy/support service staff, The iterative characteristic can improve the efficiency of the development process by providing a means of identifying and fine tuning improved technologies for a specific location-that is, climatic situation, soil type, and/or farmer resource base.

Fsd with A 'Pre-determined Focus' on Specific Technology

Researchers in station-based programmes often want FSD workers to evaluate specific technologies they have developed to see how they fit in with constraints of the farming system currently used by farmers. This type of work is referred to as FSD with a 'pre-determined focus' or 'predetermined technology focus,' In Botswana, FSD teams and participating farmers have assisted station programmes in testing cowpea varieties, evaluating the desirability of hilling groundnuts, ascertaining whether the Dutch hoe can help with the weeding operation, and so forth.

This type of testing in FSD is not analogous to typical technical transfer. Rather than testing fully finished products, FSD assists technology generating programmes, usually based on research stations. Often, the types of technology tested in this manner have not been identified as having high leverage power in the system. Leverage power refers to the ability of a technology change to have a major impact on the performance of the farming system. However, these technologies do fit possible needs and asking farmers to participate in their development is worthwhile, It is desirable to test technologies as early as possible in their development, This is to facilitate making modifications and to minimize waste of research resources. It also may accelerate the dissemination of suitable technology, Such technology, even if not fully appropriate, may sometimes stimulate thinking among farmers if they are not accustomed to these types of interventions. Feedback in these instances can contribute to new innovations in technology generation.

FSD with a 'Pre-determined Focus' on a Subject

A great deal of FSD work, particularly in its earlier years, was not focused on testing specific technology as much as a pre-determined focus on a subject matter area. FSD within the CGIAR institutions such as IRRI, CIMMYT, and ICRISAT have focused on issues related to production of their respective commodity crops. ILCA focused on issues in livestock production, ICRAF in agroforestry' etc. Some FSD teams within the national agricultural research systems (NARS) also worked under limited mandates to deal with particular enterprises; commodities; or other subject areas. The judgements of farmers, the end-users of products developed by these organizations, were foremost in these FSD activities. However, the parent institution with its technologygeneration activities would be a primary client in these instances.

FSD with a 'Whole-farm Problem Perspective, or 'FSD in the Small'

An FSD interdisciplinary team conducted a diagnostic survey and determined that labour for satisfactory weed control in field crops is a major constraint to farm production and farm family well-being. The survey also showed that due to population pressures, farmers were no longer able to shift field sites as in the past and that deteriorating soil fertility also was becoming a major issue. However, during the course of farmers participating in testing possible solutions to these problems, discussions between the FSD team and farmers identified the potential for intensified poultry production as an additional means of improving farm income. This resulted, in addition, in collaborative testing in the area of intensive poultry production and, as a result, exemplifies FSD with a 'whole-farm problem perspective.'

The major breakthrough that has occurred is through supplementing the rapid rural appraisal (RRA) techniques, which have been available for some time with the participatory rural appraisal (PRA) techniques, which have developed rapidly in recent years. To date, the first two approaches listed above have been used. Although FSD 'in the small' arrives at a focus within the system in the course of diagnosis, FSD 'with. pre determined focus' moves into the system to research an enterprise or one facet of an enterprise looking for improvements within that focus that are compatible with the whole farming system.

SMALL FARMS AND FARMING

The nature, social bases, and socioeconomic consequences of "small" farms and part-time farming are, in a sense, among the most longstanding foci in rural sociological research on agriculture. Nevertheless, sociological analysis of small-scale agriculture and part-time farming received substantially increased attention and exhibited considerable theoretical reformulation in the 1970s and 1980s for several reasons. First, spearheaded by the writings of Hightower and others on the demise of "small" farms, there was a startling renaissance of concern with the status and prospects of small family farms, which had greatly subsided a few decades earlier in the aftermath of the apparent resolution of the depression-induced increase in tenancy.

Second, quite ironically, several observers began to document the fact that the 1970s had witnessed an unprecedented trend toward the stability and persistence of small farms; in most regions the absolute number of farms with fewer than 50 acres has increased over the past one to two decades. Third, it became apparent that much of the reason for the enhanced viability of small farms was the growing access of farm households to off-farm jobs and income; indeed, the U.S. Department of Agriculture began to note that, in most years, the majority of farm households' aggregate income came from nonfarm sources and that there had been a slow but steady increase in the prevalence of off-farm labour market participation among farm families since World War II.

Small and part-time farms are treated here as a particular focal point in the new sociology of agriculture, a tack that has been taken by many others. Whereas there is a key connection between small-scale agriculture and part-time farming — the fact that part-time farming has in recent decades been much more common among "small" than among "medium-size" or "large" farms — the two phenomena, however, are by no means equivalent. In particular, some "small farms"-which provisionally can be referred to here as low-sales-volume farms, with about $50,000 or less in gross annual sales volume-have little or no off-farm income and, accordingly, have long been overrepresented among rural poverty groups in the United States and elsewhere. Moreover, in all advanced capitalist societies, especially the UK, part-time farming occurs across the whole spectrum of sizes of family-proprietor farms; in the UK it has been quite common for the operators of very large farms to have off-farm occupations, especially the operation of small businesses related to agriculture. Thus small-scale and part-time farming, although tending to be significantly related, are by no means coterminous. Further, the analysis of neither small nor part-time farming has been dominated or influenced by neo-Marxist postures nearly as much as other areas in the new sociology of agriculture. In the huge literature on these topics there have been theoretical and empirical neo-Marxist works, but these approaches have been a distinct minority component of the literature.

It is useful to begin with a very generalized model of the recent evolution of U.S. agriculture to situate the theoretical and empirical issues relating to small and part-time farms. Over the past decade or so there has been growing acceptance of the notion that U.S. agriculture has followed a dualistic or bimodal pattern of development since the early 1970s. There is evidence that prior to the late 1960s (and thus preceding the phase of dualistic development) there was a singular pattern of increased concentration and differentiation in U.S. agriculture in which there was a direct association between size of farm and its persistence or viability; thus large farms increased in numbers, acreage, assets, and so on at the expense of small farms, particularly small tenant farms without the means to acquire off-farm income. Farm operators with access to off-farm income were thought to be largely in a transitional status, either into or out of agriculture.

It is now suggested that a more complicated or dualistic pattern of concentration and differentiation has emerged. The stylized facts of this association are, first, that large farms — including large family-type, larger-than-family, and industrial farms in Rodefeld's terms — are, much as in previous decades, exhibiting a demonstrable increase in their shares of aggregate production, sales, assets, profits, use of hired labour, and so on.

Second, however, it is argued that small farms are increasing in numbers and holding their own in terms of acreage and aggregate production. This bolstered status of small farms is considered to be made possible largely by access of their operators and other family members to off-farm wage income. U.S. small farmers have been noted to have, on average, total family incomes in excess of those of medium-sized farmers and the U.S. median family income because of the higher than average educational attainments of members of small farm households and their consequent ability to acquire highly remunerative off-farm work. While outside of the South small farmers have tended to have higher educational levels than larger, full-time operators for several decades, the educational disparities between small, part-time and larger, full-time operators are suggested to have increased. This has often been explained in terms of the 1970s "population turnaround" in which well-educated persons, many of urban origin, were attracted to farm residential situations for lifestyle reasons.

Finally, it is asserted that there has been a "disappearing middle" — that medium-size (traditional family-type) farms have decreased in numbers and in shares of aggregate output, assets, profits, and the like. It has been suggested that these medium-size farms enjoy neither the advantages of bigness or of smallness. In particular, "middle farmers," because they have operations too large to be farmed on a part-time basis, tend to lack the labour resources (and often the educational qualifications) to participate effectively in off-farm labour markets. Thus "the middle" is tending to "disappear," albeit slowly. Approximately 5, 70, and 25 percent of U.S. farms are seen to lie, respectively, in the "large," "small," and "medium" categories.

These stylized facts appear to be largely accurate, although Ehrensaft *et al.*, and Barlett have raised empirical and theoretical objections to the disappearing middle thesis. Salamon and Davis-Brown's findings suggest that the "disappearing middle" may have come more from certain segments of the farm population (the more risk-accepting, entrepreneurial farmers) and less from other segments (the "yeoman" farmers whose key goals are preserving their family farms). Similarly, this conclusion may not hold for black farmers whose limited access to credit for land purchases and mechanization led to rapid declines in the total number of black farmers. Nonetheless, there has been general confirmation of this empirical model for the U.S. farm sector, which provides a preliminary framework for placing the phenomena of small and part-time farming in historical context.

As Coughenour and Wimberley have noted, "most American farm families live on small and part-time farms." Thus in the sense of understanding the composition of the U.S. farm sector, it is essential to account for the status of and trends in small-scale and part-time agriculture. A dramatically different context of research on small farms emerged in the mid- to late 1970s. This context can now be seen as contradictory; at the same time that there had been a rekindling of concern with the status and prospects of the small-scale farm that had been unprecedented since the Great Depression, there was a correspondingly unprecedented increase in the numbers of small farms in most regions of the country.

The changing context of the decline and persistence of smallscale agriculture strongly influenced the conduct of research on small farms. There emerged a number of studies examining the factors leading to small farm decline, persistence, survival, adaptation, and so on. The dominant finding in this research was that small farmers pursue multiple strategies for survival — access to off-farm income; minimizing debt; direct marketing of high-value, specialty crops; and so on — and that the choice of strategies was related to the socioeconomic characteristics of the farm, farm operator, and farm household members.

Recognition of the diversity of small farmers and their survival strategies set forth a related line of inquiry: documenting the internal differentiation among small farmers and generating typol ogies of small farms and small farming households. Major studies in this genre include Buttel and Gertler, Schulman *et al.*, Mufioz, Schroeder *et al.*, and Heffernan *et al.*. These studies have typically found that gross sales (or other indicators of scale), age, education, motivations for farming, household composition, off-farm occupation, and off-farm income are important dimensions of internal stratification among small farmers.

A third cluster of research activities relating to small farms has been to explore small farmers' sociopolitical attitudes. One major area of inquiry has been in documenting the motivations of small scale farm operators and their family members.

Another focus of attitudinal research has been on the political orientations of small farmers. Research to date has generally not reported that small farmers have dramatically different political orientations than large farmers on either general dimensions of political ideology or orientations toward agricultural politics. It has been demonstrated, however, that there are major differences in political orientations among small farmers that are primarily accounted for by off-farm occupation, total family income, and race.

Finally, there has been a nagging concern among rural sociological researchers as to how small farms should be defined. Gross farm sales has for a very long time been the principal scale variable used by the Census of Agriculture in collecting and reporting data, and our macro-sociological knowledge base on trends in the structure of small-scale farming is largely based on gross sales data. Nonetheless, it is widely recognized that gross farm sales have significant limitations as a measure of scale or "smallness."

For example, many economists would prefer value-added over gross sales as a measure of scale. But in the absence of a convenient measure of value-added (or some other concept preferable to gross annual sales) for census and survey research purposes, some arbitrary criterion based on gross sales, sometimes combined with a total family income criterion, has continued to be the measurement convention in small-farm research. Nonetheless, as Carlin and Crecink have shown, alternative definitions of small farms have significant implications for the numbers and characteristics of farmers and farm households included in that category.

Part-time farming research in the United States was initiated during the Great Depression of the 1930s, mainly by agricultural economists and to a lesser extent by rural sociologists, and was part of the concern about "agricultural adjustment" that prevailed in USDA and land-grant university circles. It was observed that many farmers were attempting to survive depression conditions and avoid (or extricate themselves from) tenancy by acquiring off-farm income. As was also to become the case in the 1980s, part-time farming came to be widely advocated as a partial solution to the difficult problems of the agricultural depression of the 1930s.

In the decades up until the early and mid-1970s, however, it was recognized that "agricultural recovery," which typically was conceptualized in terms of increased owner-operation of farms, did not correspond to a decrease in part-time farming. Indeed, coincident with agricultural recovery from the late 1930s to the 1960s was an unprecedented exodus of farmers from the land — a new "agricultural adjustment" problem born of expansion, dynamism, and what latter-day neo-Marxists would refer to as differentiation, rather than depression and decline — that again called for part-time farming as an adjustment mechanism.

Whereas in the 1930s farm size and part-time farming did not appear to be strongly correlated, by the 1950s it was widely recognized that there had emerged a strong inverse association of gross farm sales and various indicators of the involvement of the operator in off-farm work. Among many researchers, especially agricultural economists, the principal research concern, on which there was a decade of conflicting results, was with whether part-time farmers were as efficient as full-time farmers.

Whereas rural sociologists were a minority among researchers on part-time farming during this period, Fuguitt in particular made several important contributions to the literature that continue to be drawn upon today. Fuguitt first demonstrated that part-time farming represented a combination of "push-pull" factors — that the prevalence of part-time farming was positively related to off-farm employment opportunities and inversely related to opportunities in agriculture. Fuguitt then established that there were categories of part-time farmers (based on farm and nonfarm origin and on farmers' aspirations) with distinctly different social characteristics.

Fuller has noted that after 1975 research attention to part-time farming increased dramatically and that several new issues occupied the attention of rural sociologists and kindred social scientists. Definitional matters received major attention after 1975; in particular, there came to be dissatisfaction with conceptualizing part-time farming in terms of the presence or absence, or the number of days, of off-farm work by the operator — the way that off-farm employment data have traditionally been collected and reported in census sources. A consensus began to emerge on conceptualizing part-time farming in terms of the activities of the farming household rather than the farm operator. There also continued to be typological research, following Fuguitt's efforts.

Further, there was a significant amount of research on mobility into and out of part-time farming, on the significance of differences in the off-farm occupations that are held, on the implications of part-time farming for rural communities, and especially on the motivations of part-time farmers. Finally, there has emerged a significant political economy tradition in small and part-time farming research that has explored issues such as the rooting of part-time farming in the concentration and differentiation process, the roles that part-time farming plays in accumulation and legitimation in the agricultural and larger economy, the role of off-farm work in the sexual division of farm household labour, and the "embourgeoisiement" thesis — that part-time farming, by enabling smaller farmers to avoid proletarianization, may blunt the emergence of radicalism by subordinate groups in the agrarian class structure.

In light of excellent review articles on the status of part-time farming research, there is a less pressing need to summarize what has been a voluminous literature. Instead, we will comment on some "metathemes" in this literature that have not yet been the subject of thorough reviews.

First, there is a general tendency for the literature on part-time farming to be polarized into highly structural (if not structuralist in the Althusserian-Marxist sense) and highly voluntarist accounts. Structuralist accounts see part-time farming in almost teleological fashion — that the dynamics of capitalism lead inexorably to agrarian proletarianization in (among others) the form of part-time farming, which provides a mechanism for sustaining profitable capital accumulation in farm, and especially nonfarm, industry by providing a low-wage labour reserve. The subjective motivations and experiences of part-time farming households are thus ignored and viewed to be epiphenomena of larger laws of motion of capitalist development. This approach has historically been prominent in Italian rural sociology circles.

Voluntarist accounts, on the other hand, have largely focused on the motivations of members of part-time farming households — especially their desire to remain in farming and to fulfill lifestyle aspirations. To the degree that nonsubjective factors are considered, they are treated as an "opportunity structure" and not as a social structure that is rooted in the historical and material dynamics of agriculture. Fuller has made a related diagnosis of the state of research on part-time farming — namely that despite widespread social science research on part-time farming, as yet there exists no satisfactory explanatory theory.

Fuller attributes this to the fact that there has been disproportionate attention to the "farming" aspects of part-time farming and too little attention to the family and other nonfarming aspects of rural life. We would suggest that the frontier of research in part-time farming should lie in integrating objective and subjective aspects of these phenomena, which should help to stimulate the theoretical thinking that Fuller has found lacking.

A second metatheme that can be distilled from the part-time farming literature concerns the need for a truly comparative approach to part-time farming research. This observation is, in a sense, ironic, since there has probably been more cross-national comparative observation of part-time farming than in any other subject matter in the sociology of agriculture. But while cross-national variations in part-time farming are widely recognized, these divergent systems of parttime farming have yet to be explained theoretically.

This suggests the need for a comparative, integrative perspective on agricultural and rural development, which, like de Janvry's rendering of Lenin's classical schema of "roads of agricultural development," would examine the interrelations between agriculture and the larger, nonfarm rural economy and society. A major step in this theoretical direction has been taken by Bonanno, though his focus has been more on the persistence and role of small farms than on part-time farming per se. Nonetheless, much work remains to be done to explore part-time farming and other aspects of agrarian structure from the vantage point of the dynamics of agrarian and regional-development dualism.

GENDER AND AGRICULTURE

The role of women in U.S. agriculture was by no means a new area of inquiry when interest in this topic soared in the late 1970s and early 1980s. Wilkening, in particular, had sustained a major research programme on the work roles of farm women and men since the early 1950s. And for several decades there had been a substantial amount of attention paid to rural and agricultural family structures. Nonetheless, rural sociological research on gender and agriculture was to embrace distinctly different themes in the era of the "new sociology of agriculture". Whereas there have been a number of foci in this new tradition of research, the dominant cast of roughly 20 recent studies has been to explore the "invisibility" of women in U.S. agriculture, to paraphrase the title of Sachs' recent book-length work in this area.

Research on gender and agriculture has fallen into four major categories, which basically have represented stages in the development of knowledge on the role of women on American farms. The first category is that of cross-sectional descriptive studies that have sought to document the nature and extent of farm women's involvement in on-farm, household, and off-farm tasks. The second is a set of studies that have attempted to assess changes in women's involvement in on-farm, household, and off-farm tasks by utilizing longitudinal research designs. The third category is that of studies exploring the interrelations between the on-farm, household, and off-farm roles of farm men and women. The final category is that of comprehensive theoretical and empirical approaches that attempt to understand how the institutions of capitalism and patriarchy interact to shape the involvement of men and women in various work roles.

The first descriptive phase of research on gender relations in agriculture was initiated by Joyce and Leadley, who summarized the existing literature on rural women, both farm and nonfarm. Joyce and Leadley lamented the lack of research on rural farm and nonfarm women and set forth an agenda for research. Soon thereafter a number of studies would be conducted in which many of the basic parameters of women's participation in farm, off-farm, and household activities were clarified.

Because census data have included little information on farm women and thus have seriously underestimated the economic contribution of women to farming operations, the knowledge base on the productive activities of farm women has been greatest and most longstanding in terms of off-farm employment. It had been recognized for quite some time that farm women were increasing their participation in off-farm labour markets, and that women's off-farm earnings were often crucial in ensuring the viability of farms. It was also recognized, however, that farm women's off-farm jobs typically were concentrated in low-skill, low-wage sectors of the economy and tended to involve less-than-full-time or otherwise intermittent employment.

Subsequent descriptive studies on farm women's work roles focused primarily on their on-farm and household work and its relationships with off-farm employment. One of the most extensive studies was by Jones and Rosenfeld, who used 1980 national survey data to profile farm women's involvement in farm work, off-farm employment, community activities, and several USDA programmes. They found a high incidence of women's involvement in most agricultural production tasks, including those typically considered to be "men's work". Over three-quarters of farm women were found to play a role in bookkeeping and other financial management tasks on the farm. Slightly over 30 percent of the farm women (60 percent of whom worked less than 40 hours per week) reported off-farm employment.

A further major descriptive study was Salant's survey of farm households in 29 rural counties in Mississippi and Tennessee. She concluded that farm women are extensively involved in onand off-farm work; over half of the women reported employment either on-, off-, or both on- and off-farm, about 75 percent of whom worked off-farm. Women's off-farm labour force participation was found to be related to age and family life cycle. The off-farm income earned by farm women made a major contribution to the income of households in the survey — in many cases enabling the family to rise above poverty-level status — despite the fact that farm women's employment was largely in low-wage jobs. Women who worked on-farm primarily did so as unpaid family laborers. Only 3 percent of the women in the sample were principal farm operators. Salant did not, however, find any support for the hypothesis that women who work on-farm tend to do so to free the labour of men in order to acquire more lucrative off-farm jobs.

Another important descriptive study is that of Fassinger and Schwarzweller, which was based on a 1979 sample survey of 124 farm households in Michigan. Fassinger and Schwarzweller found that farm size was positively related to the amount and proportion of time spent by farm men and women in on-farm work. Farm size was inversely related to men's off-farm employment but had no association with women's off-farm employment.

They found that men's time devoted to on-farm work was positively associated with women's on-farm work. Fassinger and Schwarzweller found a very high degree of division of labour by gender for both farm and household tasks. On the farm, women were more likely to assist with financial and recordkeeping aspects of the business and to plant and maintain a vegetable garden. Men, on the other hand, were primarily involved in field operations, the purchase of farm inputs and machinery, and the repair of machinery and farm buildings. The lion's share of household tasks was performed by women regardless of the scale of the farm.

Kalbacher's study was the first comprehensive report profiling the role of women farmers in American agriculture. Kalbacher used 1978 census results and data from a 1979 Farm Finance Survey to present a statistical profile of women farm operators. About 5.2 percent of all U.S. farm operators, or 128,000 farmers, were women in 1978 — representing a near-doubling of the number of female farmers since 1970 and a tripling since 1950. Femaleoperated farms were located disproportionately in the South, and tended more often to be owner-operated, part-time, and smaller than male-operated farms.

Female farm operators were also older and more likely to be in a racial-minority group than their male counterparts. But whereas women farmers were less likely than men to report fanning as their principal occupation, their average off-farm income was significantly less, resulting in a much lower average household income for women farmers, though femaleoperated farms typically had lower debt and more favorable debtasset ratios than did male-operated farms.

A further examination of the role of farm women in the United States was conducted by Tigges and Rosenfeld in their study of differences between men and women among "independent" farmers using a national sample. The independent farmer was defined as "a woman or man who has managerial responsibility for the farm operation and lacks a spouse's regular farm labour". They found that, unlike independent farming men who were similar to men whose wives participated in farming, independent farming women tended to have lower net farm incomes and total family incomes than did other women. Part of these differences were explained by differences in type of farm, class, and life-cycle situation variables. Tigges and Rosenfeld concluded that independent women farmers tended to enter farming by default (as upon the death of a spouse) rather than seeking to enter farming on their own.

After several years of descriptive studies documenting the variety of contributions of women to U.S. farm households, the attention of several researchers shifted to changes in these roles over time. The most important study in this regard has been by Wilkening in which he presented data from a 1978 mail survey of Wisconsin farm families, personal interviews with farm couples completed in 1979, and data from a 1962 study carried out in the same region. The data reveal that families remain the most important source of labour on Wisconsin farms, particularly on dairy farms that dominate in Wisconsin.

Nonetheless, Wilkening found that both farm men and women now contribute less time to the farm operation than they did in 1962. Off-farm employment by one of the spouses was significant in the late 1970s on about one-quarter of the farms, and off-farm work was associated with low levels of gross farm and high levels of total family income, but was unrelated to stage in the family life cycle. While Wilkening found that farm women have not played an increased role in farm production activities since 1962, the data did show that farm women have begun to assume major responsibilities for recordkeeping and other financial aspects of Wisconsin farm operations. Women were reported to be equal partners in making important farm decisions on about half of the farms, a slight increase since 1962. Farm women remained largely responsible for the vast majority of household work during the study period.

Testing part of a theory about the sexual division of labour and authority in farm households, Bokemeier and Garkovich found that farm women's self-concepts relating to their identification with their farms' operations were complex in their sources, including scale of farm, farm background, women's human capital, and family characteristics. They identified four roles for farm women, ranging from little or no involvement and authority to full partnership: homemaker, "ag" helper, business manager, and full "ag" partner.

A quite different study oriented toward assessing change in women's and men's work roles was conducted by Gladwin in North Florida. Gladwin was concerned with exploring an extension of the Boserup hypothesis that the capitalization and intensification of agriculture are inversely associated with the level of participation of women in farm activities. Specifically, Gladwin examined the reverse hypothesis — that a decrease in intensification, or marginalization, of agriculture will be associated with an increase in female labour with survey and census data. Gladwin reported that women are now involved to a greater extent in farm production than in the 1920s and 1930s and that the majority of farm women think of themselves as "farmers" rather than farm wives.

But she also reported that off-farm work has become more important for both men and women on North Florida farms. About 88 percent of the farms were part-time farms with significant off-farm income supporting the farm family. These part-time farms tended to produce crops or livestock requiring relatively little labour. Gladwin concluded that there is an association between labour intensification and the growing participation of women in farm work, since the few highly capitalized farms in the region received comparably less labour from women in the farm family than did the more numerous part-time, less-capitalized farms.

The third category of research on gender and agriculture has been that of assessing the mutual interrelations of men's and women's on-farm and off-farm work and how these associations are affected by the scale of the farm. One such study, by Fassinger and Schwarzweller, found that men's and women's onfarm work were positively associated, and each was positively related to farm size. Also, men's off-farm work was found to be positively associated with farm scale, while women's off-farm work was not related to the size of the farm.

Coughenour and Swanson examined 1979 Kentucky farm survey data to explore the associations between farm size and the work status of farm men and women. They did so, however, in a manner different from the previous literature; whereas previous studies have seen men's and women's work statuses as a function of farm size, Coughenour and Swanson suggest that the opposite direction of causality may be important. They explored this hypothesis by developing a four-category farm-type typology based on men's and women's work statuses. They found that there was a modest association between farm types and size of farm (measured by acreage and gross sales). They explored this association further in terms of "exchanges" between men's and women's onand off-farm labour.

They found, for example, that farms in which the woman has off-farm employment forfeited some of the woman's on-farm labour (and hence its contribution to farm size and sales) but gained in income relative to other farms. They also found that the allocation of men's labour — either on-farm, which was positively associated with farm size and sales, or off-farm, which was inversely associated with farm size and sales — was a stronger determinant of the size structure of farms than was the allocation pattern of women's labour. Nonetheless, they did find that the women's role in farming had significant effects — in particular, that women's off-farm work, which reduced or eliminated her on-farm labour input, tended to be inversely associated with the size and sales position of the farm despite the fact that women's off-farm work provided additional capital for the farm operation.

Buttel and Gillespie examined the interdependencies of average weekly hours of on-farm and off-farm work by farm men and women, the use of hired labour, and the size of farm with 1979 New York farm survey data. They found that farm men and women tend to specialize mutually in either on-farm or off-farm work, with these interdependencies being greatest among small farm households. It was also found that hired labour tends to be substituted for women's on-farm labour input, with this inverse association between hired labour and women's on-farm work being greatest among larger, commercial-scale farms. The results of the ButtelGillespie study were largely corroborated by a replication and extension using data from tobacco-, peanut-, and soybean-producing farms in North Carolina by Simpson *et al.*

The extension pertained to distinguishing between production of commodities that is "organized around" individual units ("small batch") and those whose production is coordinated as a "continuous process". Dairy and tobacco were presented as examples of the former type and soybeans and peanuts as examples of the latter type. Simpson *et al.* found that crop production organized as continuous processes tended to make farm women's work more independent of farm men's than was the case for crop production organized around small batches, and thus technological change in the production of "continuous process" crops had different effects on men and women.

Rigorous analysis of gender relations on U.S. farms is a sufficiently new focus of research that, save for the pioneering book by Sachs, there has been little effort to incorporate gender relations phenomena into a comprehensive theory of farm structural change. Sachs' book thus is the single representative in the fourth category of research on farm gender relations and the gender division of farm household labour.

Sachs' book is a wide-ranging treatment of research on farm women from a variety of disciplines and in both First World and Third World contexts. For present purposes, Sachs' treatment of U.S. farm women receives principal emphasis. As indicated in the title of her book, one of Sachs' key arguments is that while women have historically made significant contributions to agricultural (and, increasingly, off-farm) labour, these contributions have been overlooked and undervalued.

Whereas Sachs roots her analysis in the new political economy tradition referred to at the beginning of this chapter, she contends that the invisibility of the women's role in agriculture, and the subordination of women in the agricultural labour process, cannot be explained with reference to the dynamics of capitalist development alone. She argues that the simultaneous contribution of women to agriculture along with their invisibility and subordination must be viewed through the dynamics of capitalism and patriarchy, which is defined as a "set of social relations which has a material base and in which there are hierarchical relations between men, and solidarity among them, which enable them to control women". She emphasizes, moreover, that patriarchical relations preceded (although they have been transformed by) capitalism and thus cannot be seen as a direct consequence of capitalist development.

Sachs begins by demonstrating that at each major stage of agricultural development from the seventeenth century to the present women have played important roles in facilitating the viability of U.S. farms. Until the twentieth century the importance of women was typically seen to be in producing and processing food for home consumption. There was a traditional division of labour between men and women in which men did the majority of the field work, but this division of labour would often break down during busy seasons when women would do "male" tasks. With the industrialization of U.S. society and further commercialization of agriculture, women's subsistence production activities diminished in importance. Women accordingly tended to increase their on-farm and/or off-farm work. These patterns, however, varied substantially by region, with women's direct involvement in farming being greatest in the South and West. Sachs notes that one constant historical feature of women's roles in farm households has been in their performing the bulk of household tasks. Women have become somewhat more involved in farm decision making, but Sachs suggests that their decision-making roles have been secondary to their labour contributions to the farm operation and household.

Sachs' analysis then turns to the historical development of a "domestic ideology" — that women are naturally "domestic" beings and more suited to household work than men — in U.S. society in general and in agriculture in particular. She argues that keeping women in domestic roles has facilitated the development of capitalism by maintaining a reserve labour force and creating a consumptive domestic class. Sachs suggests that the family farm, particularly because it is based on the nuclear family, has been one of the most pervasive mechanisms through which patriarchal relations have been maintained on U.S. farms.

She then explores how this patriarchal order has been reflected in the sexual division of labour on farms and in the social relations of women-operated farms. Sachs' book has been exemplary in identifying how gender relations on U.S. farms have been shaped by the twin forces of patriarchy and capitalist development. Her work remains the most comprehensive theoretical statement and overview of empirical evidence on farm gender relations that has yet appeared.

Sachs' book is wide-ranging but brief, suggesting that further elaborations and critiques will increase our theoretical and empirical understanding of farm gender relations and the gender division of farm and rural labour. In particular, several ambiguities remain in the literature. At an empirical level, while the past decade of research has established a number of frequently supported generalizations about the structuring of farm women's work roles, there are a number of conflicting results.

Some observers, for example, argue that women make major contributions to farm work, whereas others see a clear division of labour such that women participate very little in farm tasks. Other empirical controversies may well arise from the fact that most of the major studies in the field have been based on data from a single state.

Thus further research is needed to clarify these empirical patterns and to incorporate them in a larger theoretical perspective — a major component of which will likely revolve around establishing and explaining differential patterns of women's work roles across farming regions. Some of this theorizing and comparative research is, however, taking place in Europe, for example, has analyzed in detail the gender relations on farms in France and has made some initial notes toward a theory of variations in gender differences in the household, child-rearing, and farm work.

LAND OWNERSHIP AND FARM STRUCTURE

OWNERSHIP OF U.S. LAND

The land surface of the United States covers 2.3 billion acres. Private owners held 61 per cent in 2002, the Federal Government 28 per cent, State and local governments 9 per cent, and Indian reservations 3 per cent. Virtually all cropland is privately owned, as is three-fifths of grassland pasture and range and over half of forestland. Federal, State, and local government holdings consist primarily of forestland, rangeland, and other land. Most land in Federal ownership—largely in the West—is managed by the Department of the Interior (68 per cent) and the Department of Agriculture (28 per cent).

Farm operators do not own all the land used in agriculture. According to the 1999 Agricultural Economics and Land Ownership Survey (AELOS), farmers held 58 per cent of the land in farms in 1999. These landowning farmers also made up 58 per cent of the 3.4 million farmland owners. Non-operator landlords accounted for the remaining 42 per cent of land in farms. Ninety-five per cent of non-farm landlords were individuals/families or partnerships. Of these unincorporated landlords, 55 per cent were at least 65 years old. Many non-farm landlords have a historic connection to farming. Among the people who have exited farming or inherited farmland since the number of farms peaked during the Great Depression, a number have retained ownership of some or all their land.

FARM NUMBERS, FARM TYPES, AND CONSERVATION PROGRAMMES

The number of farms has declined dramatically since its peak of 6.8 million in 1935, with most of the decline occurring during the 1940s, 1950s, and 1960s. The decline in farm numbers has levelled off since the 1970s. By 2002, 2.1 million farms remained. The remaining farms have a much larger average acreage, but averages mask differences among farms. Today's farms range from very small retirement and residential farms to industrialized operations with sales in the millions. Part of this diversity stems from the very low sales threshold ($1,000) necessary for an operation to qualify as a farm for statistical purposes.

One way to address the diversity of farms is to categorize them into more homogeneous groups. The farm typology developed by ERS identifies five groups of small family farms (sales less than $250,000): limited-resource, retirement, residential/lifestyle, farming-occupation/low-sales, and farming-occupation/high-sales.

The typology also includes large family farms, very large family farms, and non-family farms. In addition, very small farms (sales less than $10,000) make up more than half of all farms. Very small farms account for a particularly large share of farms in the limited-resource (72 per cent), retirement (76 per cent), and residential/lifestyle (76 per cent) groups. Production, however, is concentrated among larger farms; small farms account for only 27 per cent of the total value of production.

The smallness of most farms has implications for conservation and the environment. An ERS study found that smaller corn farms are less likely to use conservation tillage than are larger farms. The practice is more practical for larger farms because they have more acres over which to spread the cost of new or retrofitted equipment necessary to adopt conservation tillage. Small farms whose operators are retired or farm part-time are also less likely to adopt conservation tillage, possibly because of hesitancy to change familiar production practices. Small farms, however, participate widely in the Conservation Reserve Programme (CRP) and the Wetlands Reserve Programme (WRP).

DISTRIBUTION OF CONSERVATION PROGRAMME PAYMENTS BY TYPE OF FARM

These farms harvest most of the land planted to programme commodities and therefore receive three-quarters of commodity programme payments. However, CRP and WRP—the two major conservation programmes—are targeted at particular types of land, not commodities. Since small farms own 70 per cent of the land held by farms, they play a large role in natural resource and environmental policy. Retirement, residential/lifestyle, and low-sales farms account for nearly two-thirds of conservation payments and a similar share of the land farmers enrolled in the CRP and WRP. Participating farmers in each of the three groups tend to enroll large shares of their land in these programmes: 46 per cent of the land operated on retirement farms, 28 per cent on residential/lifestyle farms, and 23 per cent on low-sales farms. In contrast, enrollment ranges from 5 to 9 per cent for participating high-sales, large, and very large farms.

Because their main job is off-farm, residential/lifestyle operators are limited in the amount of time they can spend farming. As a result, residential/lifestyle farmers find CRP and WRP attractive, since these programmes require little time. Given their life-cycle position, many retired farmers have land available to put into conservation uses. The same forces may also be acting on low-sales operators, who average 57 years of age and may be scaling down their operations. If an off-farm job and advanced age are major determinants of land going into conservation uses, it may be relatively easy to get smaller farms to enrol land in the programmes. Getting larger farms to enrol more of their land might require higher payments, if the opportunity cost of idling their land is higher.

LAND TENURE

Farm operators leased 38 per cent of their total farmland in 2002, down from 40 per cent in 1997 and 43 per cent in 1992, according to the census of agriculture. This decline may reflect increasing rental costs as parcels of land become smaller. Parcels of farmland available to rent tend to become subdivided with time due to division among heirs. Smaller parcels increase transaction costs to operators assembling land to expand their operations. Still, rented land as a share of total farmland is higher than the 35-per cent rate that prevailed in the 1950s and 1960s.

About 38 per cent of all farms rented land in 2003, 32 per cent as part owners and 6 per cent as tenants. Land leasing has changed from a way for beginning farmers to enter agriculture to a way for established farmers to access additional land. Renting allows farms to expand without the debt and commitment of capital associated with ownership. In fact, about 17 per cent of very large family farms are tenants, a larger percentage than in any other group. Conventional wisdom holds that farmland owners have a long-term interest in their land and thus are more likely than renters to adopt conservation practices. Soule and others (1999 and 2000) found this to be true among corn farmers, at least in the adoption of conservation practices that provide only long-term benefits, such as grassed waterways and strip cropping.

The situation was different for conservation tillage, which can increase profits in the short run by maintaining or increasing yields while reducing machinery, fuel, and labour costs. Cash-renters are less likely than owner-operators to use conservation tillage, but share-renters appear to act like owner-operators in adopting conservation tillage. Share-renters may have an incentive to adopt conservation tillage, if the landlord bears some of the costs that may increase under conservation tillage, such as herbicide expenditures. Share-landlords are also more likely to be involved in management decisions than cash-landlords, which may make share-renters act more like owners.

NEW HEALTH PRODUCTS FOR FARM ANIMALS

Animal diseases cost U.S. agriculture $17 billion each year, according to the Congress Office of Technology Assessment. To help prevent these losses, animal scientists are using biotechnology to develop an array of products to diagnose, treat, and prevent disease in farm animals. An area where animal biotechnology has already had a profound impact is diagnostic testing. Many animal diseases are difficult to diagnose. A veterinarian often has to wait hours or days for laboratory results to confirm a diagnosis. In the meantime, the veterinarian must either withhold treatment or risk using an inappropriate therapy.

Using biotechnology, scientists are developing fast, accurate diagnostic tests for many of the most common farm animal diseases. In some cases, a veterinarian can conduct the diagnostic tests right on the farm and immediately start the best therapy. Many of these new tests use monoclonal antibodies. Antibodies are proteins an animal's immune system produces in response to invasion by bacteria, viruses, or parasites. Each kind of antibody binds specifically with a characteristic part of a particular kind of invader. Scientists make monoclonal antibodies by fusing two kinds of cells. One is an immune system cell that produces antibodies, which bind to part of a particular disease-causing microbe; the other is a cancer cell. The cell that results from the fusion inherits from the immune system cell the ability to produce antibodies, and from the cancer cell the ability to reproduce indefinitely.

This new hybrid cell, or hybridoma, acts as an antibody factory. The antibodies are called monoclonal because they are produced by identical "clones" of the original hybridoma-all will bind with the identifying structure on the surface of the microbe in question. One application of monoclonal antibodies is as a test for brucellosis in cattle. This bacterial disease often causes cows to abort pregnancies, and it can be spread to farmers and people who drink milk from infected cows. Vaccines protect pregnant animals from abortion, but vaccinated cows may become carriers of the disease. Conventional diagnostic tests cannot distinguish between the disease-causing microbe and the vaccine, which is made from the microbe. Thus, these tests cannot help the farmer determine which animals are carriers of the disease and which are not. But diagnostic tests using monoclonal antibodies are so specific, they distinguish between cattle that carry the disease-causing bacterium and those that carry only the vaccine. With the aid of these new tests, farmers can isolate carriers to prevent the spread of infection. Diagnostic tests are not just for diseases.

A monoclonal antibody test to detect pregnancy in animals is so simple that a farmer can read the test right on the farm. Monoclonal antibodies are also used in a test to detect a disease called scours in piglets and calves. Scours is a bacterial disease of newborn animals that causes diarrhea. Many of the afflicted animals die of dehydration. By providing quick, accurate diagnosis, the monoclonal antibody test for scours permits immediate administration of therapy. The therapy for scours in calves is also based on monoclonal antibodies. Infected calves are fed monoclonal antibodies that coat the offending bacteria, preventing the microbes from causing diarrhea.

Some other therapies for animal diseases are based on recombinant DNA technology. Genetic engineering allows scientists to insert one or more new genes into an animal, plant, or microbial cell. In developing treatments for an animal disease, for instance, an animal gene is inserted into a microbe or an animal cell that can be grown in a laboratory culture. The inserted gene is one that produces a natural disease-fighting protein. The microbes or animal cells become minute factories, producing large quantities of the therapeutic protein. For example, genetic engineering is used to produce interferons and interleukin-2, natural proteins that fight viruses.

Veterinarians long have used antibiotics against bacterial diseases, but they have had few drugs to treat viral infections. Interferons and interleukin-2 not only kill viruses directly, they also stimulate the animal's immune system, improving the immune system's ability to fight disease naturally. One target of research regarding interferons and interleukin-2 is shipping fever, a disease of cattle caused by several viruses infecting an animal simultaneously.

Although the animal's immune system normally keeps the viruses in check, outbreaks of the disease may result from stress during transport to feedlots or market. Shipping fever costs the beef industry more than $250 million a year. Injections of interferons or interleukin-2 before shipping may help to augment the cattle's natural immunity and prevent shipping fever.

BETTER VACCINES TO PREVENT DISEASE

In addition to producing new diagnostic tests and therapeutic proteins, animal scientists are using biotechnology to develop vaccines to prevent disease. A safe, effective vaccine for swine pseudorabies, a fatal herpes-virus, is already in use. An advantage of the swine pseudorabies vaccine, and other vaccines made through recombinant DNA technology, is that they use only a small portion of the original microbe. Conventional vaccines use killed or weakened forms of the disease-causing microbe.

Sometimes, when the microbes are not completely killed or sufficiently weakened, a conventional vaccine may cause the disease it was supposed to prevent. Disease-causing genes are not included in the genetically engineered vaccines. Therefore, the recombinant vaccines build up the body's immunity without the risk of causing disease. Another advantage of recombinant vaccines is the speed with which they can be developed.

"Conventional vaccine development can take 20 or 30 years, maybe even 100 years," says Karen Jacobsen, D.V.M., University of Georgia College of Veterinary Medicine. "This is one reason why we don't have vaccines for a lot of animal diseases today. " Recombinant vaccines are being developed for foot-and-mouth disease, a highly contagious viral disease that infects cattle, sheep, and other animals. Although this disease has been eradicated in North America, it still causes substantial livestock productivity losses elsewhere, particularly in developing countries. Infected herds must be slaughtered, and contaminated ranches must be left idle for months to prevent new outbreaks of the disease. Conventional foot-and-mouth vaccines are made by weakening the virus that causes the disease. These vaccines sometimes revert to the virulent state, and they have caused outbreaks of the disease in Europe. The foot-and-mouth vaccine made with biotechnology cannot cause the disease because, as in other biotech vaccines, the disease-causing genes have been removed.

The biotech vaccine does not need to be kept cold and, therefore, can be used in developing countries. In addition to being used in treating swine pseudorabies and foot-and-mouth disease, recombinant DNA technology is being used to develop vaccines against dozens of livestock and poultry diseases, such as scours and a variety of respiratory disorders. Some vaccines may be injected into eggs so that chickens are immunized before they hatch. Most vaccines protect against viral or bacterial infections, but using genetic engineering, researchers in New Zealand and Australia have developed a vaccine against a parasitic disease. Farmers must destroy the meat from sheep infested with a species of tapeworm common in those two countries. The new vaccine, which protects sheep from the tapeworm, may save farmers millions of dollars annually.

PROTEINS FOR ANIMAL NUTRITION

Natural proteins called somatotropins, or growth hormones, help animals convert feed into muscle or milk. Using recombinant DNA technology, scientists have been able to develop bacteria that produce commercial quantities of somatotropins.

They have found that administering small additional amounts of these proteins to animals helps to increase the animals' efficiency of feed conversion, thereby saving the farmer money. Porcine somatotropin (PST) may help improve human health, as well as lower the farmer's cost of production. PST improves feed efficiency in hogs by 15 to 20 per cent. It also reduces fat deposition, allowing PST-treated hogs to provide consumers with leaner cuts of pork. The ability to produce lean pork has enormous implications for improving human health by reducing dietary fat and cholesterol. Another growth protein, bovine somatotropin (BST), is being administered to make dairy cows more efficient, thereby lowering the dairy farmer's costs. Regular administration of small amounts of BST increases a cow's milk yield by 10 to 25 per cent. Although feed intake also increases, this feed increase is proportionately less than the increase in milk production.

Studies indicate that BST can improve the milk-to-feed ratio by 5 to 15 per cent. Extensive laboratory testing has shown that the meat and milk of treated animals contain no more PST or BST than the meat or milk from untreated animals. Other tests have shown that these proteins do not act on the human body, and therefore, they are safe to eat. Nevertheless, the use of BST remains controversial. Testing is under way to make sure that animals are not adversely affected by treatment with somatotropins. Another protein that can be administered to animals to increase feed efficiency is called growth hormone releasing factor (GHRF). While not a growth protein itself, GHRF causes the animal to increase production of growth proteins.

IMPROVED ANIMAL BREEDS

Long before anyone knew about genes or DNA, people were selectively breeding animals to produce desirable traits, such as strength, feed efficiency, and disease resistance. Selective breeding is a crude technology, because each cross between two animals mixes hundreds of thousands of genes, both desirable and undesirable. Recombinant DNA technology is now making it possible for scientists to breed animals with great precision. One or more new genes can be inserted into an animal embryo without disturbing the rest of the animal's genes. The resulting animal with the new gene or genes is referred to as a transgenic animal. Despite the differences in technique, the scientist using recombinant DNA has the same objectives as the selective breeder: improving efficiency, disease resistance, and other desirable traits. But recombinant DNA provides a wider array of possible new traits, since recombinant techniques allow animal scientists to insert genes not only from the same or closely related species, but also from distantly related animals, plants, and even microbes. One possible application of gene transfer in animals is to develop dairy cows that produce more nutritious milk. Bottle feeding human infants with cow's milk is not an ideal substitute for breast feeding, because cow's milk differs from human milk in some important components. Scientists are looking for ways to transfer genes into dairy cows that will enable the cows to produce milk that more closely resembles human milk.

Advances in Control of Parasitic Diseases Continue to Rely on Biotechnology Parasitic disease continues to be a major health problem worldwide in humans and domestic animals. Applying biotechnology has led to progress in prevention and control in many areas, particularly in regard to diagnosis. Parasites, like most organisms, carry repetitive sequences in their genomes. These are sometimes tandem repeats (back-to-back in a chromosome), or they may be located at random throughout the genome. Often, repetitive sequences are intergenic and noncoding, and are thus far more subject to evolution and change, increasing the likelihood that they will be genus or even species specific. Since detection of these sequences is not dependent upon gene expression, which may vary over time, repetitive sequence-based assays allow detection of the parasite at any life stage.

The effectiveness of repetitive sequence-based polymerase chain reaction (PCR) methods of identification has been demonstrated for parasites of several clinically important genera (1). A PCR assay has also been developed to differentiate two closely related nematode species that are virtually indistinguishable by morphologic examination (2), as well as for diagnosis of Toxoplasma and Cryptosporidium infections (3). Attempts to develop antiparasite vaccines have met with multiple failures. The problems often result from lack of post-translational modification by currently used vector-host systems and the inability to regenerate the secondary and tertiary structure of the native immunogens. Antigen presentation in recombinant parasite vaccines has also been thought to be a problem resulting in failed efficacy. There is now some evidence to indicate that cloning of genes for antigens of interest to yield chimeric molecules expressed in the BCG strain of Mycobacterium bovis may provide improved antigen presentation to stimulate both humoural and cellular responses (4); however, it is unlikely that future applications of this technology will be simple. Nonetheless, work is proceeding on the application of biotechnology to the development of vaccines against parasites in most food animals. A particularly striking exception to the marginal successes reported in other systems has been a recombinant vaccine for a parasitic helminth of sheep (5). This product yielded greater than 94% protection in experimental work and is now available as a commercial product. A further possibility is the application of anti-idiotypic monoclonal antibodies. When monoclonal antibodies are prepared against a monoclonal antibody, it is possible to obtain, in a very small number of cases, an antibody which mimics the original antigen. Results of work with Cryptosporidium perfringens epsilon toxin suggest that animals immunized with anti-idiotype monoclonals can be protected against toxin challenge.

4

Animal Reproduction

Reproduction is one of the most characteristic features of living organisms. Life would not exist on Earth if plants and animals did not reproduce to make their offspring. By reproducing, a living organism can be sure that there is another individual of its kind to take its place when it dies. In this way a species of organism guarantees its survival.

A species is a particular type of organism. For example, a horse is a species and a zebra is another species. A species which cannot reproduce enough offspring will disappear for ever from the face of the Earth - it will become extinct. This has happened many times in the past. The best known example of animals which have become extinct is the dinosaurs.

The dinosaurs were a group of reptiles which mysteriously became extinct 60 million years ago.

Fortunately there has always been another type of living organism to replace those that become extinct. In the case of the dinosaurs they left the Earth to the group of animals that we belong to, the mammals.

WHAT IS REPRODUCTION?

In a nutshell, reproduction is the creation of a new individual or individuals from previously existing individuals. In animals, this can occur in two primary ways: through asexual reproduction and through sexual reproduction. Let's look at asexual reproduction.

Asexual Reproduction

In asexual reproduction, one individual produces offspring that are genetically identical to itself. These offspring are produced by mitosis. There are many invertebrates, including sea stars and sea anemones for example, that produce by asexual reproduction. Common forms of asexual reproduction include:

Budding

- In this form of asexual reproduction, an offspring grows out of the body of the parent.

- Hydras exhibit this type of reproduction.

Gemmules (Internal Buds)

- In this form of asexual reproduction, a parent releases a specialized mass of cells that can develop into offspring.
- Sponges exhibit this type of reproduction.

Fragmentation

- In this type of reproduction, the body of the parent breaks into distinct pieces, each of which can produce an offspring.
- Planarians exhibit this type of reproduction.

Regeneration

- In regeneration, if a piece of a parent is detached, it can grow and develop into a completely new individual.
- Echinoderms exhibit this type of reproduction.

Parthenogenesis

This type of reproduction involves the development of an egg that has not been fertilized into an individual.

Animals like most kinds of wasps, bees, and ants that have no sex chromosomes reproduce by this process. Some reptiles and fish are also capable of reproducing in this manner.

Advantages and Disadvantages of Asexual Reproduction

Asexual reproduction can be very advantageous to certain animals. Animals that remain in one particular place and are unable to look for mates would need to reproduce asexually. Another advantage of asexual reproduction is that numerous offspring can be produced without "costing" the parent a great amount of energy or time.

Environments that are stable and experience very little change are the best places for organisms that reproduce asexually. A disadvantage of this type of reproduction is the lack of genetic variation. All of the organisms are genetically identical and therefore share the same weaknesses. If the stable environment changes, the consequences could be deadly to all of the individuals.

Asexual Reproduction in Other Organisms

Animals are not the only organisms that reproduce asexually. Yeasts, plants and bacteria are capable of asexual reproduction as well. Bacterial asexual reproduction most commonly occurs by a kind of cell division called binary fission. Since the cells produced through this type of reproduction are identical, they are all susceptible to the same types of antibiotics.

HOW ANIMALS REPRODUCE

Animals can be grouped into those which give birth to living offspring and those which lay eggs that eventually hatch into offspring. Those animals which give birth to live offspring are called live-bearing or viviparous.

Those animals which lay eggs are called egg-laying or oviparous. The difference is in the place where the offspring develops before it is born. Below is a table of the vertebrates which shows which groups are viviparous and which groups are oviparous.

Group	*Oviparous (Egg-layers)*	*Viviparous* (Live-bearers)
Mammals	Only a few primitive egg-laying mammals exist. They live in Australia and New Guinea. e.g. spiny anteater and duck-billed platypus.	Nearly all mammals e.g. mouse, human, cat, dog, bear, kangaroo and dolphin.
Birds	All birds. e.g. robin, penguin, parrot, sparrow and eagle.	None.
Reptiles	Most reptiles are egg-layers. e.g. crocodile, turtle and cobra.	Some lizards and snakes are live-bearing. e.g. vipers.
amphibians	Nearly all. e.g. frog, toad and	A few species of frogs living in South America and West Africa are live-bearers.
Fish	Most species. e.g. herring, salmon and trout.	Quite a few species are live-bearers. e.g. sharks and guppies.

Amongst the invertebrates there are many which are oviparous but a few are viviparous such as sea anemones and aphids.

Sexes and Sex Cells

A species of animal usually exists in two types or sexes called males and females. Each sex has its role to play in reproduction.

When an animal gives birth or lays eggs we notice that it is always the female which does this. In some cases, such as the sea horse and the midwife toad, the male appears to give birth to the offspring.

In these exceptions the male has been given the eggs to look after but it is still the female animal that lays the eggs in the first place. To produce an offspring two special cells are needed. A cell is a microscopic part of an organism's body. The bodies of animals are made of millions of cells. There are many different sorts of cells in a body, each having specialized functions.

Some are found in the blood transporting oxygen, others are found in the walls of our stomach producing juices to digest our food. To reproduce, animals make special sex cells. In the male animal these sex cells are called sperm cells or sperms and in the female animal they are called egg cells or eggs.

Sperm Cells

Sperm cells are very small but they are very specialized too.

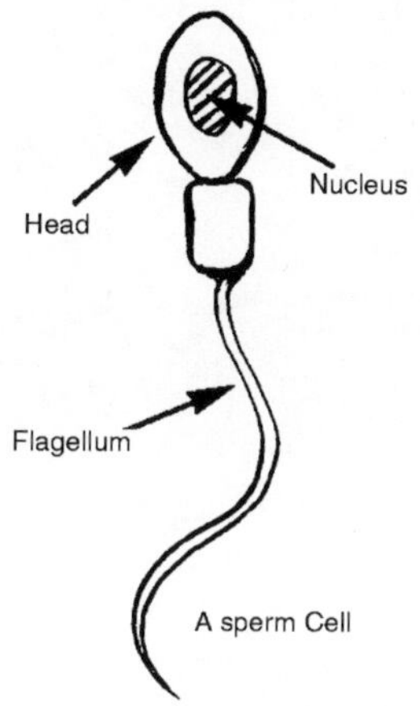

Fig. Sperm Cell

The sperm cells of different species of animal are all about the same size, about 60μm long (μm = micrometres; This is a thousandth of a millimetre). Over 11 000 sperm cells could fit on a pin head! All sperms have a head and a tail called a flagellum. They use the flagellum to swim through liquids. The head of the sperm is very important because it contains the nucleus. The nucleus of the cell is the control centre. In sex cells the nucleus carries half of the information needed for reproduction (like the plans needed to construct a house).

Sperm cells are made in very large numbers by special organs in the male's body called testes (sing. testis). In most animals the testes are carried inside the male's abdomen but in mammals they are kept in a sack of skin called sperm duct the scrotum outside the abdomen.

Egg Cells

Egg cells are produced by female animals in special organs called ovaries (sing. ovary). These are found inside the abdomen of all female animals. Egg cells are much bigger and simpler than sperm cells in their structure.

They do, however, vary a lot from one species of animal to another. For example the egg of a human is only 0,1 mm wide but the egg of a chicken is 20 mm wide.

Animal	*Number of Eggs Produced At One Time*
Cod	5000000
Frog	1500
Viper	12
Human	1

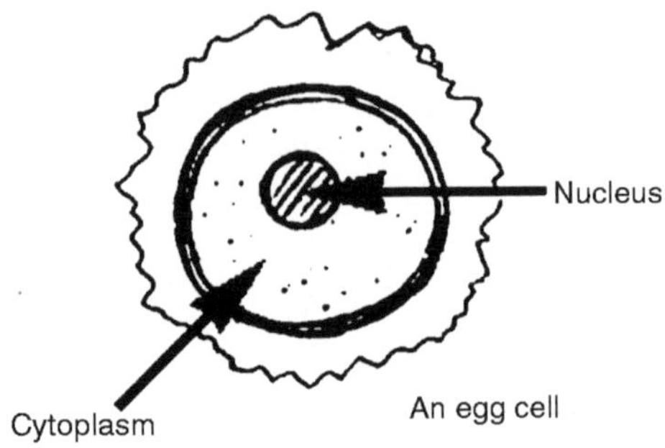

Fig. Egg Cell

Even so the human egg cell is still nearly 50 times wider than a sperm cell. The reason for the difference in size is that the egg has a large food supply stored inside it called yolk. Yolk is the yellow part of a chicken's egg.

The egg also has its own nucleus which carries the other half of the information needed for reproduction. Because eggs are bigger than sperms they are not produced in such large numbers. Even so animals which do not take care of their offspring, such as fishes and frogs, will lay a large number of eggs. Animals which take care of their offspring, such as mammals and birds, will produce less eggs.

MATING AND FERTILIZATION

The sperm cell carries half of the information needed for reproduction and the egg cell carries the other half. To produce an offspring the sperm and egg have to meet. This means that the male and female animals have to come together. This is called mating. Because the sperms can swim on their own, aquatic animals such as fish can simply release the eggs and sperms into the water. Amphibians, which return to water to reproduce, also release their eggs and sperms directly into the water.

The sperm cell is microscopic and only lives for a short time, so the eggs and sperms need to be released close together. Fish, such as herring, which live in the open sea will shoal at certain times of the year. Shoaling occurs when the males and females gather close together, usually near the surface of the sea. The females release their eggs and the males release their sperms at the same time and in the same place. Therefore, the sperms do not have far to swim.

Fish, such as salmon, trout and sticklebacks, build their nests in the mud and gravel on the bottom of shallow rivers and lakes. The male usually builds the nest and then he guards it.

When a female enters his territory he will try to encourage her to lay her eggs in his nest. This sometimes involves quite a complicated mating display. If the female decides to lay her eggs in his nest he will release his sperms over the eggs.

Amphibians

Animals which live on land have a problem sperm cells cannot walk or fly! For the amphibians the answer to this problem is to return to the water to mate. Frogs and toads will migrate to the ponds and lakes where they were born. They do this usually in Spring. Sometimes they cross roads on their migration route in such large numbers that they can cause accidents! In France the motorway which runs through the Sologne, south of Orleans, has special tunnels constructed underneath it. These tunnels allow the amphibians, which are very numerous in this region, to cross underneath the road safely. These tunnels are called crapauducts! When the frogs arrive at the water the males and females call to one another. This helps the amphibians to identify one another, as many different species may be mating at the same time and in the same place.

The male holds the female round the waist so that as she releases her eggs he will release his sperms close to them in the water. The mating calls are very important for some species because the males will try to mate with any object floating in the water, including fish or pieces of wood!

Mating on Land

Vertebrate animals such as reptiles, birds and mammals, and invertebrates, such as spiders and insects, do not return to water to mate, so the male must place the sperm cells directly inside the female's body. The male often has a special organ to place the sperms inside the female and the female has a special opening to receive the sperms. This kind of mating is called copulation and it has two advantages. First, the sperms are released in a liquid called semen which is made by the male. The sperms can swim in this liquid towards the eggs inside the female.

The second advantage is that the sperms cells inside the female are more certain to meet the egg cells. Animals which mate in this way do not produce so many eggs because only a few of them are lost. Copulation is so effective that many aquatic animals use it too, even though they are surrounded by water in which the sperms could swim. Getting closer to the female can have its disadvantages. If the animal is a carnivore, such as a spider, the female may think that the male is a meal to be eaten rather than a mate! The result is that land animals have developed complicated mating behaviour which involves dances, smells and often very colourful mating displays. It is usually the male which is colourful in the animal world, even amongst the herbivores. This is very apparent amongst birds; the peacock, for example, has brilliant tail feathers. The female (peahen) by comparison is quite drab.

Fertilization

When a sperm cell meets an egg cell, the head of the sperm fuses with the egg. The flagellum of the sperm is lost, its work is finished. The information in the nucleus of the male's sperm can combine with the information in the nucleus of the female's egg to produce an offspring. This important event is called fertilization. If fertilization takes place inside the female's body - as it does in mammals or birds - it is called internal fertilization. Only one sperm cell will fertilize one egg cell successfully. It is possible, however, for two sperms to fertilize two eggs.

Species	*Litter Size*
Bat	1
Dolphin	1
Chimpanzee	1
Lion	3
Hedgehog	5
Red Fox	6
Rabbit	6
Black rat	11

This will result in two offspring. This is not at all unusual in the animal world. Foxes give birth to litters of between four and eight puppies. A litter is a group of babies born from the same mother at the same time. The table below gives some examples of different mammals and their average litter size

Reproduction in Birds

The egg cell of a bird is really what we call the yolk. In the chicken's egg this yellow ball of yolk is about 20 mm in diameter. The yolk is the food supply for the developing embryo. The nucleus of the bird's egg can be found in the middle of a white spot which is found on top of the yolk. You need a microscope to see the nucleus itself but the white spot is visible to the naked eye. The egg that a hen (a female bird) lays is covered in layers of albumen (a material like jelly), two membranes and a hard shell. The shell and the membranes protect and support the egg.

The albumen feeds the developing embryo with water and protein. The albumen, membranes and shell are added to the yolk layer-by-layer as it passes down a tube called the oviduct inside the hen's body. The oviduct carries the egg from the ovary out of the hen's body. The last layer to be added is the shell. If a bird's egg is to be fertilized, mating must take place so that the egg will meet a sperm cell in the oviduct before these layers are added. The shell of a bird's egg is often coloured or patterned to help camouflage it from predators.

This is especially important for birds which nest on the ground, such as plovers. A hen bird may lay from 1 to 19 eggs in her nest. A group of eggs laid by a hen is called a clutch. It does not matter whether the hen bird lays 2 or 20 eggs in her clutch, she will only lay one egg per day. This is important because birds' eggs are very large. If the hen bird were to lay all her eggs at once she would have to produce smaller eggs or fewer eggs. The hen bird usually lays her egg for the day early in the morning. This is so that she can be free to hunt for food during the morning and in the middle of the day.

Taking Care of the Eggs

Birds take great care of their eggs. Usually the eggs are laid in a nest. Sometimes the nest can be a simple scrape in the ground or it can be a very elaborate nest. For example the mallee fowl of eastern Australia builds a huge nest of decomposing vegetation which helps to keep the eggs warm.In most cases it is the body heat of the parent bird which keeps the eggs warm. Keeping the eggs warm is called incubation or brooding. If the eggs get cold the embryo developing inside will die.

In some types of birds the hen develops a patch on her breast without feathers. The eggs that she is incubating will be in contact with the skin and so they will be close to her body temperature. The hen must go and feed from time to time, so the cock (the male bird) may take over. This is particularly important for one species of bird, the emperor penguin. This is the largest species of penguin, it has a mass of 16 kilogrammes and it lives on the continent of Antarctica. Antarctica is mostly covered in solid ice but it is there that the emperor penguin breeds.

It is the male bird that looks after the egg to begin with. The hen lays only one large egg per year. She uses up a lot of her food reserves to make this egg, so she must go to sea to fish for food to replace these reserves. The male stays on the ice and keeps the egg on his feet covered by a fold of skin. Obviously the emperor penguins cannot build a nest from the ice, so the male keeps the egg on his feet for eight weeks through the fierce Antarctic winter.

The temperatures can drop to - 60° C. The chick hatches at about the time when the hen returns with food. At last the cock can go and hunt for food. He will return about two weeks later with more food for the chick.

Reproduction of Horse

Pregnancy lasts for approximately 335-340 days and usually results in one foal (male: colt, female: filly). Twins are rare. Colts are usually carried 2-7 days longer than fillies. Females 4 years and over are called mares and males are stallions. A castrated male is a gelding. Horses, particularly colts, may sometimes be physically capable of reproduction at approximately 18 months but in practice are rarely allowed to breed until a minimum age of 3 years, especially females. Horses four years old are considered mature, though the skeleton usually finishes developing at the age of six, and the precise time of completion of development also depends on the horse's size (therefore a connection to breed exists), gender, and the quality of care provided by its owner.

Also, if the horse is larger, its bones are larger; therefore, not only do the bones take longer to actually form bone tissue (bones are made of cartilage in earlier stages of bone formation), but the epiphyseal plates (plates that fuse a bone into one piece by connecting the bone shaft to the bone ends) are also larger and take longer to convert from cartilage to bone as well.

These plates convert after the other parts of the bones do but are crucial to development. Depending on maturity, breed and the tasks expected, young horses are usually put under saddle and trained to be ridden between the ages of two and four.

Although Thoroughbred and American Quarter Horse race horses are put on the track at as young as two years old in some countries, horses specifically bred for sports such as show jumping and dressage are generally not entered into top-level competition until a minimum age of four years old, because their bones and muscles are not solidly developed, nor is their advanced training complete. For endurance riding competition, horses may not compete until they are a full 60 calendar months old. Horses have, on average, a skeleton of 205 bones.

A significant difference in the bones contained in the horse skeleton, as compared to that of a human, is the lack of a collarbone—their front limb system is attached to the spinal column by a powerful set of muscles, tendons and ligaments that attach the shoulder blade to the torso. The horse's legs and hooves are also unique, interesting structures.

Their leg bones are proportioned differently from those of a human. For example, the body part that is called a horse's "knee" is actually the carpal bones that correspond to the human wrist. Similarly, the hock, contains the bones equivalent to those in the human ankle and heel.

The lower leg bones of a horse correspond to the bones of the human hand or foot, and the fetlock (incorrectly called the "ankle") is actually the proximal sesamoid bones between the cannon bones (a single equivalent to the human metacarpal or metatarsal bones) and the proximal phalanges, located where one finds the "knuckles" of a human.

A horse also has no muscles in its legs below the knees and hocks, only skin and hair, bone, tendons, ligaments, cartilage, and the assorted specialized tissues that make up the hoof.

Reproduction of Dog

In domestic dogs, sexual maturity (puberty) begins to happen around age 6 to 12 months for both males and females, although this can be delayed until up to two years old for some large breeds. Adolescence for most domestic dogs is around 12 to 15 months, beyond which they are for the most part more adult than puppy. As with other domesticated species, domestication has selectively bred for higher libido and earlier and more frequent breeding cycles in dogs, than in their wild ancestors. Dogs remain reproductively active until old age. Most female dogs have their first estrous cycle between 6 and 12 months, although some larger breeds delay until as late as 2 years. Females experience estrous cycles biannually, during which her body prepares for pregnancy, and at the peak she will come into estrus, during which time she will be mentally and physically receptive to copulation.

Dogs bear their litters roughly 56 to 72 days after fertilization, although the length of gestation can vary. An average litter consists of about six puppies, though this number may vary widely based on the breed of dog. Toy dogs generally produce from one to four puppies in each litter, while much larger breeds may average as many as 12 pups in each litter.

Spaying and Neutering

Neutering (spaying females and castrating males) refers to the sterilization of animals, usually by removal of the male's testicles or the female's ovaries and uterus, in order to eliminate the ability to procreate, and reduce sex drive. Neutering has also been known to reduce aggression in male dogs, but has been shown to occasionally increase aggression in female dogs.

Animal control agencies in the United States and the ASPCA advise that dogs not intended for further breeding should be neutered so that they do not have undesired puppies. Because of the overpopulation of dogs in some countries, puppies born to strays or as the result of accidental breedings often end up being killed in animal shelters.

Neutering can also decrease or eliminate the risk of hormone-driven diseases such as mammary cancer, as well as undesired hormone-driven behaviours. However, certain medical problems are more likely after neutering, such as urinary incontinence in females and prostate cancer in males.

The hormonal changes involved with sterilization are likely to somewhat change the animal's personality, however, and some object to neutering as the sterilization could be carried out without the excision of organs.

It is not essential for a female dog to either experience a heat cycle or have puppies before spaying, and likewise, a male dog does not need the experience of mating before castration.

Female cats and dogs are seven times more likely to develop mammary tumors if they are not spayed before their first heat cycle. The high dietary estrogen content of the average commercial pet food as well as the estrogenic activity of topical pesticides may be contributing factors in the development of mammary cancer, especially when these exogenous sources are added to those normal estrogens produced by the body.

Dog food containing soybeans or soybean fractions have been found to contain phytoestrogens in levels that could have biological effects when ingested longterm.

Gender-preservative surgeries such as vasectomy and tubal ligation are possible, but do not appear to be popular due to the continuation of gender-specific behaviours and disease risks.

According to the Humane Society of the United States, 3-4 million dogs and cats are euthanized each year in the United States and many more are confined to cages in shelters because there are many more animals than there are homes.

Spaying or castrating dogs helps keep overpopulation down. Local humane societies, SPCAs and other animal protection organizations urge people to neuter their pets and to adopt animals from shelters instead of purchasing them. Several notable public figures have spoken out against animal over population, including Bob Barker.

On his game show, The Price Is Right, Barker stressed the issue at the end of every episode, saying: "Help control the pet population. Have your pets spayed or neutered." Dogs are susceptible to various diseases, ailments, and poisons, some of which affect humans in the same way, others of which are unique to dogs.

Dogs, like all mammals, are also susceptible to heat exhaustion when dealing with high levels of humidity and/or extreme temperatures. Infectious diseases commonly associated with dogs include rabies (hydrophobia), canine parvovirus, and canine distemper.

Inherited diseases of dogs can include a wide range from elbow or hip dysplasia and medial patellar luxation to epilepsy and pulmonic stenosis.

Canines can get just about anything a human can get (excluding many infections which are species specific) like hypothyroidism, cancer, dental disease, heart disease, etc. Two serious medical conditions affecting dogs are pyometra, affecting unspayed females of all types and ages, and bloat, which affects the larger breeds or deep chested dogs.

Both of these are acute conditions, and can kill rapidly; owners of dogs which may be at risk should learn about such conditions as part of good animal care. First generation hybrids (such as this terrier mix) often are healthier than either parent due to the genetic phenomenon of heterosis or "hybrid vigor". Common external parasites are various species of fleas, ticks, and mites.

Internal parasites include hookworms, tapeworms, roundworms, and heartworms. See also CVBD (Canine Vector-Borne Diseases). Some breeds of dogs are also prone to certain genetic ailments, such as hip dysplasia, luxating patellas, cleft palate, blindness, or deafness.

Dogs are also susceptible to the same ailments that humans are, including diabetes, epilepsy, cancer, and arthritis. Gastric torsion and bloat is a dangerous problem in some large-chested breeds.

The typical lifespan of dogs varies widely among breeds. Based on questionnaire surveys of owners in the UK, Denmark, and the USA/Canada, the median longevity of most dog breeds is between 10 and 13 years. The breed with the dubious distinction of the shortest lifespan (among breeds for which there is a questionnaire survey with a reasonable sample size) is the Dogue de Bordeaux with a median longevity of about 5.2 years, but several breeds, including Miniature Bull Terrier, Bulldog, Nova Scotia Duck Tolling Retriever, Bloodhound, Irish Wolfhound, Greater Swiss Mountain Dog, Great Dane, and Mastiff, are nearly as short-lived, with median longevities between 6 and 7 years.

On the other end of the spectrum, the longest-lived breeds, including Toy Poodle, Border Terrier, Miniature Dachshund, Miniature Poodle, and Tibetan Spaniel, have median longevities between 14 and 15 years. The median longevity of mixed breed dogs (average of all sizes) is one or more years longer than that of purebred dogs (all breeds averaged). As a rule of thumb, small breeds are longer-lived than large breeds, but some of the longest lived large breeds have median longevities nearly as large as those of the shortest lived small breeds, and some of the breeds with the shortest longevities are medium-sized.

"Median longevity" refers to the age at which half the dogs in a population have died and half are still alive. Individual dogs, even in breeds with low median longevities, may live well beyond the median. The dog widely reported to be the longest-lived on record is "Bluey," purportedly born in 1910 in Australia. He died in 1939 at the age of 29.5 years.

Bluey is usually identified as an Australian Cattle Dog, but the first Australian Cattle Dog breed standard was written in 1902, only eight years before Bluey's birth. It is unclear how closely Bluey was related to the breed as it exists today. The Bluey record is anecdotal and unverified. The longest verified records are of dogs living to 24 years.

ANIMAL REPRODUCTIVE SYSTEM

Animal reproductive system, any of the organ systems by which animals reproduce. The role of reproduction is to provide for the continued existence of a species; it is the process by which living organisms duplicate themselves. Animals compete with other individuals in the environment to maintain themselves for a period of time sufficient to enable them to produce tissue nonessential to their own survival, but indispensable to the maintenance of the species. The additional tissue, reproductive tissue, usually becomes separated from the individual to form a new, independent organism.

The reproductive systems in metazoans (multicelled animals) from sponges to mammals, exclusive of humans. It focuses on the gonads (sex organs), associated ducts and glands, and adaptations that aid in the union of gametes—*i.e.*, reproductive cells, male or female, that are capable of producing a new individual by union with a gamete of the opposite sex. Brief mention is made of how the organism provides for the development of embryos and of the regulatory role of gonads in vertebrate cycles.

Unlike most other organ systems, the reproductive systems of higher animals have not generally become more complex than those of lower forms. Asexual reproduction (*i.e.*, reproduction not involving the union of gametes), however, occurs only in the invertebrates, in which it is common, occurring in animals as highly evolved as the sea squirts, which are closely related to the vertebrates. Temporary gonads are common among lower animals; in higher animals, however, gonads are permanent organs. Hermaphroditism, in which one individual contains functional reproductive organs of both sexes, is common among lower invertebrates; yet separate sexes occur in such primitive animals as sponges, and hermaphroditism occurs in animals more highly evolved—*e.g.*, the lower fishes. Gonads located on or near the animal surface are common in the lowest invertebrates, but in higher animals they tend to be more deeply situated and often involve intricate duct systems. In echinoderms, which are among the highest invertebrates, the gonads hang directly into the sea and spill their gametes into the water. In protochordates, gametes are released into a stream of respiratory water that passes directly into the sea. Duct systems of the invertebrate flatworms (Platyhelminthes) are relatively complex, and those of specialized arthropods (*e.g.*, insects, spiders, crabs) are more complex than those of any vertebrate. Copulatory organs occur in flatworms, but copulatory organs are not ubiquitous among vertebrates other than reptiles and mammals. The trend towards fewer eggs and increased parental care in higher animals may account for the relative lack of complexity in the reproductive systems of some advanced forms. Whereas trends towards increasing structural complexity have often been reversed during evolution, reproductive behaviour patterns in many phylogenetic (*i.e.*, evolutionary) lines have become more complicated in order to enhance the opportunity for fertilization of eggs and maximum survival of offspring.

A direct relationship exists between behaviour and the functional state of gonads. Reproductive behaviour induced principally but not exclusively by organic substances called hormones promotes the union of sperm (spermatozoa) and eggs, as well as any parental care accorded the young. There are a number of reasons why behaviour must be synchronized with gonadal activity. Chief among these are the following:

Individuals of a species must congregate at the time the gonads contain mature gametes. This often entails migration, and some members of all major vertebrate groups migrate long distances to gather at spawning grounds or rookeries. Individuals with gametes ready to be shed must recognize members of the opposite sex. Recognition is sometimes by external appearance or by chemical substances (pheromones), but sex-linked behaviour is often the only signal.

Geographical territories frequently must be established and aggressively defended. The building of nests, however simple, is essential reproductive behaviour in many species. When fertilization of aquatic forms is external, sperm and eggs must be discharged at approximately the same time into the water, since gametes may be quickly dispersed by currents. Courtship, often involving highly intricate behaviour patterns, serves to release the gametes of both mating individuals simultaneously.

When fertilization is internal, willingness of the female to mate is often essential. Female mammals not in a state of willingness to mate not only will not mate but may injure or even kill an aggressive male. The unwillingness of a female mammal to mate when mature eggs are not present prevents loss of sperm needed to preserve the species. Parental care of fertilized eggs by one parent or the other has evolved in many species. Parental behaviour includes fanning the water or air around the eggs, thereby maintaining appropriate temperature and oxygen levels; secretion of oxygen from a parent's gills; transport of eggs on or in the parental body (including the mouth of some male parents); and brooding, or incubation, of eggs.

Some species extend parental care into the postnatal period, feeding and protecting the offspring. Such behaviour patterns are adaptations for survival and thus are essential; all are induced by the nervous and endocrine systems and are typically cyclical, because gonadal activity is cyclical.

REPRODUCTIVE SYSTEMS OF INVERTEBRATES

GONADS, ASSOCIATED STRUCTURES, AND PRODUCTS

Although asexual reproduction occurs in many invertebrate species, most reproduce sexually. The basic unit of sexual reproduction is a gamete (sperm or egg), produced by specialized tissues or organs called gonads. Sexual reproduction does not necessarily imply copulation or even a union of gametes. As might be expected of such a large and diverse group as the invertebrates, many variations have evolved to ensure survival of species. In many lower invertebrates, gonads are temporary organs; in higher forms, however, they are permanent. Some invertebrates have coexistent female and male gonads; in others the same gonad produces both sperm and eggs. Animals in which both sperm and eggs are produced by the same individual (hermaphroditism) are termed monoecious. In dioecious species, the sexes are separate. Generally, the male gonads ripen first in hermaphroditic animals (protandry); this tends to ensure cross-fertilization. Self-fertilization is normal, however, in many species, and some species undergo sex reversal.

SPONGES, COELENTERATES, FLATWORMS, AND ASCHELMINTHS

Sponges are at a cellular level of organization and thus do not have organs or even well-developed tissues; nevertheless, they produce sperm and eggs and also reproduce asexually. Some species of sponge are monoecious, others are dioecious. Sperm and eggs are formed by aggregations of cells called amoebocytes in the body wall; these are not considered gonads because of their origin and transitory nature.

In hydrozoan coelenterates, temporary gonads are formed by groups of cells in either the epidermis (outer cell layer) or gastrodermis (gut lining), depending on the species; scyphozoan and anthozoan coelenterates generally have gonads in the gastrodermis. The origin and development of gonads in coelenterates, particularly freshwater species, are often associated with the seasons. Freshwater hydrozoans, for example, reproduce asexually until the onset of cold weather, which stimulates them to form testes and ovaries. Colonial hydrozoans asexually produce individuals known as polyps. Polyps, in turn, give rise to free swimming stages (medusae), in which gonads develop. The body organization of sponges and coelenterates is such that most of their cells are in intimate contact with the environment; consequently, gametes are shed into the water, and no ducts are necessary to convey them to the outside.

In contrast to sponges and coelenterates, platyhelminths generally have well-developed organ systems of a permanent nature and, in addition, have evolved secondary reproductive structures to convey sex products. One exception is the acoels, a group of primitive turbellarians; they lack permanent gonads, and germinal cells develop from amoebocytes in much the same manner as in sponges. The majority of flatworms, however, are monoecious, the primary sex organs consisting of one or more ovaries and testes. The tube from the ovary to the outside is called the oviduct; it often has an outpocketing (seminal receptacle) for the storage of sperm received during copulation. In many species the oviduct receives a duct from yolk (vitelline) glands, whose cells nourish the fertilized egg. Beyond the entrance of the duct from the yolk glands the oviduct may be modified to secrete a protective capsule around the egg before it is discharged to the outside. The male organs consist of testes, from which extend numerous tubules (vasa efferentia) that unite to form a sperm duct (vas deferens); the latter becomes an ejaculatory duct through which sperm are released to the outside. The sperm duct may exhibit expanded areas that store sperm (seminal vesicles), and it may be surrounded by prostatic cells that contribute to the seminal fluid. The sperm duct eventually passes through a copulatory organ. The same basic structural pattern, somewhat modified, is found in most higher invertebrates.

Aschelminthes (roundworms) are mostly dioecious; frequently there are external differences between males and females (sexual dimorphism). The males are generally smaller and often have copulatory spicules. Nematodes have relatively simple reproductive organs, a tubular testis or ovary being located at the end of a twisted tube. The portion of the female tract nearest the ovary forms a uterus for temporary storage of fertilized eggs. Some species lay eggs, but others retain the egg in the uterus until the larva hatches. The sperm are released into a cavity called the cloaca. A number of free-living nematodes are capable of sex reversal—if the sex ratio in a given population is not optimal or if environmental conditions are not ideal, the ratio of males to females can be altered. This sometimes results in intersexes; *i.e.*, females with some male characteristics. Hermaphroditism occurs in nematodes, and self-fertilization in such species is common.

ANNELIDS AND MOLLUSKS

Annelids have a well-developed body cavity (coelom), a part of the lining of which gives rise to gonads. In some annelids, gonads occur in several successive body segments. This is true, for example, in polychaetes, most of which are dioecious. Testes and ovaries usually develop, though not invariably, in many body segments; and the sperm and eggs, often in enormous numbers, are stored in the coelom. Fertilization is external. In oligochaetes (all of which are monoecious) on the other hand, the gonads develop in a few specific segments. Sperm are stored in a seminal vesicle and eggs in an egg sac, rather than in the coelom. A portion of the peritoneum, the membrane lining the coelom, becomes a saclike seminal receptacle that stores sperm received from the mate. The earthworm, Lumbricus terrestris, is an example of a specialized annelid reproductive system.

Female organs consist of a pair of ovaries in segment 13; a pair of oviducts that open via a ciliated funnel (*i.e.*, with hairlike structures) into segment 13 but open to the exterior in segment 14; an egg sac near each funnel; and a pair of seminal receptacles in segment 9 and another in segment 10. Male organs consist of two pairs of minute testes in segments 10 and 11, each associated with a ciliated sperm funnel leading to a tiny duct, the vas efferens. The two ducts on each side lead to a vas deferens that opens in segment 15. Testes and funnels are contained within two of three pairs of large seminal vesicles that occupy six body segments. Leeches (Hirudinea), also monoecious, have one pair of ovaries and a segmentally arranged series of testes with duct systems basically similar to those of earthworms.

Although mollusks have a close evolutionary kinship to annelids, they have reduced or lost many structures characteristic of segmented worms. The coelom persists only as three regional cavities: gonadal, nephridial (kidney), and pericardial (heart). In ancestral forms these were interconnected so that gametes from the gonad passed through the pericardial cavity, the nephridial cavity, then to the outside through a nephridial pore. The various groups of mollusks have tended to modify this arrangement, with the result that gonads have their own pore; among amphineurans, for example, the sexes are usually separate, and there is one gonad with an associated pore. Gastropods show considerable variability, but generally one gonad (ovary, testis, or ovotestis—a structure combining the functional gonads of both sexes) is located in the visceral hump and connected to the outside by a remnant of the right kidney. In hermaphroditic forms, one duct carries sperm as well as eggs.

The gonadal ducts of gastropod females often secrete a protective capsule around the fertilized eggs; in males, the terminal portion of the duct is sometimes contained in a copulatory organ. Pelecypods may be either monoecious or dioecious, but the gonads are usually paired. In mussels and oysters, the gonads open through the nephridial pore, but in clams the reproductive system opens independently. The cephalopods are all dioecious. The single testis or ovary releases its products into the pericardial cavity and this, in turn, leads to a gonopore, the external opening. The oviduct of the squid is terminally modified to form a shell gland. The male system is more complex—the gonoduct leads into a seminal vesicle where a complicated torpedo-shaped sperm case (spermatophore) is secreted and contains the sperm. Spermatophores are then stored in a special structure (Needham's sac) until copulation occurs.

A remarkable characteristic of some mollusks is the ability to alter their sex. Some species are clearly dioecious; however, among the monoecious species there is considerable variability in their hermaphroditic condition. In some species, male and female gonads, although in the same individual, are independent functionally and structurally. In others, an ovotestis produces both sperm and eggs. Oysters display a third condition; young oysters have a tendency towards maleness, but, if water temperature or food availability is altered, some individuals develop into females. Later, a reversal to the male condition may occur. The sexual makeup of an entire oyster population also has a seasonal aspect; in harmony with the group, an individual may undergo several alterations in the course of a year. A similar phenomenon, called consecutive sexuality, occurs in limpets. These gastropods stack themselves in piles, with the younger animals on top. The animals on top are males with well-developed testes and copulatory organs; those in the middle are hermaphroditic; those on the bottom are females, having lost the testes and copulatory organ (penis) by degeneration. A decrease in the number of females in a stack induces males to assume female characteristics, but the transition is retarded when an excess of females is present. The degree of maleness or femaleness is probably controlled in part by environmental and internal factors.

ARTHROPODS

The phylum Arthropoda includes a vast number of organisms of great diversity. Most arthropods are dioecious, but many are hermaphroditic, and some reproduce parthenogenetically (*i.e.*, without fertilization). The primary reproductive organs are much the same as in other higher invertebrates, but the secondary structures are often greatly modified. Such modifications depend on whether fertilization is internal or external, whether the egg or zygote (*i.e.*, the fertilized egg) is retained or immediately released, and whether eggs are provided some means of protection after they have left the body of the female. The mandibulate arthropods (*e.g.*, crustaceans, insects) include more species than any other group and have invaded most habitats, a fact reflected in their reproductive processes.

Crustaceans (*e.g.*, crabs, crayfish, barnacles) are for the most part dioecious. The primary reproductive organs generally consist of paired gonads that open through paired ventral (bottom side) gonopores. Females often have a seminal receptacle (spermatheca) in the form of an outpocketing of the lower part of the female tract or as an invagination (inpocketing) of the body near the gonopore. Males have appendages modified for clasping the female during copulation or for guiding sperm. A number of groups have members that reproduce parthenogenetically. Branchiopods (*e.g.*, water fleas, fairy shrimp) have simple paired gonads. The female gonopore often opens dorsally (on the back side) into a brood chamber; the male gonopore opens near the anus. Males have appendages for clasping females during copulation. Ostracods, or seed shrimp, have paired, tubular gonads. The eggs may be brooded by the female, or they may be released into the water via a gonoduct and gonopore. The terminal portion of the male gonoduct is enclosed in a single or paired penis. Many species reproduce parthenogenetically. Some experts contend that this is the only method employed, even though functional males may be present in the population. Copepods (*e.g.*, Cyclops) have paired ovaries and an unpaired testis. The terminal portion of the oviduct constitutes an ovisac for storage of eggs. The male deposits sperm in a spermatophore that is transferred to the female. Sexual dimorphism is particularly evident among parasitic copepods. Frequently, parasitic females can hardly be recognized as copepods except for the distinctive ovisacs. Males, on the other hand, are free-living and are recognizable as copepods.

The hermaphroditic Cirripedia (*e.g.*, barnacles) are among the exceptions to the generalization that crustaceans are dioecious. It has been suggested that hermaphroditism in barnacles is an adaptation to their sessile, or stationary, existence, but cross-fertilization is more common than self-fertilization. The ovaries lie either in the base or in the stalk of the animal, and the female gonopore is near the base of the first pair of middle appendages (cirri). The testes empty into a seminal vesicle through a series of ducts; from the vesicle extends a long sperm duct within a penis that may be extended to deposit sperm in the mantle cavity of an adjacent barnacle. The terminal portion of the oviduct secretes a substance that forms a kind of ovisac within the mantle cavity, where fertilized eggs undergo early development. Although most barnacles are hermaphrodites, some display a peculiar adaptation in that they contain parasitic dwarf or accessory males. Dwarf males are much smaller than the host barnacle in which they live and are degenerate, except for the testes. In some species they live in the mantle cavity of hermaphroditic forms and produce accessory sperm; in other species only the female organs persist in the host animal, and the accessory male is a necessity.

Amphipods and isopods (*e.g.*, pill bugs, sow bugs), like most crustaceans, are dioecious and have paired gonads. Females of both groups have a ventral brood chamber (marsupium) formed by a series of medially directed (*i.e.*, towards the body midline) plates (oostegites) in the region of the thorax, the region between head and abdomen. Many isopods are parasitic and have developed unusual sex-related activities. Certain species are parasitic on other crustaceans. After a series of molts (*i.e.*, shedding of the body covering) a parasitic larval (immature) isopod attaches to the shell of a crab. If it is the only larva to do so, it increases in size and develops into an adult female. If another larva subsequently attaches, the new arrival becomes a male. It has been demonstrated that the testes of the functioning male larva will change to ovaries if the larva is removed to a new, uninfected host. Thus, the larvae of these species apparently are intersexual and can develop into either sex. This phenomenon, reminiscent of that in mollusks, demonstrates the way in which similar adaptations have evolved in diverse groups of organisms.

The gonads of crabs and lobsters are paired, as are the gonopores. The females of many species have external seminal receptacles on the ventral part of the thorax; those of other species have internal receptacles in the same region. In some species, seminal receptacles are absent, and the male simply attaches a spermatophore to the female. Thus, males either have appendages (gonopods) by which sperm are inserted in the body of the female or produce spermatophores for sperm transfer. The sexual dimorphism of many decapods can be altered by parasitism. An example of this is the crab that is parasitized by a barnacle. A barnacle infection in male crabs induces the secondary sex characters of the crab to resemble those of a female; however, masculinization does not occur in parasitized females. At each molt a parasitized male crab increasingly resembles a female even though the testes may be completely unaffected. Feminization results from a hormonal alteration of the parasitized crab. Insects are rarely hermaphroditic, but many species reproduce parthenogenetically (without fertilization). The insect ovary is composed of clusters of tubules (ovarioles) with no lumen, or cavity.

The upper portion of each ovariole gives rise to oocytes (immature eggs) that mature and are nourished by yolk from the lower portion. The oviduct leads to a genital chamber (copulatory bursa, or vagina), with which are often associated accessory glands and a seminal receptacle. Some accessory glands form secretions by which eggs become attached to a hard surface; others secrete a protective envelope around the egg. The eighth and ninth body segments are often modified for egg-laying. The paired testes consist of a series of seminal tubules that form primary spermatogonia (immature spermatozoa) at their upper ends. As the spermatogonia mature a covering is secreted around them. Eventually they enter a storage area (seminal vesicle). The terminal portion of the male system is an ejaculatory duct that passes through a copulatory organ. A pair of accessory glands, often associated with the ejaculatory duct, contributes to the semen (fluid containing sperm) or participates in spermatophore formation.

The ninth body segment and sometimes the tenth bear appendages for sperm transfer. Scorpions and spiders have tubular or saclike gonads; the female system is equipped to receive and store sperm, and, in some species, the female retains the eggs long after fertilization has occurred. Male spiders may have a cluster of accessory glands associated with the terminal portion of the reproductive system for the manufacture of spermatophores, or they may have expanded seminal vesicles for the retention of sperm until copulation takes place. Often specific appendages are adapted for sperm transfer.

REPRODUCTIVE SYSTEMS OF VERTEBRATES

GONADS, ASSOCIATED STRUCTURES, AND PRODUCTS

The reproductive organs of vertebrates consist of gonads and associated ducts and glands. In addition, some vertebrates, including some of the more primitive fishes, have organs for sperm transfer or ovipository (egg-laying) organs. Gonads produce the gametes and hormones essential for reproduction. Associated ducts and glands store and transport the gametes and secrete necessary substances. In addition to these structures, most male and female vertebrates have a cloaca, a cavity that serves as a common terminal chamber for the digestive, urinary, and reproductive tracts and empties to the outside. In lampreys and most ray-finned fishes in which the cloaca is small or absent, the alimentary canal has a separate external opening, the anus.

In some teleosts the alimentary, genital, and urinary tracts open independently. Hagfishes, which are closely related to the lampreys, have a short cloaca. In many vertebrates other than mammals, especially reptiles and birds, the cephalic, or head, end of the cloaca is partitioned by folds into a urinogenital chamber (urodeum) and an alimentary chamber (coprodeum) that open into a common terminal chamber (proctodeum). Above monotremes (*e.g.*, platypus, echidna) the embryonic cloaca becomes completely partitioned into a urinogenital sinus conveying urine and the products of the gonads, and an alimentary pathway; the two open independently to the exterior.

Gonads arise as a pair of longitudinal thickenings of the coelomic epithelium and underlying mesenchyme (unspecialized tissue) on either side of the attachment of a supporting membrane, the dorsal mesentery, to the body wall. At first, gonadal ridges bulge into the coelom and are continuous with the embryonic kidney. The germinal epithelium covering the gonadal ridges gives rise to primary sex cords (medullary cords) that invade the underlying mesenchyme. These cords establish within the gonadal blastema (a tissue mass that gives rise to an organ) a potentially male component, the medulla. Secondary sex cords grow inward, spreading just beneath the germinal epithelium to form a cortex. If the gonad is to become a testis, only the medullary component differentiates. If the gonad is to become an ovary, only the cortex differentiates.

The length of an adult gonad depends, in part, upon the extent of gonadal-ridge differentiation. In cyclostomes (lampreys and hagfish), elasmobranchs (sharks, skates, and rays), and teleosts most of it differentiates, and the gonads extend nearly the length of the body trunk. In tetrapods (amphibians, reptiles, birds, and mammals), the cranial portion, at the anterior end, generally does not differentiate; in toads only the more caudal, or posterior, portion does so. The middle segment in toads of both sexes gives rise to a Bidder's organ containing immature eggs. In anurans (frogs and toads) and some lizards of both sexes, one segment of the gonadal ridge gives rise to yellow fat bodies that, especially in anurans, diminish in size just prior to the breeding season. In mammals, only the middle portion of the gonadal ridge differentiates.

Some vertebrate species have only one gonad, which may lie in the midline or on one side; the condition is more common among females. Adult cyclostomes of both sexes have one gonad. In lampreys it is in the middle of the body; in hagfishes it is on the right side. Birds are the only other major group of vertebrates in which most females have one gonad, the right ovary being typically absent. Male birds have a pair of testes, however. Exceptions to the condition of single ovaries among birds include members of the falcon family, in which more than 50 percent of mature hawks have two well-developed ovaries. In all bird species a small percentage of females probably have two ovaries; reported instances include owls, parrots, sparrows, and doves, with estimates for doves ranging from 5 percent to 25 percent. A few teleosts and viviparous elasmobranchs have only one ovary; in sharks the right one is usually present, in rays, the left. In amniotes (*i.e.*, reptiles, birds, and mammals) unpaired gonads are unusual. Some lizards have one testis, and some female crocodiles have one ovary. Among mammals, the platypus usually has only a left ovary, and some bat species (family Vespertilionidae) have only the right.

One of two explanations may account for unpaired gonads: the paired embryonic gonadal ridges may fuse to form a median gonad—as in lampreys and the perch—or only one gonadal ridge may receive immigrating primordial germ cells (immature sperm or eggs), with the result that the opposite gonad does not develop—as in chickens and ducks. Both gonadal ridges have been reported to exhibit an equal number of primordial germ cells in embryonic hawks, and these typically have two ovaries. Among lower vertebrates, mature gonads sometimes produce both sperm and eggs. Hermaphroditism is more general in cyclostomes and teleosts than in other fishes. A teleost may function as a male during the early part of its sexual life and as a female later. In some teleost families sperm and eggs mature simultaneously but in different regions of the gonad. These fish normally function as males during one season and as females the next. Cyclostomes generally are ambisexual during juvenile life—*i.e.*, immature male and female sex cells exist side by side, or, as in Myxine, the anterior part of the immature gonad may be ovary and the caudal part, the testis. It is thought that cyclostomes normally become unisexual at maturity. Hermaphroditism is uncommon among amphibians, although it frequently occurs as an anomaly. In vertebrates above amphibians, true hermaphroditism probably does not exist. Both male and female duct systems are occasionally absent. In cyclostomes, a few elasmobranchs, and some teleosts, such as salmon, trout, and eels, the gametes are propelled towards the posterior within the coelom, often by cilia (minute hairlike structures), and exit via a pair of funnel-like genital pores near the base of the tail. In cyclostomes, the pores lead to a sinus, or cavity, within a median papilla (*i.e.*, a fingerlike structure) and are open only during breeding seasons.

MALE SYSTEMS

Tests

In anurans, amniotes (reptiles, birds, and mammals), and even some teleosts, testes are composed largely of seminiferous tubules—coiled tubes, the walls of which contain cells that produce sperm—and are surrounded by a capsule, the tunica albuginea. Seminiferous tubules may constitute up to 90 percent of the testis. The tubule walls consist of a multilayered germinal epithelium containing spermatogenic cells and Sertoli cells, nutritive cells that have the heads of maturing sperm embedded in them. Seminiferous tubules may begin blindly at the tunic, or outermost tissue layer, and pass towards the centre, becoming tortuous before emptying into a system of collecting tubules, the rete testis.

Such an arrangement is characteristic of frogs. In certain amniotes—the rat, for example—the tubules may be open ended, running a zigzag course from the rete to the periphery and back again. The average length of such tubules is 30 centimetres (12 inches), and they seldom communicate with each other. In many mammals the tubules are grouped into lobules separated by connective-tissue septa, or walls. The arrangement permits the packing of an extensive amount of germinal epithelium into a small space. In immature males and in adult males between breeding seasons, the tubules are inconspicuous and the epithelium is inactive; in some species, however, spermatogenesis, or production of sperm, proceeds at a variable pace throughout the year. An active epithelium may exhibit all stages of developing sperm. The lumen, or tubule cavity, contains the tails of many sperm (the heads of which are embedded in

Sertoli cells), free sperm, and fluid that is probably resorbed. In mammals, in any single zone along a tubule, all sperm are at the same stage of maturation; adjacent zones contain different generations of sperm, and a period of sperm formation and discharge is followed by an interval of inactivity.

In cyclostomes, most fishes, and tailed amphibians the germinal epithelium is arranged differently. Instead of seminiferous tubules there are large numbers of spermatogonial cysts (also called spermatocysts, sperm follicles, ampullae, crypts, sacs, acini, and capsules) in which sperm develop, but in which the epithelium is not germinal. Spermatogenic cells migrate into the cysts from a permanent germinal layer, which, depending on the species, may lie among cysts at the periphery of the testes or in a ridge along one margin of the testis. After invading the thin non-germinal epithelium of a cyst, spermatogenic cells multiply, producing enormous numbers of sperm. The cysts become greatly swollen and whitish in colour; the entire testis also swells and has a granular appearance. As sperm mature, they separate from the epithelium and move freely in the cystic fluid. Finally, the cysts burst, and the sperm are shed into ducts. In the case of cyclostomes and a few teleosts the sperm are shed into the coelom. The cysts, totally emptied, collapse. Then either they are replaced by new ones, or they become repopulated by additional spermatogenic cells. It is not yet known which of these processes occurs.

Testicular stroma, which fills the spaces between seminiferous tubules or spermatogenic cysts, consists chiefly of connective tissue, blood and lymphatic vessels, and nerves; it is more abundant in some vertebrates than in others. Glandular Leydig (interstitial) cells are also present in most, if not all, vertebrates. Thought to be a primary source of androgens, or male hormones, Leydig cells are not always readily distinguishable, and, in some bird species, they may be seen only with the electron microscope. The capillary system of the rat testis, and probably that of many other vertebrates, is such that blood that has bathed the Leydig cells flows to the tubules; it is thus probable that Leydig cell hormones have an immediate effect on the germinal epithelium.

Testes in vertebrates below mammals lie within the body. This is also true of many, sometimes all, members of the mammalian orders Monotremata, Insectivora, Hyracoidea, Edentata, Sirenia, Cetacea, and Proboscidea. Some male mammals—most marsupials, ungulates, carnivores, and primates after infancy—have a special pouch (scrotum) that the testes occupy permanently. A few mammals have a pouch into which the testes descend and from which they can be retracted by muscular action. These include a few rodents such as ground squirrels; most, if not all, bats; and some primitive primates (loris, potto). The scrotum consists of two scrotal sacs, each connected to the abdominal cavity by an inguinal canal lined with the peritoneal membrane. The canals are the path of descent (and retraction) of the testes to the sacs. In descending, the testes carry along a spermatic duct, blood and lymphatic vessels, and a nerve supply wrapped in peritoneum and constituting, collectively, the spermatic cord. Rabbits, most rodents, and some insectivores, which lack scrotal sacs, have instead a wide inguinal canal into which the testes may be drawn and from which they are retracted when in danger of injury. In these mammals, descended testes cause a temporary bulge in the perineal region (*i.e.*, between the anus and the urinogenital opening). In a small number of mammals, the testes permanently occupy the perineal location. The scrotum is a temperature-regulating device. Warm blood approaching the testis comes close to the vessels carrying cool blood leaving the testis, so that the blood approaching the testis is cooled; the vessels form an intricate vascular network (pampiniform plexus) within the spermatic cord. Failure of both testes to enter the scrotal sacs (cryptorchidism) results in permanent sterility. In cold weather two sets of muscles, the dartos and cremasteric, pull the testes close to the body. The dartos lies between the two scrotal sacs and is attached to the scrotal skin. The cremaster, wrapped around the spermatic cord, is an extension of the abdominal wall musculature. It retracts the testis. Birds, like mammals, are homoiothermic (warm-blooded), and their testes are near air sacs (extensions of incurrent respiratory tubes). Air in the sacs may help regulate the temperature of the testes.

Ducts

The male duct system begins as the rete testis, a network within the testis of thin-walled ductules, or minute ducts, that collects sperm from the seminiferous tubules. The rete is drained by a number of small ducts—usually fewer than ten—called the vasa efferentia, which are modified kidney tubules. In some fishes and amphibians the vasa efferentia connect the testes with the cranial (anterior) end of the kidneys. In anamniotes (*e.g.*, fish and amphibians), therefore, except teleosts, the ducts that drain the kidneys usually drain the testes also. In most amphibians these ducts pass caudad, or posteriorly, to empty independently into the cloaca; in some fishes they pass through a median urinogenital papilla.

Although drainage of the testis and the kidney by the same duct is a basic pattern, there has been a tendency in many vertebrates towards separate spermatic and urinary ducts. This tendency is manifested in one of two ways among anamniotes. In many sharks and in some amphibians (Plethodontidae, Salamandridae, Ambystomatidae), the embryonic kidney duct ultimately drains the testis, and one or more new ducts (ureters) drain the adult kidney. On the other hand, in the primitive fish Polypterus and in most teleosts, the embryonic kidney duct drains the adult kidney, and a new duct arises to drain the testis. Many degrees of separation of the two ducts occur in anamniotes, from the condition of the sturgeon, in which the spermatic duct unites with the urinary duct far towards the head, to the condition in Esox (a pike), in which spermatic and urinary ducts empty independently to the exterior. In amniotes, the mesonephric kidney is a temporary structure confined to the embryo, but the mesonephric duct persists in the adult male as a sperm duct. A separate ureter drains the adult kidney. The spermatic and urinary ducts empty independently into the cloaca except in mammals above monotremes, in which they are confluent with the urethra. The epididymis of amniotes, a highly tortuous duct draining the vasa efferentia, usually serves as a temporary storage place for sperm; it is small in birds and large in turtles. In mammals, the first part of the epididymis consists of a head, body, and tail that wrap around the testis; it gradually straightens to become the spermatic duct. The epididymis secretes substances that prolong the life of stored sperm and increase their capacity for motility.

In all vertebrates certain regions of the spermatic duct are lined by cilia and a variety of secretory epithelial cells. One end may enlarge to form a sperm reservoir or secrete seminal fluid. In the catfish Trachycorystes mirabilis secretions of the spermatic duct form a gelatinous plug in the female similar to the vaginal plug of mammals. A seminal glomulus in birds functions as a sperm reservoir. In some mammals an enlargement of the spermatic duct called the ampulla contributes to the seminal fluid and stores sperm. Small mucous glands (of Littré) and other glandular structures open into the urethra along its length. Cloacal glands in basking sharks and many salamanders form a jelly that encloses sperm in a spermatophore. Cloacal glands of some lizards produce secretions called pheromones. The siphon sac of elasmobranchs is one of the few accessory sex glands that is a separate organ in animals below mammals. It extends as an elongated pocket into the pelvic fin and secretes a nutritive mucus that enters the female reproductive tract with sperm.

ACCESSORY GLANDS

Accessory sex glands that are conspicuous outgrowths of the genital tract are almost uniquely mammalian. The major mammalian sex glands include the prostate, the bulbourethral, and the ampullary glands, and the seminal vesicles. All are outgrowths of the spermatic duct or of the urethra and all four occur in elephants and horses and in most moles, bats, rodents, rabbits, cattle, and primates. A few members of these groups lack ampullary glands, or ampullary glands and seminal vesicles. Cetaceans (whales, porpoises) have only the prostate, as do some carnivores, including dogs, weasels, ferrets, and bears.

The prostate, the most widely distributed mammalian accessory sex gland, is absent only in Echidna (a marsupial) and a few carnivores. It empties into the urethra by multiple ducts. Many rodents, insectivores, and lagomorphs have three separate prostatic lobes; in a few mammals (some primates and carnivores) the prostate is a single mass with lobules and encircles the urethra at the base of the bladder. In a few mammals (*e.g.*, opossum), the prostate is not a compact mass but a partly diffuse gland. In many rodents (*e.g.*, rat, guinea pig, mouse, hamster) and some other mammals, the semen coagulates quickly after ejaculation as a result of a secretion from a male coagulating gland, which is usually considered part of the prostatic mass. Coagulated semen forms a vaginal plug that temporarily prevents copulation.

Bulbourethral (Cowper's) glands arise from the urethra near the penis and are surrounded by the muscle of the urethra or penis. Typically, there is one pair, but as many as three (marsupials) may be found. The glands, small in man, large in rodents, elephants, and some ungulates including pigs, camels, and horses, are absent in cetaceans, mustelids (*e.g.*, mink, weasel), sirenians (manatees, dugongs), pholidotans (pangolins), some edentates, and carnivores such as walrus, sea lion, bear, and dog.

Although many mammals have an ampullary swelling on the spermatic duct near the urethra, only a small number form a separate ampullary gland as an outgrowth of the duct. It is very large in some bats, absent in many mammalian orders, and variable in the rest. Although common in rodents, it is absent in guinea pigs and some strains of mice.

Seminal vesicles are paired, typically elongated and coiled fibromuscular sacs that empty into either the spermatic duct or the urethra. Absent in monotremes, marsupials, carnivores, cetaceans, and in some insectivores, chiropterans, and primates, seminal vesicles are exceptionally large in rhesus monkeys and small in man. They are absent in domesticated rabbits, small or rudimentary in cottontails, large in armadillos, and variable in sloths. They contribute the sugar fructose and citric acid to the semen but do not serve as sperm reservoirs.

FEMALE SYSTEMS

Ovaries

Ovaries lie within the body cavity and are suspended by a dorsal mesentery (mesovarium), through which pass blood and lymph vessels and nerves. Primitive vertebrate ovaries occur in the hagfish, in which a mesentery-like fold of gonadal tissue stretches nearly the length of the body cavity. Unique in the hagfish is the fact that functional ovarian tissue occupies only the forward half of the gonadal mass, the rear part containing rudimentary testicular tissue. In most fishes except very primitive forms, the ovaries are similarly elongated. In tetrapods other than mammals, the ovaries are usually confined to the middle third or half of the body cavity, particularly during non-breeding seasons. The ovaries of mammals undergo moderate caudal displacement, finally coming to lie between the kidney and the pelvis.

The appearance of an ovary depends on many factors—*e.g.*, whether one egg or thousands are discharged (ovulated); whether the eggs are immature or ripe; whether mature eggs are small or large; or whether pigments occur in the egg cytoplasm, such as those responsible for yellow yolk. Other factors also affect the appearance of the ovary: the season of the year in seasonal breeders (the ovary enlarges during breeding seasons, diminishes in size between seasons); the age of the animal (whether juvenile, reproductively active, or senile, particularly in birds and mammals); and the fate of ovulated, or discharged, egg follicles, or sacs. The ovaries are covered with a germinal epithelium that is continuous with the peritoneum lining the body cavity. The term germinal epithelium is inappropriate because in most adults it contains no germ cells, these having moved deeper into the ovary. In hagfishes and amphibians, cells that give rise to eggs are known to occur in the germinal epithelium, and it may be that the germinal epithelium in a few other vertebrates contains similar cells. The germinal epithelium undergoes cell division, however. This is particularly true of species in which enormous expansion of the ovary occurs each breeding season. Beneath the epithelium is a layer of connective tissue, the tunica albuginea, which is much thinner than that surrounding the testes.

A typical vertebrate ovary consists of cortex and medulla. The cortex, immediately internal to the tunica albuginea, contains future eggs and, at one time or another, eggs in ovarian follicles (*i.e.*, developing eggs); it undergoes fluctuations in size and appearance that correlate with stages of the reproductive cycle. The cortex also contains remnants of ovulated follicles and, in mammals, clusters of interstitial cells that, in some species, are glandular. The cortical components are embedded in a supportive framework of connective, vascular, and neural tissue constituting the stroma. Internal to the cortex is the medulla, consisting of blood and lymph vessels, nerves, and connective tissue. The medulla, which contains no germinal elements, exhibits no significant cyclical activity, is usually inconspicuous, is continuous with the dorsal mesentery, and, in cyclostomes, is hardly distinguishable from the latter. The mammalian medulla, on the contrary, is almost completely surrounded by cortex and converges on the mesovarium (*i.e.*, the part of the peritoneum that supports the ovary) at a narrow hilus, at which nerves and vessels enter the ovary. In the medulla of the mammalian ovary near the hilus are small masses of blind tubules or solid cords—the rete ovarii—which are homologous (*i.e.*, of the same embryonic origin) with the rete testis in the male. The microscopic right ovary of birds usually consists only of medullary tissue.

Ovaries are characterized as saccular, hollow, lacunate (*i.e.*, compartmented), or compact. The ovary of many teleosts, especially viviparous ones, contains a permanent cavity, which is formed during ovarian development when an invagination of the ovarian surface traps a portion of the coelom. The cavity is therefore unique in that it is lined by germinal epithelium. The lining develops numerous ovigerous folds that project into the lumen and greatly increase the surface area for proliferation of eggs. In most other teleosts, a temporary ovarian cavity develops after each ovulation, when the shrinking cortex withdraws from the outside ovarian wall along one side of the ovary. The resulting cavity is obliterated as eggs of the next generation enlarge. The permanent and temporary cavities of teleost ovaries and a similar cavity in garfish ovaries are continuous with the lumen of the oviduct, and eggs are shed into them. The ovaries of other fishes lack cavities and are characterized as compact. The amphibian ovary, which contains six or more central, hollow sacs that give it a lobed appearance, is characterized as saccular. The sacs are formed when the embryonic medullary and rete cords become hollow and coalesce. Maturing eggs bulge into the sacs but are not shed into them. The ovaries of reptiles, birds, and monotremes have cavities homologous to those in amphibians; the number of medullary spaces in the adults is considerably larger, however, so that the ovaries contain an extensive network of fluid-filled cavities (lacunae). Such ovaries are characterized as lacunate. The ovaries of mammals above monotremes are compact, having no medullary cavities.

An ovarian follicle consists of an oocyte, or immature egg, surrounded by an epithelium, the cells of which are referred to variously as follicular, nurse, or granulosa cells. In cyclostomes, teleosts, and amphibians, the epithelium is one layer thick. In the hagfish and those vertebrates in which the oocyte receives heavy deposits of yolk (elasmobranchs, reptiles, birds, and monotremes), the epithelium appears to be two cells thick, apparently the result of layering of nuclei in a simple columnar epithelium (*i.e.*, epithelium consisting of relatively "tall" cells). Above monotremes the follicular epithelium appears to be many cells thick; in at least one species, however, this is considered an artifact, and all granulosa cells are said to extend between the outer boundary of the epithelium and the oocyte.

The follicular epithelium originates as a few flattened cells derived from the germinal epithelium. Primary follicles are usually situated just under the tunica albuginea; secondary follicles lie deeper in the cortex. The primitive role of the follicular cells appears to be the secretion of the yolk-forming material onto or into the oocyte. Evidence from mammals indicates that the follicular cells may also have a role in converting substances produced elsewhere into female hormones, or estrogens. In some hibernating bats the granulosa cells are filled with glycogen, or animal starch, which may be a source of energy. Mammalian follicles above monotremes are unique in that they develop a fluid-filled cavity (antrum) within the granulosa layer. During antrum formation cell division of the granulosa cells increases, and fluid-filled spaces develop among the cells.

The spaces coalesce to form the antrum. Under the influence of pituitary gonadotropic hormones, many antral follicles thereafter continue to grow, forming large so-called Graafian follicles—less than 400 microns, or 0.4 millimetre (0.16 inch), in diameter in large mammals, 150–200 microns, or 0.15–

0.2 millimetre (0.006–0.008 inch), in small ones. Graafian follicles contain mature eggs and appear as large blisters on the ovary. At this stage the ovum, suspended within the fluid of the antrum (liquor folliculi) by a slender stalk of granulosa cells, is surrounded by a cluster of these cells, the cumulus oophorus, or discus proligerus. The remaining follicular cells form a thin wall surrounding the antrum. Antra are lacking in a few insectivores (Hemicentetes, Euriculus) because the granulosa cells swell and multiply to form corpora lutea, masses of yellow tissue. In the bat Myotis the antrum is likewise compressed and disappears just before discharge of the egg, or ovulation. In all vertebrates, oocytes that have begun to grow and mature may, at any time until just before ovulation, cease development and undergo atresia, or degeneration. This is a normal process that reduces the number of eggs ovulated. In small laboratory rodents, atresia takes place in 50 percent of the Graafian follicles in each ovary one or two days before ovulation, thus reducing the number of ovulatable eggs by 50 percent. A similar reduction takes place in hagfish prior to ovulation. Atretic follicles eventually become lost in the stroma of the cortex of the ovary. In mammals especially, follicles lacking oocytes and antra, called anovular follicles, as well as polyovular follicles (*i.e.*, containing more than one oocyte), occasionally occur.

The ovarian follicle of vertebrates, commencing with hagfish, is surrounded by a theca, or sheath, composed of two concentric layers of stromal cells. The outer layer (theca externa) is chiefly connective tissue but may contain smooth muscle fibres. The inner layer (theca interna) has more blood vessels and, in vertebrates that produce heavily yolked eggs, the largest vessels carry venous blood. In these species the cell membranes of the oocyte and granulosa cells have many microvilli (*i.e.*, fingerlike projections), which probably facilitate transport of substances important in yolk formation from the blood vessels to the egg. Mature follicles in the marsupial Dasyatus are said to lack theca, and in some bats only one thecal layer has been described.

During the growth phase, eggs in species with massive amounts of yolk may increase in size 106 (1,000,000) or more times as a result of vitellogenesis (deposit of yolk). In goldfish, on the other hand, when vitellogenesis commences, the egg has a diameter of 150 microns (0.15 millimetre [0.006 inch]); that of the mature egg is only 500 microns (0.5 millimetre [0.02 inch]). Mammalian eggs contain little yolk and vary little in size. Oogonia (*i.e.*, cells that form oocytes) of the golden hamster average 15 microns (0.015 millimetre [0.0006 inch]) in diameter, and eggs in Graafian follicles average 70 microns (0.07 millimetre [0.003 inch]). The mature eggs of horses and humans are approximately the same size—somewhat less than 150 microns. In seasonally breeding oviparous fishes and amphibians, all eggs are usually in the same stage of development, and the ovary grows to a mature state quite rapidly as a result of growth of the eggs, which frequently number more than 1,000,000. Such ovaries distend the body wall when mature; following spawning, the ovaries shrink rapidly to inconspicuous bodies consisting mainly of oogonia, immature oocytes, and a few stromal cells.

In reptiles and birds, ovarian weight also is high in proportion to body weight during egg-laying seasons. The weight of the ovary of the starling, for example, may increase from eight milligrams in early winter to 1,400 milligrams immediately before ovulation. The mature eggs of reptiles and birds are unique in that they are suspended from the ovary by a short stalk (pedicle). The stalk contains a cortex with additional oocytes in various stages of development and extensions of vessels and nerves. Full growth of the follicle in reptiles and birds requires only a few days or weeks (nine days in the domestic hen). In mammals, the ratio of ovarian weight to body weight varies insignificantly throughout the reproductive life of the female, and follicles in many stages of development are constantly present.

Vertebrate eggs are almost universally shed into the coelom or into a subdivision thereof, from which they enter the female reproductive tract. Even in those teleosts in which the eggs are shed into an ovarian cavity, the latter is often of coelomic origin. In many mammals a membranous sac of peritoneum, the ovarian bursa, traps part of the coelom in a chamber along with the ovary. The bursal cavity (periovarian space) may be broadly open to the main coelom, completely closed off from the coelom, or in communication with the coelom by a narrow, slitlike passage. The bursa, moderately developed in lower primates and catarrhines (Old World monkeys), is poorly developed in man. In horses, one edge of the ovary contains a long groove (ovulation fossa) into which all eggs are shed; the groove is found in a cleftlike ovarian bursa. The ovarian bursa increases the probability that all ovulated eggs will enter the oviduct.

The process of ovulation has been described for all vertebrate classes. Elasmobranchs, reptiles, and birds have massively yolked eggs. As ovulation approaches, the fimbria (*i.e.*, frills, or fringes) of the membranous and muscular funnel surrounding the entrance to the oviduct wave in a gentle, undulating motion. An egg that is nearly free of the ovary is grasped and partially encompassed by the fimbria; when the egg is freed, the fimbria draw the egg into the funnel. At this time, the egg has little shape and is partly squirted and partly flows into the oviduct; never completely free in the coelom, its chances of not entering the oviduct are small. In the case of moderately or poorly yolked eggs cilia help to sweep the eggs into the ostium, or opening, of the oviduct. During ovulation in Japanese rice fish, Oryzias latipes, a tiny papilla, or fingerlike process, develops on the surface of a bulging mature follicle in the centre (stigma) of the follicle. The follicle becomes thin at the stigma, an aperture appears, and the egg rolls out. In rabbits this process differs only in detail. During the final 20 minutes before ovulation in rabbits, some of the tiny blood vessels surrounding the stigma rupture, and a small pool of blood forms under the apex of the cone-shaped papilla. The follicular wall shortly gives way at the apex, and follicular fluid oozes from the opening, followed soon after by the egg. The ovulated mammalian egg typically is surrounded by a layer of columnar follicular cells, the corona radiata; but it is naked in some insectivores and some marsupials. Following ovulation in all vertebrates, the ovary may become smaller, become modified for maintenance of pregnancy, or proceed to form additional eggs.

The process of ovulation in vertebrates has been documented, but the immediate causes remain to be clarified. It is almost certain that an ovulatory hormone is secreted by the pituitary gland (*i.e.*, the so-called master endocrine gland) of all vertebrates. It is highly probable that breakdown of very small fibres that bind the follicular cells together may occur at the stigma, weakening the follicular wall at that location. Hormones from the ovary and other sources may play a role, as may neurohormones, which are hormones released at nerve endings. Rhythmic contractions of the entire ovary occur at ovulation in many vertebrates and have been described in rabbits. The role of mechanical pressure within the follicle, however, is not understood. Ovulation in most mammals (spontaneous ovulators) occurs cyclically as a result of the spontaneous release of the ovulatory hormone. In a few mammals (reflex ovulators) the stimulus of copulation is essential for release of the ovulatory hormone.

Striking postovulatory changes take place in the follicles of mammals and, to lesser degrees, of lower vertebrates. Blood vessels from the theca interna invade the ovulated follicles; the granulosa cells divide, enlarge, accumulate fats, and obliterate any remnants of the collapsed antra. Thereafter, they are known as lutein cells. Theca interna cells undergo changes identical to those of the granulosa cells.

The result in mammals is the formation of solid masses called corpora lutea, recognizable as prominent reddish-yellow bulges on the ovary. Corpora lutea produce the hormone progesterone, which is essential for the maintenance of pregnancy. The conversion of postovulatory follicles into structures more or less resembling mammalian corpora lutea has been demonstrated in numerous viviparous reptiles, amphibians, and elasmobranchs; in certain other fishes, including cyclostomes; and in some oviparous amphibians and reptiles. In birds, the postovulatory follicle shrinks, and identifiable corpora lutea do not develop, although some granulosa cells accumulate lipids of unknown significance.

TRACTS

The female reproductive tract consists of a pair of tubes (gonoducts) extending from anterior, funnel-like openings (ostia) to the cloaca, except as noted below. The gonoducts are specialized along their length for secretion of substances added to the eggs; for transport, storage, nutrition, and expulsion of eggs or the products of conception; and, in species with internal fertilization, for receipt, transport, storage, and nutrition of inseminated sperm. The predominately muscular tracts are lined by a secretory epithelium and ciliated over at least part of their length. Fusion of the caudal (tail) ends of the paired ducts may occur. Gonoducts are absent in cyclostomes and a few gnathostome fishes that have abdominal pores. A few vertebrates have only one functional gonoduct.

Gonoducts in lungfishes and amphibians are coiled muscular tubes that are ciliated over most of their length. Only occasionally do they unite caudally in a genital papilla before opening into the cloaca. During breeding seasons their diameter increases severalfold because of the highly active secretory epithelium. Between breeding seasons they are small. In some anurans (frogs, toads), such as Rana, the lower end of each gonoduct is expanded to form an ovisac, in which ovulated eggs are stored until spawning; the tube between the ostium (funnel-like opening) and ovisac is the oviduct. In viviparous amphibians the young develop in the ovisac. In amphibians, numerous multicellular glands extend deep into the lining of the female tract. Six successive glandular zones have been described in some urodeles, and these secrete six different gelatinous substances upon the egg. Female urodeles often have convoluted tubular outpocketings of the cloaca called spermatheca; they temporarily store sperm liberated from the male spermatophore.

The two gonoducts of elasmobranchs share a single ostium, a trait found only in Chondrichthyes. The ostium is a wide caudally directed funnel supported in the falciform ligament, which is attached to the liver. The role of the fimbria of the ostium at ovulation has been described. Two oviducts pass forward from the ostium to the septum transversum (*i.e.*, between the heart and abdominal cavities), curve around one end of the liver, then pass posteriorly on each side. Approximately midway between ostium and uterus each oviduct has a shell (nidamental) gland.

Fertilization takes place above the shell gland, which may be immense or almost undifferentiated. Half of the shell gland secretes a substance high in protein content (albumen), and the other half secretes the shell—delicate in viviparous forms, thick and horny in most oviparous species. Horny shells may have spiral ridges and many long tendrils, which entwine about an appropriate surface after the egg is deposited. In the viviparous shark Squalus acanthias several eggs pass one after the other through the shell gland, where they are enclosed in one long delicate membranous shell that soon disintegrates. Beyond the shell gland the oviducts terminate in an enlargement, which, in viviparous species, serves as a uterus. An oviducal valve may be found at the junction of oviduct and uterus. Although the two uteri usually open independently into the cloaca, they occasionally unite to form a bicornuate (two-horned) structure. In immature females, the uterus may be separated from the cloaca by a hymen, or membrane. The tract enlarges enormously during the first pregnancy and does not thereafter fully regress to its original size.

The gonoducts of most lower ray-finned fishes resemble those of lungfish, but those of gars and teleosts are exceptional in that the oviducts are usually continuous with the ovarian cavities. A median genital papilla receives the oviducts in teleosts, and the papilla is sometimes elongated to form an ovipositor. European bitterlings deposit their eggs in a mussel by means of the ovipositor, and female pipefish and sea horses deposit them in the brood pouch of a male. With certain modifications, the gonoducts of reptiles and birds are comparable to those of lower vertebrates. Crocodilians, some lizards, and nearly all birds have one gonoduct; the other is not well developed. Even in birds of prey having two functional ovaries, the right oviduct is sometimes undeveloped. The tracts of reptiles generally show less regional differentiation than do those of birds.

The oviduct funnel (ostium) in birds forms the chalazae—two coiled, springlike cords extending from the yolk to the ends of the egg. In both reptiles and birds, much of the length of the female tract is oviduct. This region, called the magnum in birds, secretes albumen; lizards and snakes do not form albumen. Behind the albumen-secreting region is a shell gland. In lizards, the gland is midway along the tract. In birds, the shell gland is at the posterior end, has thick muscular walls, and is often inappropriately called a uterus. It is preceded by a narrow region, or isthmus, which secretes the non-calcareous, or soft, membranes of the shell. The shell gland leads to a narrow muscular vagina that empties into the cloaca. The vagina secretes mucus that seals the pores of the shell before the egg is expelled. Special vaginal tubules (spermatheca) store sperm over winter in some snakes and lizards; seminal receptacles have been described in the oviduct funnel in some snakes. In birds, sperm storage glands (sperm nests) often occur in the funnel and at the uterovaginal junction. In lizards and birds, ovulation does not usually occur into a tract already containing an egg. Some lizards shed very few eggs per season; the gecko, for example, sheds only two.

The female reproductive tracts of monotremes, the egg-laying mammals, consist of two oviducts, the lower ends of which are shell glands. These open into a urinogenital sinus, which, in turn, empties into a cloaca. Marsupials have two oviducts, two uteri (duplex uterus), and two vaginas. The upper parts of the vaginas unite to form a median vagina that may or may not be paired internally. Beyond the median vagina, the vaginas are again paired (lateral vaginas) and lead to a urinogenital sinus. The posterior end of the pouchlike median vagina is separated from the forward end of the urinogenital sinus by a partition. When the female is delivering young, the fetuses are usually forced through the partition and into the urinogenital sinus, bypassing the lateral vaginas.

The ruptured partition may remain open thereafter, resulting in a pseudovagina. It closes in opossums, and in kangaroos both the median and lateral routes may serve as birth canals. The lateral vaginas in marsupials receive the forked tips of the male penis. Fertilization in all mammals takes place in the oviducts (Fallopian tubes). In eutherian mammals (*i.e.*, all mammals except monotremes and marsupials), with exceptions noted below, female reproductive tracts beyond the ostia (oviduct funnels) consist of two narrow and somewhat tortuous Fallopian tubes, two large uterine horns (each of which receives a Fallopian tube), a uterine body, and one vagina. Fallopian tubes often have a short dilated ampulla, or saclike swelling, just beyond the ostium. Implantation of the egg occurs only in the uterine horns; the embryos become spaced equidistant from one another in both horns even if only one ovary has ovulated. In some species one horn is rudimentary—the left in the impala (an African antelope)—and the embryos become implanted in the other horn, even though both ovaries ovulate. The body of the uterus in some mammals (*e.g.*, rabbits, elephants, aardvarks; some rodents, bats, insectivores) contains two separate canals (bipartite uterus).

In other mammals (ungulates, many cetaceans, most carnivores and bats) the body of the uterus has one chamber into which the two horns empty (bicornuate uterus). There are numerous intermediate conditions between the bipartite and bicornuate condition. Apes, monkeys, and man have no horns, and the Fallopian tubes empty directly into the body of the uterus (simplex uterus). In all mammals, the uterine body tapers to a narrow neck (cervix). The opening (os uteri) into the vagina is guarded by fleshy folds (lips of the cervix). The vagina in eutherian mammals other than rodents and primates terminates in a urinogenital sinus that opens to the exterior by a urinogenital aperture. In some rodents and in higher primates the vagina opens directly to the exterior. In the young of many species a membrane, the hymen, closes the vaginal opening. In guinea pigs the hymen reseals the opening after each reproductive period. Sperm are stored over winter in the uterus of some bats and in vaginal pouches in others.

MECHANISMS OF REPRODUCTION

Death by accident, disease, or aging awaits all organisms. So, if a population is to continue, those that die must be replaced. Reproductive mechanisms are as diverse as species, but even though we usually associate reproduction and sex, some species can reproduce asexually. In *asexual* reproduction a single parent produces offspring with characteristics identical to itself. This method has important advantages. For one, it isn't dependent on the presence of other individuals. One organism can start a whole new population. Furthermore, asexual reproduction may be simpler and more efficient. Asexual reproduction does not occur in vertebrates, however, it occurs in some chordates specially the Urochordates. By comparison, sexual reproduction may seem needlessly complicated.

Sexual reproduction is almost universal in vertebrates; only a very few lizards normally reproduce *parthenogenetically*—without needing (or having) mates. In vertebrates, the sexes are functionally separate in the overwhelming majority of cases. The sexual reproduction consists of ovaries and testes, which produce and nurture the gametes, and ducts that transport the eggs or sperms.

Fertilization may take place either internally or externally. If fertilization is external, the female deposits her eggs in the water (*spawns*) and the male fertilizes them during the process. Although the sperm remains viable for several hours, the egg soon loses its ability to become fertilized, perhaps because of the absorption of water by its jelly coats, which then act as a barrier to sperm penetration. *Intermittent organs* develop in those animals (mainly terrestrial) with internal fertilization and aid in guiding the sperm to the female reproductive trace. *Accessory sexual structures* present in the male compensate for the lack of an aquatic environment by secreting a fluid medium for the sperms. Since the sperms are hardly more than sacks of chromosomes, this material, the semen, also provides nourishment for them and induces their motility. The type of fertilization practiced by a few vertebrates, such as newts and salamanders, does not fall into either of the above categories. The male of these animals deposits immobile sperms in small gelatinous packets called *spermatophores,* on twigs and debris near the female, and she picks them up with her cloacal lips and stores them. Fertilization occurs internally as the eggs leave the oviducts.

The number of eggs released by the vertebrate animal during a breeding period shows a definite relationship to the type of fertilization practiced. Fishes and amphibians ovulate large number of eggs. Since the adults abandon the eggs after they are spawned and show no further interest in their fate, the prodigious numbers ensure species survival. For example, it is estimated that a population of Pacific herring releases from 8 to 75 billion eggs each year. Most of them hatch (95 per cent), but only 0.1 per cent survive long enough to reach maturity. On the other hand, in those species in which the young ones are nurtured within the body of the mother until the time of birth, only a few eggs are usually released during a reproductive cycle. Normally, a human female ovulates only one.

The vertebrate reproductive system, like all systems in the body, is under the control of chemical agents or *hormones.* Ductless or endocrine glands of the body generally produce the hormones, which basically function by modifying cellular metabolism. As a result, they coordinate and integrate the adjustments made by the body to living conditions. Changes in structure, function, or development of the individual organs reflect hormonal control and result in the maintenance of a stable state. The vertebrate reproductive system is an excellent example of a system operating under endocrine control. The cyclic changes in structure, function, and behaviour associated with the female reproductive cycle can be correlated with fluctuations in hormone secretion.

MALE REPRODUCTIVE ORGANS

The reproductive system in most males consists of a pair of *testes, sperm-conducting ducts,* and certain *accessory tissues.* In all vertebrates the testes perform two essential functions necessary for reproduction. They act both as an exocrine organ and an endocrine gland, since they produce sperms and also secrete the male hormone. The testes of most vertebrates are located in the upper part of the abdominal cavity lateral to or just anterior to the kidneys. A double fold of the peritoneal mesentery, the *mesorchium,* usually suspends then from the dorsal wall. In the majority of mammals, however, the testes do not remain in this position within the abdominal cavity, but descend posteriorly into the *scrotal sacs.*

Each testis is covered by a this layer of peritoneal epithelium under which lies a tough connective tissue capsule, the *tunica albuginea.* In some animals septa extend radially to the capsule, thus partitioning the gland into smaller areas; in other cases the testis is lobular or segmented. *Seminiferous tubules,* which produce the sex cells, and *interstitial tissue,* which secretes *androgen* (the general term of the male hormone), compose the testis regardless of its shape, size, or location. The tubules may be short (ampulla) or long and tortuous, but in either case their walls usually consist of an outer connective tissue capsule, a thin basement membrane, and a lining of stratified epithelium. *Spermatogenesis* occurs in these seminiferous tubules. By a series of rapid cell divisions the spermatogonia, undifferentiated sex cells located beneath the basement membrane, transform into mature spermatozoa and pass into the lumen of the tubules. A cross section through the tubules reveals many stages of differentiating sperm, usually in more or less distinct layers, located between the basement membrane and the cavity of the tubule. It is possible by repeated cell division of a small number of spermatogonial cells to produce several million sperms each reproductive cycle.

Interspersed among the spermatogonia of the seminiferous tubule is a second type of cell, the sustentacular or *Sertoli cell.* These cells, wedged between the many layers of cells undergoing division, extend from the basement membrane to the lumen. They help provide structure to the tubule and function as nurse cells for the sperms. The mature sperm has no way or manufacturing its food and must depend upon some external source for its nutritive materials. The heads of several mature sperms embed in membrane folds of the *Sertoli cells,* where they apparently receive nourishment and remain until released into the genital tract of the male. In general, most male vertebrates have mature sperms only during breeding periods, whereas during periods of sexual quiescence the tubules undergo changes in which large amounts of lipid material accumulate. There are exceptions to this general statement, however, since some vertebrates have mature sperms the year round. Amphibia and certain male mammals, such as man, remain spermatogenically active. Not all spermatogonia within a tubule mature at one time, since some must remain as reserve cells for future sperms. In some anuran testes, for example, one lobe of the testis may contain active sperms, whereas in an adjacent lobe, all the spermatogonia remain in a quiescent state. In man, one area of seminiferous tubules may be actively producing mature sperms whereas another area may be inactive.

The seminiferous tubules usually anastomose (join together) at their distal ends and empty into straight tubules. These tubules empty into a network of sinuses of channels known as the *rete tubules.* In man, the rete tubules converge to form 12 to 14 *efferent ductules,* which transport the sperms from the testis to the sperms duct. The efferent ductules are located in the mesorchium of lower animals. The ductules appear as this threads leading from the testis to the anterior end of the sperm duct and may be seen even in preserved animal if the mesentery is held up to the light.

The *interstitial cells,* or *Leydig cells,* of the testes produce *andorgen,* the masculinizing hormone. These cells, located between the seminiferous tubules, are obvious in the mammal but may exist in less typical forms in other vertebrates. Although the testes are the main androgen-secreting organs, other structures contain the necessary enzymatic machinery for producing the hormone. For example, the ovary, adrenal cortex, and probably the placenta also are capable of synthesizing the male steroid androgen. *Testosterone* and *androstenedione* are the main androgens secreted by the testes. These compounds contain 19 carbon atoms and possess methyl groups at C-10 and C-13.

FEMALE REPRODUCTIVE ORGANS

The vertebrate *ovary* also functions both as an exocrine and as an endocrine organ. It produces ova and the steroid sex hormones *estrogen* and *progesterone,* which regulate the reproductive trace, secondary sex characteristics, and mating behaviour of the female. The ovary is suspended from the dorsal body wall just behind the kidneys by a section of the peritoneum called the *mesovarium* and may be classified morphologically into either a solid or saccular type. Most vertebrates have a pair of ovaries, but a single structure is found in many fishes, probably representing a fusing of the two original ovaries since it has a typical vertebrate structure. In some animals, birds for example, only one side of the reproductive tract develops and the ovary on the other side remains inactive or degenerates. The external shape of the ovary varies according to the species and to the reproductive state of the individual. In the frog, during the breeding months the multilobed organ fills the body cavity, actually stretching the body wall. After ovulation, the ovary involutes the looks like a ruffled piece of mesentery. The ovary of birds is an enormous structure because of both the accumulation of yolk in individual eggs and of the fact that the female bird ovulates one egg per day for several months.

The saccular type of ovary, represented by the frog consists primarily of an outer area where the eggs are located and a lymph-filled central space, connecting with the cavities of other lobes. The solid ovary can be divided into two areas: an outer region, the *cortex,* containing many ova in various stages of development, and an inner *medulla,* consisting of irregularly arranged connective tissue, blood, and lymph vessels. *Interstitial cells,* similar to those found in the male, are embedded in the framework or *stroma* of the ovary, and are prominent in some species. A dense connective tissue layer, the *tunica albuginea,* covers the cortex. External to the runic, the peritoneal epithelium surrounds the ovary. This epithelium was unfortunately named the germinal epithelium by the early biologists who believed that these cells were the source of the primordial germ cells, the future eggs and sperms.

Regardless of the morphology of the ovary (saccular or solid) possessed by the vertebrate, the germ cells take up their position within the cortical region where clusters of epithelial cells surround each ovum and form a *follicle.* The follicular cells are believed to arise from the germinal epithelium. In electron micrographs protoplasmic connections extend from the follicle cells to the surface of the egg and evidently aid in providing nutrition for the growing oocyte. For this reason, they are often referred to as "*nurse cells.*" These follicular cells (along possibly with the *theca interna cells*) synthesize the hormone *estrogens.* A young follicle, one whose wall is only one cell thick, is initially located in the cortex of the ovary embedded in the connective tissue stroma. The stroma contributes additional connective tissue cells to the follicle in the form of sheaths around the follicular epithelium. These sheaths differentiate into an external fibrous layer, the *theca externa,* and an inner vascular layer, the *theca interna.* As the number of connective tissue sheaths increase, the follicles move away from the periphery of the ovary into the deeper areas of the cortex. When the egg and follicle mature, surface of the ovary thins because of the increased internal pressure. The wall becomes necrotic, ruptures, and releases the egg into the body cavity. This process of freeing the ovum from the ovary is called *ovulation* and is under the control of pituitary, hypothalamic, and ovarian hormones.

The number of eggs released from the ovary at any one time is determined genetically for each species. In those vertebrates that lay their eggs and then desert them, species survival depends on the production of large numbers of eggs, since the mortality is high. Certain fishes spawn several million eggs each season. Frogs shed as many as 2000 to 3000 eggs at one time.

5

Livestock Farming

BRIEF DESCRIPTION OF THE SECTOR

The use of cattle, sheep, goats, pigs, poultry, and other livestock offers many benefits to millions of farmers in the developing world. These animals are integral to rural livelihoods and culture, providing food (meat, blood, eggs and dairy products), materials (wool, hide, horns, etc.), income, and mechanical power for pulling carts or plowing fields. Asia has been identified as the developing region in which the demand for livestock products is expected to rise most rapidly.

Livestock manure can serve as a source of fertilizer. Grazing can help sustain vegetation and promote biodiversity by dispersing seeds, controlling shrub growth, breaking soil crusts, stimulating grass growth and improving seed germination. Livestock also may represent savings and currency, and have cultural value as well. For example, gifts of livestock may serve to resolve conflicts or cement marriages.

Livestock production can be categorized under three main systems: grazing, mixed farming and industrial:

1. Grazing systems generally rely on native grassland or forests for fodder, with little or no use of crops or imported inputs, and are traditionally managed by pastoralist communities.
2. Mixed farming systems integrate livestock and crop production. Adding livestock to their farms helps farmers diversify risk and extract value from otherwise valueless or low-value by-products of each activity: crop residue becomes feed, manure becomes fertilizer. Soil nutrients can be further replenished by rotating leguminous (nitrogenfixing) fodder crops with food crops. These systems are managed by settled farmers.
3. Industrial production systems concentrate livestock populations in special facilities and separate their feeding and waste processing from the land on which they live. Feed is provided directly instead of being acquired through grazing, and manure is transported off-site.

Generally, these systems are owned by relatively wealthy individuals and managed by local employees.

Grazing systems are most favoured in arid, semi-arid, or other areas of marginal value for crop-based agricultural production, while mixed farming systems flourish in temperate, subhumid, humid, and some highland climates. Industrial production, because it does not depend on local fodder supplies, can be conducted in any climate and generally occurs near the urban centers it supplies, sometimes even in peri-urban areas.

Although more slowly in sub-Saharan Africa than in most other regions. This increase is driven by growing population, increasing urbanization and rising incomes. This situation is expected to continue throughout the next decade. A shift towards industrial production—farming of monogastric species (pigs, poultry) fed with grain—may be an unavoidable trend in areas with rapidly growing demand for animal food products. Properly managed, livestock production can enhance land and water quality, biodiversity, and social and economic well-being.

However, when improperly managed, livestock production may cause significant economic, social and environmental damage. Increasing livestock production has the potential to increase environmental harm. This guideline is intended to help identify potential adverse environmental impacts and to suggest mitigation and monitoring options, as well as "best management practices" to address them.

EVALUATE CURRENT AND PROPOSED LIVESTOCK MANAGEMENT PRACTICES

To maximize forage productivity, it is best to combine or alternate various livestock breeds on a range. Their differing food preferences can help to keep plants productive by minimizing overgrazing of a particular favoured area and allowing less preferred plant species time to mature. It is prudent to make superior forage available to those animals with the highest needs. When forage is limited, livestock managers may decide that young and milk-producing animals must have first access to new pastures and ranges with a wide variety of abundant forage.

Managers should investigate the value of various systems of rotating livestock. Rotation allows land to be grazed continuously throughout the year. Livestock can be rotated between fields or ranges to prevent the buildup of disease and to vary grazing pressures. Through either fencing or herding, they can be relocated into croplands to consume crop residues. Assessment of seasonal grazing patterns should include potential impact on soils.

Dry-season grazing can benefit the land by breaking up crusted soil and working seeds into the ground. By contrast, grazing on moist soil can cause considerable soil compaction, which reduces the soils ability to absorb moisture and can result in increased erosion from run-off during the rainy season. Many of the environmental impacts from livestock production are associated with particular practices of livestock managers.

Thus it is critical to understand current practices and how the proposed project might alter these practices or promote new ones:

- Who are the local community's livestock managers?
- What practices do a family or community use to control the size and composition of livestock herds?
- How do livestock managers currently control livestock movement? Will the proposed project change these movements in a way that might harm the environment?
- Does the proposed project require the construction of fences? If so, will they interfere with wildlife migration or transit of livestock belonging to other communities? Could the fences lead to overgrazing and land degradation? Will the fences be built with local materials? Would living fencing be practical? Would solar-powered electric fencing be technically and economically feasible?
- Are streams and riverbanks currently protected from livestock damage? If the proposed project will open new areas to grazing, will water supplies need to be protected?
- Must steps be taken to prevent new livestock and associated animals (*e.g.*, dogs) from transmitting disease to wildlife? Is there a vaccination/animal disease control programme available for this purpose?
- Will the project involve construction of improvements (*e.g.*, boreholes or other infrastructure)? Could these lead to unplanned changes in herding patterns and overgrazing?

ASSESS DEMAND AND USE OF LIVESTOCK PRODUCTS

- Who is marketing livestock and livestock products?
- Is the demand for livestock products coming from local or outside populations? How rapidly is it increasing or decreasing? How stable is the demand?
- In preparing livestock products, are people using technologies which reduce impacts on the environment, open additional markets, or improve health and nutrition (of people and animals.)

ASSESS LIVESTOCK ECTOPARASITE MANAGEMENT

Epidemic and endemic diseases continue to be a major constraint to livestock productivity in large parts of the developing world. Although vaccines have controlled many of the epidemic diseases, they continue to cause severe economic losses through morbidity and mortality. These diseases include the infections caused by vector-borne haemoparasites and helminths. Existing technologies, such as chemotherapeutic agents and live vaccines that were previously successful in controlling these diseases, are no longer effective—because of acquired resistance or weakened delivery services. Appropriately designed alternatives are often lacking.

CONSIDER POPULATION PRESSURE AND DISEASE BURDEN

Two factors that may affect the outcome and impact of projects population growth and fatal or debilitating epidemic diseases. Population growth may lead to conflicting uses of grazing lands or reduce individual farms or rangeland to sizes that cannot sustain livestock. Fatal or debilitating epidemic diseases may weaken effective dissemination or replication of proper livestock management techniques.

- What is the current and projected population growth rate in the project area? How might this affect project sustainability in the future?
- What is the current extent of the HIV/AIDS epidemic in the region? How might this affect the composition of the population (size, ethnic makeup, age/gender distribution) and family structures necessary for project sustainability? How will development and livestock technical support services be affected?
- Are there other epidemic diseases in the region, such as sleeping sickness, that might adversely affect project implementation?

LIVESTOCK PRODUCTION

ARECA SHEATH

Areca (*Areca catechu*) is a commercial crop and the fallen areca sheath is a good source of dry roughage to animals. The nutritional value of areca sheath is similar to paddy straw (CP, NDF, ADF) with an advantage of having less lignin, silica and more calcium, sulphur and copper. Feeding of areca sheath to sheep and dairy animals replacing paddy straw could be an alternate approach.

Lignin Degradation

Activities of lignolytic enzymes produced by immobilization on areca sheath recorded temperature and pH stabilities over a wide range, thus making it desirable for supplementation in animals.

Feed Additives

Out of the 44 plant products, *viz.* seed pulp (5), leaves (20), spices (13), citrus fruit peel (3), oils (3), and 23 plant combinations, screened for their anti-methanogenic activity using *in vitro* gas production test, a mixture of plant products and yet another combination of four plants appeared to have good potential as feed additives in the ruminants. Use of nitrate reducing bacteria as probiotic exhibited 80 to 99% reduction of *in vitro* methane emission. There was no accumulation of hydrogen, nitrate or nitrite in the incubation medium, and no adverse effect on *in vitro* true digestibility of feed. Supplementation of Chicory root powder, *Lactobacillus acidophilus* as symbiotic in the diet of dogs proved beneficial in terms of fibre utilization, calcium and lipid metabolism, hind gut fermentation and cellmediated and humoural immunity. *Helianthus tuberoses* L. *Jerusalem artichoke Hathi choke*; a novel phytogenic-origin probiotic, showed immense potential for use in the diet of dogs for enhancing their digestive health, lipid profile and immunity.

Unconventional Feed

Detoxification procedures for jatropha, *karanj* and neem seed cakes were standardized in an industrial process. In growing lambs, replacement of soybean meal with detoxified jatropha cake up to 37.5% showed no adverse effect on DM intake, nutrient utilization, bloodbiochemical profile, growth rate and feed conversion efficiency. Nutritive value of *Robinia pseudoacacia* in terms of high organic matter, crude protein and low lignin content was comparable to *Grewia optiva*, and showed higher nutritional values than other tanniniferous forages. Combined treatment with curry and bel leaf powder had positive effect in restoration of cyclicity and fertility in acyclic goats and buffaloes.

Chelated Minerals

Cell-mediated immune response was significantly higher in kids given organic zinc (40 ppm zinc as zinc methionine) as compared to zinc sulphate.

Nutritional Manipulation

Feeding growing Frieswal bull calves with three different levels of rumen degradable protein and rumen undegradable protein for achieving a targeted body weight gain of 700 g and 1,000 g/day, yielded only 608 g and 772 g of average daily gain suggesting that there is a need to develop our own feeding standards for higher body weight gains.

Complete Pellets

Loin eye area increased linearly with increased level of concentrate mixture in different supplementation group. Kids fed diet without any concentrate mixture had lower yield of separated lean, fat and bone content. Different levels of supplementation improved meat bone ratio of Barbari kids under intensive system.

YAK

Two complete feed blocks (CFB) were prepared based on tree leaves (50% tree leaves, 43% concentrate and 7% molasses) and paddy straw (50% paddy straw, 43% concentrate and 7% molasses). The result indicated that tree leaves based CFB supplemented with limiting micronutrients can be fed during winter to overcome the weight loss in yaks.

POULTRY

Feeding Schedules for Poultry

In PB-1 females, ME restriction was successful in achieving the desired body weight (83 g/week and 5. 75% growth rate) at 20 weeks of age. *Karanj* cake in the diet adversely affected Krishibro chicks, but feeding of detoxified cake improved the performance. The production performance of PB-2 line was better with organic Zn supplementation at 60 ppm and hatchability at 80 ppm. Trace mineral retention in tissues of broiler chicks was linearly related to their level in diet. Herbal antifungal compounds, *viz. Piper longum*, *Ferula assafoetida* and *Cymbopogon citratus* oil were identified as potent antifungal herbal compounds against *Aspergillus parasiticus* in poultry feed under *in vitro* conditions.

Quality Protein Maize in Diet

Dietary replacement (50 to 100% w/w) of normal maize (higher in lysin and tryptophan by 44 and 33%, respectively, over normal maize) resulted in 6.93% higher body weight gain, 7.69% improvement in feed utilization efficiency and 5.15% higher breast meat yield in broiler chicken.

Amelioration of Heat Stress in Broilers

Amelioration of heat stress in broiler chickens was achieved through supplementation of vitamin C and potassium chloride in diet. Heat shock protein 70 (HSP70) was detected in muscle and blood lymphocyte in birds as potential marker of heat stress.

REPRODUCTION

Cattle

Good quality fresh and post-thaw semen of crossbred bulls had significantly higher percentage of HOST positive spermatozoa with low LDH and GOT enzyme leakage as compared to poor quality semen. Frozen semen doses (166,610) of Frieswal bulls were produced and 73,205 doses were distributed that included 50,725 to military farms, 9,000 to field progeny testing project, and 13,480 to various developmental agencies and farmers for cattle improvement.

Buffalo

Buffalo sperm membrane is highly prone for lipid peroxidation as compared to cattle, and in a study on nutrition and male fertility, it was observed that

PUFA feeding may reduce lipid peroxidation and improve frozen semen fertility. Studies on nutrition-endocrine interactions in attainment of puberty in female buffaloes showed that supplementation of higher dietary energy (through bypass fat) has a positive effect on early maturity of hypothalamo-pituitary—ovarian axis in terms of early and better LH secretion and ovarian follicular activity in female buffalo calves.

Studies on physiological reasons for low reproductive efficiency and early embryonic mortality in buffaloes showed that antioxidant enzymes, Cu and Zn-SOD are induced significantly during early pregnancy and may be important for overcoming oxidative stress that otherwise would lead to early embryonic mortality. LPS and TNFα appear to modulate prostaglandin production in endometrium and corpus luteum and are important for protecting and enhancing function of corpus luteum to prevent early embryonic mortality.

IGF-I also stimulated progesterone secretion from the corpus luteum. Five proteins, specific to uterine secretions of pregnant buffalo were recognized and it may be possible to use one of these for early pregnancy diagnosis to monitor pregnancy and prevent early embryonic mortality.

Goat

Study on effect of dextran, a new cryoprotectant, revealed that dextran alone did not prove to be an efficient cryoprotectant, but 3% glycerol dextran in combination with 6% glycerol resulted in 39.33±1.45% post-thaw recovery. Hormone delivery system for synchronization of heat in goats was standardized by using sponges of different sizes and shapes. The estrus response, onset of estrus and duration of estrus following use of intra vaginal sponge pessaries for induction of multiple birth during peak breeding activity were 100%, 29.14±4.42 h and 25.71±0.37 h, respectively, with 57.14% conception rate.

The kidding rate following induction of multiple birth was 2.0. During low breeding activity season (summer) the estrus response, onset of estrus and duration of estrus were 71.42%, 60.0±16.54 h and 26.4±11.63 h, respectively, following use of intra vaginal pessaries. Oocytes were recovered by follicle puncture technique for *in vitro* maturation, fertilization and culture from goat ovaries. The average recovery of oocytes using follicle puncture was 1.76/ovary.

The oocytes were cultured in tissue culture medium (TCM-199) supplemented with EGF (10 ng/ml) and fertilized *in vitro* with a cleavage rate of 41.55%. Embryos were bisected through microtools. Out of 25 bisected cells 3 single cells divided into double cells whereas one bunch of 3 cells developed into 4 cells. Adaptability study on goats revealed that water intake of goats kept under asbestos sheet roof sheds was significantly higher compared to the thatch roof shed during both the moderate and cool seasons. Milk production of animals kept inside the thatch roof shed was significantly higher than the animals inside the asbestos roof shed.

Camel

Physiological parameters (rectal temperature, respiration rate, pulse rate) were higher during evening as compared to morning. The morning and evening variation in the THI value was significant for all weeks in different months from October to March. The Benezara coefficient of adaptability (BCA) of camel revealed that BCA was significantly higher during evening as compared to morning. Non-pregnant female camels were divided in groups A,B,C.

Each group was tested with poll gland secretion collected in breeding season (PGS-BS), poll gland secretion collected during non-breeding season (PGS-NBS) and water. The results indicated that the PGS-BS evoked greater olfactory, contact and flehmen response than PGS-NBS and water. Nine female camels, which did not conceive during breeding season were given injection of highly purified hCG for infertility treatment at the time of mating. Pregnancy was diagnosed in three out of nine camels.

Yak

On application of the vaginal sponge containing 600 mg progesterone for synchronization of oestrus followed by AI in yaks resulted in 100% oestrus response with 60% conception rate. The yaks do not experience cold stress (environmental stress) during winter but they are subjected to heat stress (during summer) when the ambient temperature exceeds 10°C at the altitude of 2,750 m above mean sea level. The body weight gain was significantly higher during summer than winter in adult yaks.

The lactating yak cows lost about 5.84% of their body weight during winter along with reduced milk yield mainly due to shortage of feed and fodder. Estrous was displayed year round but the highest incidence (around 65%) was restricted to warmer part of the year. Physiological responses, *viz.* rectal temperature, respiration and pulse rates of yak calves, bulls and lactating cows were recorded during winter and summer. The physiological responses of calves were significantly higher than bulls and lactating cows during summer. Standardization for heat shock protein 70 (Hsp 70) assay was carried out in yaks.

Mithun

Different protocols for preservation of mithun semen indicated that it is possible to preserve mithun spermatozoa at refrigeration temperature in tris-egg yolk diluents till 108 h, and doubling the concentrations of fructose in the diluent buffer increased the progressive motility, live sperm count and decreased total sperm abnormalities. Hence, addition of extra-energy source (fructose) might increase the storage time of mithun sperm at refrigerated temperature up to 5 days.

Pig

Gloved hand method of semen collection was standardized. Semen preservation was done for Hampshire, Duroc and Ghungroo pigs for first time in Asom. The AI technology was used in the farm and in the nearby villages and highest litter size (15 piglets) at birth was recorded from a Ghungroo sow in the farm.

Poultry

Exposure of birds to 650 nm and 475 nm wavelengths of light improved egg production by 6.55 and 3.05%, respectively, compared to 450 nm (incandescent) light.

LIVESTOCK SECTOR AND THEIR CAUSES

LARGE AREAS OF LAND DEGRADED

Overgrazing of rangeland reduces the density of vegetation and the amount of organic matter generated. This, in turn, increases soil erosion from wind and water and decreases soil fertility through loss of nutrients. In arid and semi-arid areas these impacts may also contribute to desertification. Fortunately, ecosystems in these areas demonstrate considerable resilience and often recover when grazing pressure is reduced, either through traditional methods or through modern management practices.

Policy and Legal Problems

In areas that traditionally rely on grazing, the health of rangeland is generally best maintained by traditional pastoralist practices, which regulate grazing location and herd size in accordance with drought cycles and the supply of fodder. In fact, government policies or donor interventions that disrupt or discourage these practices may be a root cause of degradation. A variety of government policies may restrict the movement of livestock within a range area and prevent livestock managers from moving stock from areas that have been depleted of fodder to better supplied areas.

Two particular policy-based problems are:

1. *Land Tenure Insecurity:* Lack of confidence in secure title to rangeland (especially on communal lands) has been shown to reduce the incentive to manage the land sustainably. Many national governments have either implicitly or explicitly claimed ownership of range and wildlands and ignored traditional or customary claims.
2. *Privatization of Communal Resources:* Where national governments have privatized, or are privatizing, formerly state-owned or communal

lands, new owners may erect fencing or prevent herds from crossing or grazing on their property.

Wells and Boreholes

Traditionally, access to water on critical grazing lands has been controlled to limit livestock populations and prevent herds from outgrowing the forage supply in dry areas. Thus, new wells or boreholes, generally sponsored by donors or governments, may undermine traditional livestock management systems practices by allowing herds to grow beyond sustainable levels for surrounding areas.

Overgrazing and degradation are most noticeable in the immediate vicinity of the boreholes or wells, but their effects can extend (in gradually decreasing severity) over a considerable radius. Boreholes also reduce pressure on livestock owners to decrease herd size during drought and may discourage movement of herds to other rangelands, disrupting historic wet season/dry season grazing patterns. Larger herd sizes and reductions in pastoral movement may prove to be a recipe for severe degradation of soil and vegetation.

Wet-season Grazing

Poor timing in the use of rangeland can also damage the soil. Wet-season grazing can compact the moist earth, reducing its ability to absorb moisture This increases erosion from water run-off.

Poor Balance of Livestock Species

Each species or breed of livestock has foraging preferences and will graze favoured areas and plants while neglecting others. Browsing animals, such as goats and camels, prefer the leafy tops of shrubs. By contrast, grazers tend to consume ground-level grasses and leafy plants. A poor balance between browsers and grazers can change the mix of plants in ways that degrade the area. For example, too many grazers can diminish the number and populations of herbaceous plant species and allow woody plants to become dominant, changing, possibly irreversibly, the character and utility of the ecosystem.

DAMAGED HABITAT AND REDUCED BIODIVERSITY

Livestock production can damage habitats and reduce biodiversity in wildlife and domestic stock, in vegetation, and in aquatic and wetland ecosystems.

Harm to Wildlife and Domestic Stock and Loss of Wildlife Habitat

The loss of habitat caused by livestock production in grazing and mixed farming systems may be one of the greatest threats to wildlife. Human population growth and density, and the accompanying increase in livestock, often leads producers to expand livestock grazing ranges into wild lands and the conversion of wild lands to mixed farming use.

Slaughter of Wildlife by Livestock Managers

Another danger to wildlife is intentional slaughter by livestock managers. Fear that the wildlife will prey on livestock and damage crops is a common motivation, as is the belief that the wildlife are competing with livestock for fodder, the desire to prevent spread of disease to livestock, and concern for human safety.

Extinction of Local Livestock Breeds

Systematic livestock production may result in loss of genetic diversity in livestock species. This is unfortunate because genetic diversity is a measure of a species' robustness. Local breeds may have traits conferring resistance to emergent or future pathogens, or have other favourable adaptations to local environments. The consistent replacement of local breeds with more productive imported ones can contribute to the extinction of that breed and of all the genetic diversity harbored within its population.

HARM TO VEGETATION

Clearing of Forest and Wild Lands

Vegetation is typically altered or destroyed when forests/wild lands are cleared or are burned to promote new growth. This changes local ecosystems and may contribute to global warming. Fires to burn vegetation are dangerous and degrade air quality.

Loss of Rangeland Fertility

Ironically, mixed farming systems may reduce the fertility of rangeland while helping to solve a farmland problem. Traditional farming practices cause a net loss of nutrients in farm soils; that is, when crops are harvested and sold (or even when human waste is deposited in a latrine) nutrients that make the soil fertile may be lost.

Mixed farming reduces the extent of this loss, by transferring nutrients from the range to the farm in the form of manure. The gain in fertility for the farm is, of course, a net loss for rangeland. Over time, the altered nutrient balance can reduce the productive capacity of the range and/or lead to changes in the composition and density of plant species.

Loss of Farm Fertility

Damage from mixed farming often occurs when poverty and population growth pressure change the crop/grazing land ratio, where other nutrient sources are not available. As the land area available for each grazing animal shrinks, overgrazing becomes prevalent leading to soil erosion and nutrient loss. Harvesting livestock further reduces nutrients available for crop production.

Damage to Riparian Soil and Vegetation

Livestock in grazing and mixed farming systems often graze very heavily in riparian areas along streams and lakes. Results include trampling, loss of vegetation, soil disturbance, and soil compaction, erosion and/or sedimentation which can severely damage riparian habitats, increase siltation and adversely affect watersheds. Pesticide contamination from treatments to protect livestock from insectborne infections (*e.g.*, livestock "dipping") may ultimately reach the aquatic environment. Here it can be toxic to aquatic organisms, as well as people or animals who depend on these sources for drinking water.

Introduction of Invasive Plants

New breeds or fodder crops can introduce invasive non-native plants into a region. The manure, coats and hooves of newly introduced breeds can carry plant seeds. Most non-native plants are not invasive, but when they are, the results can be devastating.

Contamination from Manure

Livestock manure contains relatively high concentrations of nutrients, solids, enteric bacteria and other microorganisms, and organic material. The manure from industrial livestock operations is often discharged or "leaked" into lakes or streams, because it cannot be economically transported to replenish crop fields. When this occurs, the nutrients can cause eutrophication (rapid plant growth in water bodies), solids can create sedimentation, and organic material leads to oxygen depletion (BOD) of the water. Manure from mixed farming, if applied in a concentrated fashion, can lead to similar problems.

Degrade Water Quality and Reduce Water Supplies

Where water is scarce, either chronically or seasonally, the diversion of water to sustain livestock potentially limits its availability for other purposes. This is of particular concern in arid and semi-arid regions, where the construction of boreholes to supply livestock can lead to unsustainable withdrawal rates and the dangerous depletion of aquifer reserves.

Stockpiled manure can contaminate bodies of water, causing myriad adverse effects. These include eutrophication, oxygen depletion, sedimentation, contamination with enteric bacteria and possibly other pathogenic organisms, toxic pollution from pesticides, and contamination of groundwater and aquifers with both nitrates and pesticides. Moreover, high concentrations of nitrate in potable water supplies represent a potential health hazard, especially for children.

Harm to Human Health

Using water for farm animals may make it less available for the many uses that influence health, such as bathing, washing, cooking, and drinking. Moreover, excessive contamination by enteric microorganisms, toxic pesticides or nitrates in may render water unfit for human consumption and may be especially dangerous to children. Pesticides or other vector control treatments used on livestock represent threats to the health of livestock managers, their families, and others exposed directly or through water use. These substances may be toxic, cause birth defects, alter children's proper development, promote cancer, or slowly poison one or more organ systems.

Climate Change/Global Warming

Approximately 17% of global methane is produced by livestock digestion and manure. Methane is a greenhouse gas that has 56 times more global warming potential than carbon dioxide over a 20-year time horizon. Furthermore, clearing wild or forest land for new fields reduces the land's ability to act as a carbon sink. Setting fire to vegetation to clear land or promote new growth for livestock to graze on may also contribute to global warming.

Odour

Concentrated manure stored at industrial livestock facilities can generate strong and unpleasant odours, damaging the quality of life of nearby residents. This problem is most evident when facilities are located in densely populated areas.

6

Breeding of Animals

The science of Animal breeding is defined as the application of the principles of Genetics and biometry to improve the efficiency of production in farm animals. These principles were applied to change animal populations thousands of years before the sciences of genetics and biometry were formally established.

The practice of animal breeding dates back to the Neolithic period (approximately 7000 BC), when people attempted to domesticate wild species such as reindeer, goats, hogs and dogs. Domestication was performed through controlled mating and reproduction of captive animals which were selected and mated based on their behaviour and temperament. Judging from cave paintings that have survived, selection was also applied to some qualitative traits such as coat colour and the absence or presence of horns.

Without written records, there is no certain knowledge of the evolution of animal breeding practices, but written documents dating back more than 4000 years indicate that humans appreciated the significance of family resemblance in mating systems, recognized the dangers of intense inbreeding, and used castration to prevent the reproduction of undesirable males.

Progress in the performance of domesticated animals through these selection practices was very slow; improvements were mainly due to animals adapting better to their environments. Robert Bakewell, an English animal breeder of the 18th century, is considered the founder of systematized animal breeding. He was the first to emphasize the importance of accurate breeding records, introduced the concept of progeny testing to evaluate the genetic potentials of young sires, and applied inbreeding to stabilize desired qualitative traits. He also promoted concepts such as "like begets like," "prepotency is associated with inbreeding" and "breed the best to the best."

Bakewell and his contemporaries in Europe pioneered the development of diverse breeds of Beef Cattle, Dairy Cattle, Sheep, Hogs and Horses. Most livestock breeds with pedigree herd books and breed associations were established between the late 18th century and the second half of the 19th century. Colour, conformation, geographical origin and some production characteristics were the main factors that differentiated these breeds. Wide geographical redistribution of animal populations was also an important factor in the formation of new breeds, as invading armies, migrating people and traders transported livestock to new lands.

Animal breeding as a modern science belongs to the 20th century. Although numerous geneticists and biometricians have made significant contributions to the development of this science, J.L. Lush of Iowa State University is considered the father of the modern science of animal breeding. Lush and his students developed major scientific procedures applicable to the genetic improvement of farm animals.

SELECTIVE BREEDING

Selective breeding in domesticated animals is the process of a breeder developing a cultivated breed over time, and selecting qualities within individuals of the breed that will be best to pass on to the next generation. Breeding techniques such as inbreeding, linebreeding and outcrossing are utilized by breeders in the maintenance and improvement of their chosen breeds.

BREEDING STOCK

"Breeding stock" is a term used to describe a group of animals used for purpose of planned breeding. When an individual is looking to breed animals, he or she is looking for certain valuable traits in purebred stock for a certain purpose, or may intend to use some type of crossbreeding to produce a new type of stock with different, and presumably superior abilities in a given area of endeavor. For example, to breed chickens, a typical breeder intends to receive eggs, meat, and new, young birds for further reproduction.

Thus the breeder has to study different breeds and types of chickens and analyze what can be expected from a certain set of characteristics before he or she starts breeding them. Accordingly, when purchasing initial breeding stock, the breeder seeks a group of birds that will most closely fit the purpose intended.

Purebred Breeding

The idea of *breed purity* may strike an unpleasant chord with some individuals because it is reminiscent of nineteenth-century eugenics notions of a "superior strain" in humans, supposedly exemplified by members of an aristocratic social class or of certain races. The application of theories of eugenics had far-reaching consequences for human beings, particularly in cases such as The Holocaust. However, the idea of a superior strain in animals, such as that incorporated in the breeding of Thoroughbred race horses, was less controversial.

By "breeding the best to the best," employing a certain degree of careful inbreeding, considerable culling, and selection for "superior" qualities, one could develop a bloodline or "breed" superior in certain respects to the original base stock or landrace which had been produced primarily by natural selection.

This process forms the basis for a "true-breeding" line of animals, which today is generally called a "purebred" line. Once created, a purified line is "closed" and preserved from dilution and possible "debasement" by unrelated stock. When such animals are recorded with a breed registry, the closing of a bloodline group is sometimes referred to as having a closed stud book. However, this theory is not without some negative consequences.

When the ideal of purified lineage or purely aesthetic type is seen as an end in itself, without consideration of usefulness or vigor, a breed can suffer over time. Also, if a breed registry is closed with too small a gene pool, over time, degradation due to inbreeding depression may occur. For many purebred breeders, the need for a periodic outcross is recognized and incorporated into an overall plan.

Some breed registries thus maintain an "open" stud book for certain types of approved outcrossed or crossbred animals. The observable phenomenon of hybrid vigor stands in contrast to the notion of breed purity. However, on the other hand, indiscriminate breeding of crossbred or hybrid animals may also result in degradation of quality, particularly when animals are mass-produced without concern for overall health or longevity.

Backyard Breeding

The term backyard breeder is a general term, sometimes considered derisive, used to describe people who are "home breeders", and breed only a few animals for personal use or occasional sale. However, while some backyard breeders pay strict attention to quality, the term usually describes those who allow animals, particularly dogs or horses, to procreate regardless of physical, genetic, and/or emotional health.

It is also used to describe unreputable commercial breeders. In the process of careless breeding, many backyard breeders produce genetically weak animals that can be predisposed to genetic disorders or debilitating physical deformities, particularly those that result from inbreeding.

In dog breeding, these individuals focus on only a limited aspect of the dog, while ignoring the original function and temperament of the breed. When such breeding is carried out on a large scale, the venue is called a *puppy mill* or *puppy farm*. Backyard breeding is popularly blamed for the proliferation of aggressive dogs for the sports of dog baiting and dog fighting.

Dog fanciers generally believe that such ill-bred dogs are the reason for the bad public reputation of certain dog breeds, and the increasingly frequent enactment of breed-specific legislation. Because of the time and expense of feeding and caring for horses, which produce one foal per year, there are fewer horse breeders who produce animals *en masse,* though some individuals do engage in animal hoarding and breed far more animals than they can support.

In the horse world, overbreeding of grade animals that cannot be sold except for very low prices raises concerns that such animals will be slaughtered for horsemeat. Some, though not all, colour breeders make errors similar to puppy mill breeders by focusing on only one characteristic to the detriment of health, soundness, and quality conformation.

In addition, industrial breeders, such as those who produce *pregnant mare urine* for the manufacture of premarin, allow many mares to be bred indiscriminately, simply so that they are able to produce the urine laced with reproductive hormones that forms the base for the drug. Their foals are generally considered an unwanted byproduct, particularly males (colts), and sold off, frequently to the slaughter market.

IN-BREEDING AND OTHER BREEDING METHODS

One of the most hotly talked about topics with regard to pure-bred dogs is the use of in-breeding. This is a term that is often misused and is extremely misunderstood. Part of the misunderstandings come from differences in the way the terms are used within the scientific/medical field, and how it is commonly used by breeders. These are the most commonly accepted definitions used by serious dog breeders and will be the definitions used within this article.

In-breeding

This is the breeding of closely related animals. Brother-Sister, Parent-Offspring, ½ brother - ½ Sister.

Line-breeding

This is the breeding of animals that share common ancestors but are not closely related. For example the dogs may share a common great-grandparent.

Out-cross

This is generally considered the breeding of animals with no common ancestors within the first 4 or 5 generations.

COMMON MISCONCEPTIONS

In-breeding Causes Genetic Diseases

Breeding closely related animals increases the possibility that any bad genes in a line will show up. It does not 'cause' genetic disease.

Out-crossed Dogs are Healthier

This is only partly true. There is a known phenomenon called Hybrid Vigor. Two animals of unrelated strains breed and the offspring is often bigger and grows faster than it's purebred cousins.

This method is often used by farmers in order to get their animals to market sooner. But one of the biggest misconceptions of hybrid vigor is that it applies to all animals of mixed heritage. Hybrid Vigor only applies to the animals that are the direct offspring of the crossing of the unrelated strains.

In other words if you continue to breed animals of different strains there generally will not be any additional increase in hybrid vigor. If the unrelated strains share common genes for genetic disorders, hybrid vigor will not over ride the risk of the disorder showing up. Out-crossing can also cause problems if widely divergent physical types are mixed due to differences in growth rates and bone and muscle sizes.

BENEFITS OF EACH TYPE OF BREEDING

By definition, purebred dogs have a smaller gene pool to draw on than mixed breed dogs. That smaller gene pool gives the breed its individual characteristics, such as physical appearance and temperament. It is what makes a poodle a Poodle and a Golden Retriever a Golden Retriever. But there is considerable controversy with regard to whether the gene pools of the modern pure-bred have become too small.

INBREEDING

In-breeding is more likely to help "set" or "fix" a particular trait within a breed or a line by narrowing the gene pool to favour those traits. So if a breeder is looking to set a particular desirable feature of their line then in-breeding and choosing the offspring most strongly possessing that trait can be beneficial. In the first pedigree Sydney is the maternal son of Annie, thus doubling on Annie's genes. Or another way to look at it is Annie is both Sydney's mother and Grandmother. In the second example we have a full brother/sister in-breeding so we are doubling up on the genes from both Kiley and Annie. In-breeding can also help identify those bad genes that exist within a line. Dogs possessing the bad genes can be eliminated from a breeding programme and carriers also identified. Intermittent in-breeding within a line or breed is not damaging to the long term health of the animals.

However, in-breeding over successive generations can lead to reduced fitness and fertility problems among the offspring, resulting in a phenomena known as In-breeding Depression. It can take many generations to show up depending on the traits involved. To use this method responsibly a breeder would not want to in-breed on animals with known genetic disorders, temperaments not in keeping with it's given breed, or known serious structural faults, or to in-breed frequently even on healthy-superior specimens.

Line-Breeding

Line-breeding is another way to help "set" or "fix" desirable traits. With line-breeding you breed animals that are related, but you are also routinely introducing genes from other lines into the genetic mix. It takes longer to fix the desirable traits this way, but doing so lowers the risk of those problems associated with repeated in-breeding. With a tight line-breeding you might find the same 3, 4 or more dogs showing up numerous times in a 5 generation pedigree.

Here's an example of a tightly line-bred pedigree.

Loose line-breeding over successive generations will result in more variations of physical appearance than would in-breeding or tight line-breeding, but will keep the physical look and structure within the same general size and shape, it also carries fewer long term risks.

According to geneticists

Line-breeding can be carried on for many many generations without deleterious effects on the line or breed as long as the individuals involved have few hidden genetic disorders.

Out-Crossing

Out-crossing in terms of pure-bred dogs is the breeding of unrelated dogs. On a pedigree no names will be repeated within the first 5 generations. This type of breeding has both advantages and disadvantages.

Which as it turns out are flip sides of the same argument. With out-crossing you are maintaining the greatest genetic diversity, but this also leads to the least consistency in terms of physical appearance and other traits. Out-crossing does not guarantee that the animals won't develop genetic disorders, but it does tend to reduce the numbers of affected offspring.

Your best chance of getting an animal that is less prone to developing a genetic disorder comes more from finding a conscientious breeder that screens their animals for hereditary disorders and breeds for the betterment of the breed. All three methods of breeding have their place in a long term, well thought out breeding programme. Talk to the breeder, ask questions as to what their goal is in doing a particular breeding. Ask about the risks and what problems are known to that line. And all lines have some because the perfect dog and the perfect lines are still goals of the future for all breeders.

DOG

The dog is a domesticated subspecies of the wolf, a mammal of the Canidae family of the order Carnivora. The term encompasses both feral and pet varieties and is also sometimes used to describe wild canids of other subspecies or species.

The domestic dog has been one of the most widely-kept working and companion animals in human history, as well as being a food source in some cultures. There are estimated to be 400,000,000 dogs in the world. The dog has developed into hundreds of varied breeds. Height measured to the withers ranges from a few inches in the Chihuahua to a few feet in the Irish Wolfhound; colour varies from white through grays to black, and browns from light to dark in a wide variation of patterns; and, coats can be very short to many centimeters long, from coarse hair to something akin to wool, straight or curly, or smooth. In breeding circles, a male canine is referred to as a *dog*, while a female canine is called a *bitch*.

The father of a litter is called the *sire*, and the mother of a litter is called the *dam*. Offspring are generally called *pups* or *puppies* until they are about a year old. A group of offspring is a *litter*. The process of birth is *whelping*. Many terms are used for dogs that are not purebred.

Origin and Evolution

Based on DNA evidence, the wolf ancestors of modern dogs diverged from other wolves about 100,000 years ago, and dogs were domesticated from those wolf ancestors about 15,000 years ago.

This date would make dogs the first species to be domesticated by humans. Evidence suggests that dogs were first domesticated in East Asia, possibly China, and some of the peoples who entered North America took dogs with them from Asia. As humans migrated around the planet a variety of dog forms migrated with them.

The agricultural revolution and subsequent urban revolution led to an increase in the dog population and a demand for specialization.

These circumstances would provide the opportunity for selective breeding to create specialized working dogs and pets. This ancient mosaic, likely Roman, shows a large dog with a collar hunting a lion.

Molecular systematics indicate that the domestic dog descends from one or more populations of wild wolves. As reflected in the nomenclature, dogs are descended from the wolf and are able to interbreed with wolves. The relationship between human and canine has deep roots. Converging archaeological and genetic evidence indicate a time of domestication in the late Upper Paleolithic close to the Pleistocene/Holocene boundary, between 17,000 and 14,000 years ago.

Fossil bone morphologies and genetic analysis of current and ancient dog and wolf populations have not yet been able to conclusively determine whether all dogs descend from a single domestication event, or whether dogs were domesticated independently in more than one location. Domesticated dogs may have interbred with local populations of wild wolves on several occasions (a process known in genetics as introgression).

The earliest dog fossils, two crania from Russia and a mandible from Germany, date from 13,000 to 17,000 years ago. Their likely ancestor is the large northern Holarctic wolf, *Canis lupus lupus*. Remains of smaller dogs from

Mesolithic cave deposits in the Middle East, dated to around 12,000 years ago, have been interpreted as descendants of a lighter Southwest Asian wolf, *Canis lupus arabs*. Rock art and skeletal remains indicate that by 14,000 years ago, dogs were present from North Africa across Eurasia to North America.

Dog burials at the Mesolithic cemetery of Svaerdborg in Denmark suggest that in ancient Europe dogs were valued companions. Genetic analyses have so far yielded divergent results. The ancestors of dogs split off from other wolves between 75,000 and 135,000 years ago, while a subsequent analysis indicated a "common origin from a single gene pool for all dog populations" between 40,000 and 15,000 years ago in East Asia.

Both sets of dates must be reevaluated in light of recent findings showing that poorly calibrated molecular clocks have systematically overestimated the age of geologically recent events. On balance, and in agreement with the archaeological evidence, 15,000 years ago is the most likely time for the wolf-dog divergence.

Development of Dog Breeds

Dogs have been bred into a variety of shapes, colours and sizes. Variation can be wide even within a breed, as with these Cavalier King Charles Spaniels. There are numerous dog breeds, with over 800 being recognized by various kennel clubs worldwide.

Many dogs, especially outside the United States of America and Western Europe, belong to no recognized breed. A few basic breed types have evolved gradually during the domesticated dog's relationship with humans over the last 10,000 or more years, but all modern breeds are of relatively recent derivation.

Many of these are the product of a deliberate process of artificial selection. Because of this, some breeds are highly specialized, and there is extraordinary morphological diversity across different breeds. Despite these differences, dogs are able to distinguish dogs from other kinds of animal. The definition of a dog breed is a matter of some controversy.

Depending on the size of the original founding population, closed gene pool breeds can have problems with inbreeding, specifically due to the founder effect. Dog breeders are increasingly aware of the importance of population genetics and of maintaining diverse gene pools.

Health testing and new DNA tests can help avoid problems, by providing a replacement for natural selection. Without selection, inbreeding and closed gene pools can increase the risk of severe health or behavioural problems. Some organizations define a breed more loosely, such that an individual may be considered of one breed as long as 75 per cent of its parentage is of that breed.

These considerations affect both pets and the show dogs entered in dog shows. Even prize-winning purebred dogs sometimes possess crippling genetic defects due to founder effect or inbreeding. These problems are not limited to purebred dogs and can affect cross-breed populations. The behaviour and appearance of a dog of a particular breed can be predicted to a degree, while mixed-breed dogs show a broader range of innovative appearance and behaviour. Mixed-breed dogs or Mongrels (also called "mutts") are dogs that do not belong to specific breeds, being mixtures more than two in variant percentages.

Mixed breed dogs and purebred dogs are both suitable as companions, pets, working dogs, or competitors in dog sports. Sometimes different breed dogs are deliberately bred, to create cross-breeds such as the Cockapoo, a mixture of Cocker Spaniel and Miniature Poodle.

Such deliberate crosses may display some degree of hybrid vigor and other desirable traits, but may or may not inherit any of the desired traits of their parents, such as temperament or a particular colour or coat. Without genetic testing of the parents, the crosses can end up inheriting genetic defects that occur in both parental breeds.

A breed is a group of animals that possesses a set of inherited characteristics that distinguishes it from other animals within the same species. Deliberately crossing two or more breeds is also a manner of establishing new breeds, but it is only a breed when offspring will reliably demonstrate that particular set of characteristics and qualities.

Breed Popularity

Breed popularity varies widely over time and in different parts of the world and different segments of the population. Counting by American Kennel Club (AKC) registration (not by licensing registration or by United Kennel Club (UKC) registration, which could present different statistics), the Labrador Retriever has been the

United States's most commonly registered breed of dog since 1991. However, even within parts of the United States, popularity varies; for example, in 2005 the most-registered breed in New York City was the Poodle while the Yorkshire Terrier was the second-most-registered breed in Houston.

However, animal shelters in many parts of the United States report that the most-commonly available dog for adoption is the American Pit Bull Terrier or pit bull-type mixes, making up as much as 20 per cent of dogs available for adoption, none of which would be registered with the AKC. Two decades ago, in 1983, the AKC's top two registered breeds were the American Cocker Spaniel and the Poodle. In the United Kingdom, The Kennel Club reports that the most-registered breed from at least 1999 to 2005 was the Labrador Retriever. It rounds out the top three for 1999 to 2005 with the German Shepherd Dog, also popular in the States, and the English Cocker Spaniel, which is no longer even in the top ten in the States. In the United Kingdom a national dog adoption and rescue service indicates that the most common breed appearing in shelters is the Greyhound followed by the Staffordshire Bull Terrier.

Physical Characteristics

Modern dog breeds show more variation in size, appearance, and behaviour than any other domestic animal. Within the range of extremes, dogs generally share attributes with their wild ancestors, the wolves. Dogs are predators and scavengers, possessing sharp teeth and strong jaws for attacking, holding, and tearing their food.

Although selective breeding has changed the appearance of many breeds, all dogs retain basic traits from their distant ancestors. Like many other predatory mammals, the dog has powerful muscles, fused wristbones, a cardiovascular system that supports both sprinting and endurance, and teeth for catching and tearing. Compared to the bone structure of the human foot, technically, dogs walk on their toes.

Differences from Other Canids

Compared to equally sized wolves, dogs tend to have 20 per cent smaller skulls and 10 per cent smaller brains, as well as proportionately smaller teeth than other canid species. Dogs require less calories to function than wolves. Their diet of human refuse in antiquity made the large brains and jaw muscles needed for hunting unnecessary. It is thought by certain experts that the dog's limp ears are a result of atrophy of the jaw muscles.

Dogs tend to be poorer than wolves and coyotes at observational learning, being more responsive to instrumental conditioning. Dogs have thicker skins than similairly sized wolves. Inuits tend to favour dog skin over wolf skin due to the formers resistance to tear. Unlike wolves, but like coyotes, domestic dogs have sweat glands on their pawpads. The paws of a dog are half the size of those of a wolf, and their tails tend to curl upwards, another trait not found in wolves.

Feral dogs show little of the complex social structure or dominance heirarchy present in wolf packs. For dogs, other members of their kind are of no help in locating food items, and are more like competitors.

Dogs differ from wolves and most other large canid species by the fact that they do not regurgitate food for their young, nor the young of other dogs in the same territory. Many dogs, such as this American Water Spaniel, have had their natural hunting instincts suppressed or altered to suit human needs. Like most mammals, dogs are dichromats and have colour vision equivalent to red-green colour blindness in humans.

Different breeds of dogs have different eye shapes and dimensions, and they also have different retina configurations. Dogs with long noses have a "*visual streak*" which runs across the width of the retina and gives them a very wide field of excellent vision, while those with short noses have an "*area centralis*" — a central patch with up to three times the density of nerve endings as the *visual streak* — giving them detailed sight much more like a human's.

Some breeds, particularly the sighthounds, have a field of vision up to 270° (compared to 180° for humans), although broad-headed breeds with short noses have a much narrower field of vision, as low as 180°.

Dogs detect sounds as low as the 16 to 20 Hz frequency range (compared to 20 to 70 Hz for humans) and above 45 kHz (compared to 13 to 20 kHz for humans), and in addition have a degree of ear mobility that helps them to rapidly pinpoint the exact location of a sound.

Eighteen or more muscles can tilt, rotate and raise or lower a dog's ear. Additionally, a dog can identify a sound's location much faster than a human can, as well as hear sounds up to four times the distance that humans are able to. Those with more natural ear shapes, like those of wild canids like the fox, generally hear better than those with the floppier ears of many domesticated species. Dogs have nearly 220 million smell-sensitive cells over an area about the size of a pocket handkerchief (compared to 5 million over an area the size of a postage stamp for humans).

Some breeds have been selectively bred for excellence in detecting scents, even compared to their canine brethren. What information a dog actually detects when he is scenting is not perfectly understood; although once a matter of debate, it now seems to be well established that dogs can distinguish two different types of scents when trailing, an air scent from some person or thing that has recently passed by, as well as a ground scent that remains detectable for a much longer period.

The characteristics and behaviour of these two types of scent trail would seem, after some thought, to be quite different, the air scent being intermittent but perhaps less obscured by competing scents, whereas the ground scent would be relatively permanent with respect to careful and repetitive search by the dog, but would seem to be much more contaminated with other scents.

In any event, it is established by those who train tracking dogs that it is impossible to teach the dog how to track any better than it does naturally; the object instead is to motivate it properly, and teach it to maintain focus on a single track and ignore any others that might otherwise seem of greater interest to an untrained dog.

An intensive search for a scent, for instance searching a ship for contraband, can actually be very fatiguing for a dog, and the dog must be motivated to continue this hard work for a long period of time. Domestic dogs often display the remnants of counter-shading, a common natural camouflage pattern. The general theory of countershading is that an animal that is lit from above will appear lighter on its upper half and darker on its lower half where it will usually be in its own shade.

This is a pattern that predators can learn to watch for. A countershaded animal will have dark colouring on its upper surfaces and light colouring below. This reduces the general visibility of the animal. One reminder of this pattern is that many breeds will have the occasional "blaze", stripe, or "star" of white fur on their chest or undersides. Dogs can generate large amounts of energy for a short period of time. A dog's heart and lungs are oversized relative to its body and its normal everyday needs.

A dog also has relatively more red blood cells than a human. Most of the time the dog will keep the extra red blood cells stored in its spleen. When the animal enters into a situation where its full metabolism is required, such as play, catching game, or fighting other dogs, the extra cells are released into the bloodstream.

The "oversized" heart and lungs will now be running at full capacity, and the animal will have an enhanced ability to engage in aerobic activity. This activity will produce internal heating. Dogs, being covered in fur, are limited in their ability to cool down. After a short time the animal must either cease its athletic activity or risk harming itself from overheating. One can easily observe this pattern of intense activity followed by rest periods in puppies.

During the rest phase the spleen collects red blood cells and the animal may pant to cool down. Dogs are very social animals, but their personality and behaviour vary with breed as well as how they are treated by their owners and others who come in contact with them.

It is not uncommon for dogs to attack humans and other animals; however, this is usually because of lack of care or improper upbringing by its owner. Many dogs can be trained to skillfully perform tasks not natural to canines, such as in this dog agility competition. Dogs are valued for their intelligence.

This intelligence is expressed differently with different breeds and individuals, however. For example, Border Collies are noted for their ability to learn commands, while other breeds may not be so motivated towards obedience, but instead show their cleverness in devising ways to steal food or escape from a yard.

Being highly adaptable animals themselves, dogs have learned to do many jobs as required by humans over the generations. Dogs are employed in various roles across the globe, proving invaluable assets in areas such as search-and-rescue; law enforcement (including attack dogs, sniffer dogs and tracking dogs); guards for livestock, people or property; herding; Arctic exploration sled-pullers; guiding the blind and acting as a pair of ears for the deaf; assisting with hunting, and a great many other roles which they may be trained to assume. Most dogs rarely have to deal with complex tasks and are unlikely to learn relatively complicated activities (such as opening doors) unaided.

Some dogs (such as guide dogs for the visually impaired) are specially trained to recognize and avoid dangerous situations. The meaning of "intelligence" in general, not only in reference to dogs, is hard to define. Some tests measure problem-solving abilities and others test the ability to learn in comparison to others of the same age. Defining it for dogs is just as difficult. It is likely that dogs do not have the ability to premeditate an action to solve a problem. For example, the ability to learn quickly could be a sign of intelligence. Conversely it could be interpreted as a sign of a desire to please.

In contrast, some dogs who do not learn very quickly may have other talents. An example is breeds that are not particularly interested in pleasing their owners, such as Siberian Huskies. Huskies are often fascinated with the myriad of possibilities for escaping from yards, catching small animals, and often figuring out on their own numerous inventive ways of doing both. Assistance dogs are also required to be obedient at all times. This means they must learn a tremendous number of commands, understand how to act in a large variety of situations, and recognize threats to their human companion, some of which they might never before have encountered.

Many owners of livestock guardian breeds believe that breeds like the Great Pyrenees or the Kuvasz are not easily trained because their stubborn nature prevents them from seeing the point of such commands as "sit" or "down". Hounds may also suffer from this type of ranking.

These dogs are bred to have more of a "pack" mentality with other dogs and less reliance on a master's direct commands. While they may not have the same kind of intelligence as a Border Collie, they were not bred to learn and obey commands quickly, but to think for themselves while trailing game. Dogs are highly social animals.

This can account for their trainability, playfulness, and ability to fit into human households and social situations. These attributes have earned dogs a unique position in the realm of interspecies relationships despite being one of the most effective, voracious, and potentially dangerous predators. Dogs and humans at times co-operate in some of the most effective hunting in the animal world; in that context, dogs are superpredators. The loyalty and devotion that dogs demonstrate as part of their natural instincts as pack animals closely mimics the human idea of love and friendship, leading many dog owners to view their pets as full-fledged family members.

Conversely, dogs seem to view their human companions as members of their pack, and make few, if any, distinctions between their owners and fellow dogs. Dogs fill a variety of roles in human society and are often trained as working dogs. For dogs that do not have traditional jobs, a wide range of dog sports provide the opportunity to exhibit their natural skills. In many countries, the most common and perhaps most important role of dogs is as companions.

Dogs have lived and worked with humans in so many roles that their loyalty has earned them the unique sobriquet "man's best friend". However, some cultures consider dogs to be unclean. In some parts of the world, dogs are raised as livestock to produce dog meat for human consumption.

In many places, consumption of dog meat is discouraged by social convention or cultural taboo. Laughter might not be confined or unique to humans, despite Aristotle's observation that "only the human animal laughs". The differences between chimpanzee and human laughter may be the result of the adaptations that enable human speech.

However, some behavioural psychologists argue that self-awareness of one's situation, or the ability to identify with somebody else's predicament, are prerequisites for laughter, so animals are not really laughing in the same way that humans do. The dog-laugh sounds similar to a normal pant. However by analyzing the pant using a spectrograph, this pant varies with bursts of frequencies, resulting in a laugh.

When this recorded dog-laugh vocalization is played to dogs in a shelter setting, it can initiate play, promote pro-social behaviour, and decrease stress levels. In a study by Simonet, Versteeg, and Storie, one hundred and twenty subject dogs residing in a mid-size county animal shelter were observed. Dogs ranging from 4 months to 10 years of age were compared with and without exposure to a dog-laugh recording. The stress behaviours measured included panting, growling, salivating, pacing, barking, cowering, lunging, play-bows, sitting, orienting and lying down.

The study resulted in positive findings. Exposure to the dog laughing recording resulted in the following: significantly reduced stress behaviours, increased tail wagging and the display of a play-face when playing was initiated, and more frequent pro-social behaviour such as approaching and lip licking. This research suggests exposure to dog-laugh vocalizations can calm and possibly increase shelter adoptions.

HORSE BREEDING

Horse breeding refers to reproduction in horses, and particularly the human-directed process of selective breeding of animals, particularly purebred horses of a given breed. While feral and wild horses breed successfully without human assistance, planned matings can be used to produce specifically desired characteristics in domesticated horses. Furthermore, modern breeding management and technologies can increase the rate of conception, a healthy pregnancy, and successful foaling. The male parent of a horse, a stallion, is commonly known as the sire and the female parent, the mare, is called the dam.

Both are genetically important, as each parent provides half of the genetic makeup of the ensuing offspring, called a foal. (Contrary to popular misuse, the word "colt" refers to a young male horse only; "filly" is a young female). Though many amateur horse owners may simply breed a family mare to a local stallion in order to produce a companion animal, most professional breeders use selective breeding to produce individuals of a given phenotype, or breed.

Alternatively, a breeder could, using individuals of differing phenotypes, create a new breed with specific characteristics. Some breeders consider the quality of the sire to be more important than the quality of the dam. However, other breeders maintain that the mare is the most important parent. Because stallions can produce far more offspring than mares, a single stallion can have a greater overall impact on a breed. However, the mare may have a greater influence on an individual foal because its physical characteristics influence the developing foal in the womb and the foal also learns habits from its dam when young.

Foals may also learn the "language of intimidation and submission" from their dam, and this imprinting may affect the foal's status and rank within the herd. Many times, a mature horse will achieve status in a herd similar to that of its dam; the offspring of dominant mares become dominant themselves.

Estrus Cycle of the Mare

The estrous cycle controls when a mare is sexually receptive toward a stallion, and helps to physically prepare the mare for conception. It generally occurs during the spring and summer months, although some mares may be sexually receptive into the late fall, and is controlled by the photoperiod (length of the day), the cycle first triggered when the days begin to lengthen. The estrus cycle lasts about 19-22 days, with the average being 21 days.

As the days shorten, the mare returns to a period when she is not sexually receptive, known as anestrus. Anestrus prevents the mare from conceiving in the winter months, as that would result in her foaling during the harshest part of the year, a time when it would be most difficult for the foal to survive. This cycle contains 2 phases:

- Estrus, or Follicular, phase: 5-7 days in length, when the mare is sexually recepetive to a stallion. Estrogen is secreted by the follicle. Ovulation occurs in the final 24-48 hours of estrus.
- Diestrus, or Luteal, phase: 14-15 days in length, the mare is not sexually recepetive to the stallion. The corpus luteum secretes progesterone.

Although 16 per cent of mares have double ovulations (allowing them to twin), this does not affect the length of time of estrus or diestrus.

Effects on the Reproductive System During the Estrous Cycle

Changes in hormone levels can have great affects on the physical characteristics of the reproductive organs of the mare, thereby preparing, or preventing, her from conceiving.

- Uterus: increased levels of estrogen cause edema within the uterus, making it feel heavier, and the uterus loses its tone. This edema decreases following ovulation, and the muscular tone increases. High levels of progesterone do not cause edema within the uterus. The uterus becomes flaccid during anestrus.

- Cervix: the cervix relaxes right before estrus occurs, and the secretions of the cervix increase. High progesterone levels (during diestrus) cause the cervix to close.
- Vagina: the portion of the vagina near the cervix becomes engorged with blood right before estrus. The vagina becomes relaxed and secretions increase.
- Vulva: relaxes right before estrus begins. Becomes dry, and closes tightly, during diestrus.

Hormones Involved in the Estrus Cycle, During Foaling, and after Birth

The cycle is controlled by several hormones which regulate the estrous cycle, the mare's behaviour, and the reproductive system of the mare. The cycle begins when the increased day length causes the pineal gland to reduce the levels of melatonin, thereby allowing the hypothalamus to secrete GnRH.

- *GnRH (Gonadotropin releasing hormone):* Secreted by the hypothalamus, causes the pituitary to release of 2 gonadotrophins: LH and FSH.
- *LH (Luteinizing hormone):* Levels are highest 2 days following ovulation, then slowly decrease over 4-5 days, dipping to their lowest levels 5-16 days after ovulation. Stimulates the follicle to develop, which in turn causes an increase in estrogen. Unlike most mammals, the mare does not have an increase of LH right before ovulation.
- *FSH (Follicle-stimulating hormone):* Secreted by the pituitary, causes the ovarian follicle to develop. Levels of FSH rise slightly at the end of estrus, but have their highest peak about 10 days before the next ovulation. FSH is inhibited by inhibin, at the same time LH and estrogen levels rise, which prevents immature follicles from continuing their growth.
- *Estrogen:* Secreted by the developing follicle, it causes the pituitary gland to secrete more LH (therefore, these 2 hormones are in a positive feedback loop). Additionally, it causes behavioural changes in the mare, making her more recpetive toward the stallion, and causes physical changes in the cervix, uterus, and vagina to prepare the mare for conception. Estrogen peaks 1-2 days before ovulation, and decreases within 2 days following ovulation.
- *Inhibin:* Secreted by the developed follicle right before ovulation, "turns off" FSH, which is no longer needed now that the follicle is larger.
- *Progesterone:* Prevents conception and decreases sexual receptibility of the mare to the stallion. Progesterone is therefore lowest during the estrus phase, and increases during diestrus. It decreases about 15 days after ovulation, when the corpus luteum begns to decrease in size.

- *Prostaglandin:* Secreted by the endrometrium 13-15 days following ovulation, causes luteolysis and prevents the corpus luteum from secreting progesterone
- *eCG (equine chorionic gonadotropin - formerly called PMS/PMSG pregnant mare serum and pregnant mare serum gonadotropin):* Chorionic gonadotropins secreted if the mare conceives. First secreted by the endometrial cups around the 36th day of gestation, peaking around day 60, and decreasing after about 120 days of gestation. Also help to stimulate the growth of the fetal gonads.
- *Prolactin:* Stimulates lactation
- *Oxytocin:* Stimulates the uterus to contract

While horses in the wild mate and foal in mid to late spring, many horses domestically bred for competitive purposes, especially horse racing and various futurities, are usually born as close to January first as possible, so as to be at an advantage in size and maturity when competing against other horses in the same age group.

When an early foal is desired, barn managers will put the mare "under lights" by keeping the barn lights on in the winter to simulate a longer day, thus bringing the mare into estrus sooner than she would in nature. Mares signal estrus and ovulation by urination in the presence of a stallion, raising the tail and revealing the vulva. A stallion, approaching with a high head, will usually nicker, nip and nudge the mare, as well as sniff her urine to determine her readiness for mating. Once fertilized, the oocyte (egg) remains in the oviduct for approximately 5.5 more days, and then descends into the uterus. The initial single cell combination is already dividing and by the time of entry into the uterus, the egg might have already reached the blastocyst stage.

The gestation period lasts for about eleven months, or about 342 days (normal average range 320-370 days). During the early days of pregnancy, the embryo (up to about 35 days), or fetus (after 35-40 days), is mobile, moving about in the uterus until about day 16 when "fixation" occurs.

True implantation - invasion into the endometrium of any sort - does not occur until about day 35 of pregnancy with the formation of the endometrial cups, and true placentation (formation of the placenta) is not initiated until about day 40-45 and not completed until about 140 days of pregnancy. A heartbeat can be detected on day 21, and the fetus gender can be determined by day 70 of the gestation. Halfway through gestation the fetus is the size of between a rabbit and a beagle. The most dramatic fetus development occurs in the last 3 months of pregnancy when 60 per cent of fetal growth occurs.

Care of the Pregnant Mare

Domestic mares receive specific care and nutrition to ensure that they and their foals are healthy. Mares are given vaccinations against diseases such as the Rhinopneumonitis (EHV-1) virus (which can cause abortions) as well as vaccines for other conditions that may occur in a given region of the world. Mares can be used for riding or driving during most of their pregnancy, and it's healthy for them to have exercise. But only moderate exercise, especially when they become heavy in foal.

Exercise in excessively high temperatures has been suggested as being detrimental to pregnancy maintenance during the embryonic period - it should however be noted that ambient temperatures encountered during the research were in the region of 100 degrees F and the same results may not be encountered in regions with lower ambient temperatures. Mares are generally fed more than other horses and are provided supplemental vitamins and minerals because their bodies require extra nutrition to form (and later to nurse) a foal. Many feeds designed for pregnant and lactating mares also have extra fat and protein.

Overfeeding the pregnant mare should also be avoided, as excess weight may contribute to difficulties foaling or fetal/foal related problems. A special foaling stall or shed that is large and clutter free provides the mare with a safe place to give birth. Most mares foal at night or early in the morning, and prefer to give birth alone when possible. Labour is rapid, often no more than 30 minutes, and from the time the feet of the foal appear to full delivery is often only about 15 to 20 minutes.

Once the foal is born, the mare will lick the newborn foal to clean it and help blood circulation. In a very short time, the foal will attempt to stand and get milk from its mother. A foal should stand and nurse within the first hour of life. To create a bond with her foal, the mare licks and nuzzles the foal, enabling her to distinguish hers from others.

Some mares are aggressive when protecting their foals, and may attack other horses or unfamiliar humans that come near their newborns.

After birth, a foal's navel is dipped in iodine to prevent infection, it is sometimes given an enema to help clear the meconium from its digestive tract, and the newborn is monitored to ensure that it stands and nurses without difficulty.

While most horse births happen without complications, many owners have first aid supplies prepared and a veterinarian on call in case of a birthing emergency. People who supervise foaling should also watch the mare to be sure that she passes the placenta in a timely fashion, and that it is complete with no fragments remaining in the uterus, where they could cause a serious infection.

If the placenta is not removed from the stall after it is passed, a mare will often eat it, an instinct from the wild, where blood would attract predators. Foals develop rapidly, and within a few hours a wild foal can travel with the herd. In domestic breeding, the foal & dam are usually kept separate for a while, but within a few weeks are typically pastured with the other horses. A foal will begin to eat hay, grass & grain alongside the mare at about 4 weeks old; by 10-12 weeks the foal requires more nutrition than the mare's milk can supply. Foals are typically weaned at 4-8 months of age, although in the wild a foal may nurse for a year.

How Breeds Develop

Beyond the appearance and conformation of a specific type of horse, breeders aspire to improve physical performance abilities. This concept, known as matching "form to function," has led to the development of not only different breeds, but also families or bloodlines within breeds that are specialists for excelling at specific tasks.

For example, the Arabian horse of the desert naturally developed speed and endurance to travel long distances and survive in a harsh environment, and domestication by humans added a trainable disposition to the animal's natural abilities. In the meantime, in northern Europe, the locally adapted heavy horse with a thick, warm coat was domesticated and put to work as a farm animal that could pull a plow or wagon.

This animal was later adapted through selective breeding to create a strong but ridable animal suitable for the heavily-armored knight in warfare. Then, centuries later, when people in Europe wanted faster horses than could be produced from local horses through simple selective breeding, they imported Arabians and other oriental horses to breed as an outcross to the heavier, local animals. This led to the development of breeds such as the Thoroughbred, a horse taller than the Arabian and faster over the distances of a few miles required of a European race horse or light cavalry horse.

Another cross between oriental and European horses produced the Andalusian, a horse developed in Spain that was powerfully built, but extremely nimble and capable of the quick bursts of speed over short distances necessary for certain types of combat as well as for tasks such as bullfighting. Later, the people who settled the Americas needed a hardy horse that was capable of working with cattle. Thus, Arabians and Thoroughbreds were crossed on Spanish horses, both domesticated animals descended from those brought over by the Conquistadors, and feral horses such as the Mustangs, descended from the Spanish horse, but adapted by natural selection to the ecology and climate of the west.

These crosses ultimately produced new breeds such as the American quarter horse and the Criollo of Argentina. In modern times, these breeds themselves have since been selectively bred to further specialize at certain tasks. One example of this is the American quarter horse. Once a general-purpose working ranch horse, different bloodlines now specialize in different events.

For example, larger, heavier animals with a very steady attitude are bred to give competitors an advantage in events such as team roping, where a horse has to start and stop quickly, but also must calmly hold a full-grown steer at the end of a rope. On the other hand, for an event known as cutting, where the horse must separate a cow from a herd and prevent it from rejoining the group, the best horses are smaller, quick, alert, athletic and highly trainable.

They must learn quickly, have conformation that allows quick stops and fast, low turns, and the best competitors have a certain amount of independent mental ability to anticipate and counter the movement of a cow, popularly known as "cow sense." Another example is the Thoroughbred. While most representatives of this breed are bred for horse racing, there are also specialized bloodlines suitable as show hunters or show jumpers. The hunter must have a tall, smooth build that allows it to trot and canter smoothly and efficiently.

Instead of speed, value is placed on appearance and upon giving the equestrian a comfortable ride, with natural jumping ability that shows bascule and good form. A show jumper, however, is bred less for overall form and more for power over tall fences, along with speed, scope, and agility. This favours a horse with a good galloping stride, powerful hindquarters that can change speed or direction easily, plus a good shoulder angle and length of neck. A jumper has a more powerful build than either the hunter or the racehorse.

The history of horse breeding goes back millennia. Though the precise date is in dispute, humans could have domesticated the horse as far back as approximately 4500 BCE. However, evidence of planned breeding has a more blurry history. One of the earliest people known to document the breedings of their horses were the Bedouin of the Middle East, the breeders of the Arabian horse. While it is difficult to determine how far back the Bedouin passed on pedigree information via an oral tradition, there were written pedigrees of Arabian horses by A.D. 1330. The Akhal-Teke of West-Central Asia is another breed with roots in ancient times that was also bred specifically for war and racing.

The nomads of the Mongolian steppes bred horses for several thousand years as well. The types of horses bred varied with culture and with the times. The uses to which a horse was put also determined its qualities, including smooth amblers for riding, fast horses for carrying messengers, heavy horses for plowing and pulling heavy wagons, ponies for hauling cars of ore from mines, packhorses, carriage horses and many others.

Medieval Europe bred large horses specifically for war, called destriers. These horses were the ancestors of the great heavy horses of today, and their size was preferred not simply because of the weight of the armor, but also because a large horse provided more power for the knight's lance.

Weighing almost twice as much as a normal riding horse, the destrier was a powerful weapon in battle. On the other hand, during this same time, lighter horses were bred in northern Africa and the Middle East by Muslim warriors, who preferred a faster, more agile horse.

The lighter horse suited the raids and battles of the Bedouins, allowing them to outmaneuver rather than overpower the enemy. When Muslim warriors and European knights collided in warfare, the heavy knights were frequently outmaneuvered. The Europeans, however, soon made up for the lack of speed of their native breeds by incorporating genetic traits from captured oriental horses such as the Arabian, Barb to their stables.

This cross-breeding led both to a nimbler war horse, such as today's Percheron, but also to created a type of horse known as a Courser, a predecessor to the Thoroughbred, which was used as a message horse.

During the Renaissance, horses were bred not only for war, but for haute ecole riding, derived from the most athletic movements required of a war horse, and popular among the elite nobility of the time.

Breeds such as the Lipizzan were developed from Spanish-bred horses for this purpose, and also became the preferred mounts of cavalry officers, who were derived mostly from the ranks of the nobility.

It was during this time that gunpowder was developed, and so the light cavalry horse, a faster and quicker war horse, was bred for a "shoot and run" tactic rather than the close hand-to-hand fighting seen in the Middle Ages. After Charles II retook the British throne in 1660, horse racing, which had been banned by Cromwell, was revived.

The Thoroughbred was developed 40 years later, bred to be the ultimate racehorse, through the lines of 3 foundation Arabian stallions. The 17th and 18th centuries saw more of a need for fine carriage horses in Europe, bringing in the dawn of the warmblood.

The warmblood breeds have been exceptionally good at adapting to changing times, and from their carriage horse beginnings they easily transitioned during the 1900s into a sport horse type. Today's warmblood breeds, although still used for competitive driving, are more often seen competing in the show jumping or dressage arenas.

The Thoroughbred continues to dominate the horseracing world, although its lines have been more recently used to improve warmblood breeds and to develop sport horses. The predecessor of the American Quarter Horse was developed in the 1700s, mainly for quarter racing (racing ¼ of a mile). The breed was later adapted for work in the west, and "cow sense" was particularly bred for as their use for herding cattle increased. However, because there was also a need for animals suitable for sprint racing, the modern Quarter Horse has two distinct types: the sleeker racing type and the stock horse type.

The racing type most resembles the finer-boned ancestors of the first racing Quarter Horses, and the type is still used for ¼-mile races. The stock horse type, used in western events, is bred for a shorter stride, docile temperament, and cow sense. The need for horses for heavy draft and carriage work continued until the industrial revolution and the advent of the automobile and the tractor.

After this time, draft and carriage horse numbers dropped significantly, though light riding horses remained popular for recreational pursuits. Draft horses today are used on a few small farms, but today are seen mainly for pulling and plowing competitions rather than farm work.

Heavy harness horses are now used as an outcross with lighter breeds, such as the Thoroughbred, to produce the modern warmblood breeds popular in Olympic and sport horse disciplines. Breeding a horse is an endeavor where the owner, particularly of the mare, will usually need to invest considerable time and money.

For this reason, a horse owner needs to consider several factors, including:

- Does the proposed breeding animal have valuable genetic qualities to pass on?
- Is the proposed breeding animal in good physical health, fertile, and able to withstand the vigors of reproduction?
- For what purpose will the foal be used?
- Is there a market for the foal in the event that the owner does not wish to keep the foal for its entire life?
- What is the anticipated economic benefit, if any, to the owner of the ensuing foal?
- What is the anticipated economic benefit, if any, to the owner(s) of the sire and dam or the foal?
- Does the owner of the mare have the expertise to properly manage the mare through gestation and parturition?
- Does the owner of the potential foal have the expertise to properly manage and train a young animal once it is born?

There are value judgements involved in considering whether an animal is suitable breeding stock, hotly debated by breeders. Additional personal beliefs may come into play when considering a suitable level of care for the mare and ensuing foal, the potential market or use for the foal, and other tangible and intangible benefits to the owner. If the breeding endeavor is intended to make a profit, there are additional market factors to consider, which may vary considerably from year to year, from breed to breed, and by region of the world.

In many cases, the low end of the market is saturated with horses, and the law of supply and demand thus allows little or no profit to be made from breeding unregistered animals or animals of poor quality, even if registered.

The minimum cost of breeding for a mare owner includes the stud fee, and the cost of proper nutrition, management and veterinary care of the mare throughout gestation, parturition, and care of both mare and foal up to the time of weaning. Veterinary expenses may be higher if specialized reproductive technologies are used or health complications occur. Making a profit in horse breeding is often difficult.

While some owners of only a few horses may keep a foal for purely personal enjoyment, many individuals breed horses in hopes of making some money in the process. A general rule of thumb is that a foal intended for sale should be worth three times the cost of the stud fee if it were sold at the moment of birth. From birth forward, the costs of care and training are added to the value of the foal, with a sale price going up accordingly. If the foal wins awards in some form of competition, that may also enhance the price.

On the other hand, without careful thought, foals bred without a potential market for them may wind up being sold at a loss, and in a worst-case scenario, sold for "salvage" value — a euphemism for sale to slaughter as horsemeat. Therefore, a mare owner must consider their reasons for breeding, asking hard questions of themselves as to whether their motivations are based on either emotion or profit and how realistic those motivations may be.The stallion should be chosen to complement the mare, with the goal of producing a foal that has the best qualities of both animals, yet avoids having the weaker qualities of either parent.

Generally, the stallion should have proven himself in the discipline or sport the mare owner wishes for the "career" of the ensuing foal. Mares should also have a competition record showing that they also have suitable traits, though this does not happen as often. A purebred horse is usually worth more than a horse of mixed breeding, though this matters more in some disciplines than others. The breed of the horse is sometimes secondary when breeding for a sport horse, but some disciplines may prefer a certain breed or a specific phenotype of horse.

Sometimes, purebred bloodlines are an absolute requirement: For example most Racehorses in the United States must be recorded with a breed registry in order to race.Bloodlines are often considered, as some bloodlines are known to cross well with others.

If the parents have not yet proven themselves by competition or by producing quality offspring, the bloodlines of the horse are often a good indicator of quality and possible strengths and weaknesses. Some bloodlines are known not only for their athletic ability, but could also carry a conformational or genetic defect, poor temperament, or for a medical problem.

Some bloodlines are also fashionable or otherwise marketable, which is an important consideration should the mare owner wish to sell the foal. Horse breeders also consider conformation, size and temperament. All of these traits are heritable, and will determine if the foal will be a success in its chosen discipline.

The offspring, or "get," of a stallion are often excellent indicators of his ability to pass on his characteristics, and the particular traits he actually passes on. Some stallions are fantastic performers but never produce offspring of comparable quality. Others sire fillies of great abilities but not colts. At times, a horse of mediocre ability sires foals of outstanding quality. Mare owners also look into the question of if the stallion is fertile and has successfully "settled" (i.e. impregnated) mares. A stallion may not be able to breed naturally, or old age may decrease his performance. Mare care boarding fees and semen collection fees can be a major cost.

Breeding a horse can be an expensive endeavor, whether breeding a backyard competition horse or the next Olympic medalist. Costs may include:

- The stud and booking fee
- Fees for collecting, handling, and transporting semen (if AI is used and semen is shipped)
- Mare exams: to determine if she is healthy enough to breed, to determine when she ovulates, and (if AI is used) to inseminate her
- Mare transport, care, and board if the mare is bred live cover at the stallion's residence
- Veterinary bills to keep the pregnant mare healthy while in foal
- Possible veterinary bills during pregnancy or foaling should something go wrong
- Veterinary bills for the foal for its first exam a few days following foaling

Stud fees are determined by the quality of the stallion, his performance record, the performance record of his get (offspring), as well as the sport and general market that the animal is standing for. The highest stud fees are generally for racing Thoroughbreds, which may charge from two to three thousand dollars for a breeding to a new or unproven stallion, to several hundred thousand dollars for a breeding to a stakes winner.

Sport horse stallions generally range from $1000 to $3000, although the top stallions may reach $4000 for one breeding. The lowest stud fees may only be $100-$200, but there are trade-offs: the horse will probably be unproven, and probably much less athletic than a horse with a stud fee that is a bit higher. As a stallion's career, either performance or breeding, improves, his stud fee tends to increase in proportion. If one or two offspring are especially successful, winning several stakes races or an Olympic medal, the stud fee will generally greatly increase. Younger, unproven stallions will generally have a lower stud fee earlier on in their careers. To help decrease the risk of financial loss should the mare die or abort the foal while pregnant, many studs have a live foal guarantee (LFG), allowing the owner to have a free breeding to their stallion the next year. However, this is not offered for every breeding.

Covering the Mare

After the mare is bred or artificially inseminated, she is checked 16 days later to see if she "took", and is pregnant. A second check is usually performed at 28 days. If the mare is not pregnant, she may be bred again during her next cycle. It is considered safe in the horse to breed a mare to a stallion of much larger size. Because of the mare's type of placenta and its attachment and blood supply, the foal will be limited in its growth within the uterus to the size of the mare's uterus, but will grow to its genetic potential after it is born. Test breedings have been done with draft horse stallions bred to small mares with no increase in the number of difficult births.

When breeding live cover, the mare is usually boarded at the stud. She is "teased" several times with a stallion that will not breed to her, usually with the stallion being presented to the mare over a barrier. Her reaction to the teaser, whether hostile or passive, is noted. A mare that is in heat will generally tolerate a teaser (although this is not always the case), and may present herself to him, holding her tail to the side. A veterinarian may also determine if the mare is ready to be bred, by ultrasound or palpating daily to determine if ovulation has occurred. When it has been determined that the mare is ready, both the mare and intended stud will be cleaned. The mare will then be presented to the stallion, usually with one handler controlling the mare and one or more handlers in charge of the stallion. Multiple handlers are preferred, as the mare and stallion can be easily separated should there be any trouble. The Thoroughbred industry requires all registered foals to be bred through live cover. Artificial fertility treatments, listed below, are not permitted.

By contrast, the standardbred industry allows registered foals to be bred by live cover, or by artificial insemination (AI) with fresh or frozen (not dried) semen. No other artificial fertility treatment is allowed. In addition, foals bred via AI may only be registered if the stallion's sperm was collected during his lifetime, and used no later than the calendar year of his death or castration.

Artificial Insemination

Artificial insemination (AI) has several advantages over live cover, and has a very similar conception rate:

- The mare and stallion never have to come in contact with each other, which therefore reduces breeding accidents, such as the mare kicking the stallion.
- AI opens up the world to international breeding, as semen may be shipped across continents to mares that would otherwise be unable to breed to a particular stallion.
- A mare also does not have to travel to the stallion, so the process is less stressful on her, and if she already has a foal, the foal does not have to travel.
- AI allows more mares to be bred from one stallion, as the ejaculate may be split between mares.
- AI reduces the chance of spreading sexually transmitted diseases between mare and stallion.

- AI allows mares or stallions with health issues, such as sore hocks which may prevent a stallion from mounting, to continue to breed.
- Frozen semen may be stored and used to breed mares even after the stallion is dead, allowing his lines to continue. However, the semen of some stallions does not freeze well, and live cover is sometimes the only option. Also, as noted earlier with respect to standardbreds, a particular breed registry may ban the use of frozen semen in posthumous breeding.

A stallion is usually trained to mount a phantom mare, although a live mare may be used, and he is collected using an artificial vagina (AV), which is often heated to simulate the vagina of the mare. The AV has a filter and collection area at one end to collect the semen, which is then processed in a lab. The semen is then chilled or frozen and shipped to the mare owner.

When the mare is in heat, a veterinarian introduces the semen directly into her via a syringe and pipette. Often an owner does not want to take a valuable competition mare out of training to carry a foal. This presents a problem, as the mare will usually be quite old by the time she is retired from her competitive career, at which time it is more difficult to impregnate her. Other times, a mare may have physical problems that prevent or discourage breeding. However, there are now several options for breeding these mares. These options also allow a mare to produce multiple foals each breeding season, instead of the usual one. Therefore, mares may have an even greater value for breeding.

Embryo Transfer

The relatively new method involves removing the mare's fertilized embryo a few days following insemination, and transferred to a surrogate mare.

Gamete Intrafallopian Transfer (GIFT)

The mare's ovum and the stallion's sperm are deposited in the oviduct of a surrogate dam. This technique is very useful for subfertile stallions, as fewer sperm are needed, so a stallion with a low sperm count can still successfully breed.

Egg Transfer

An oocyte is removed from the mare's follicle and transferred into the oviduct of the recipient mare, who is then bred. This is best for mares with physical problems, such as an obstructed oviduct, that prevent breeding.

COW BREEDING

Different Breeds for Different Needs

Dairy cattle are breeds that have been developed to produce milk whereas beef cattle are bred and selected primarily for the production of meat. Dual-purpose breeds do exist and they have been selected for both meat and milk production. In the last 20 years the genetic make up of the UK dairy herd has changed and is still changing.

Cows are larger, earlier maturing, are producing far more milk and are more efficient at converting feed into milk. As it is the birth of the calf that causes milk production, it is critical for the farmer that the cow becomes pregnant again during lactation and produces the next calf one year after the previous one. The average length of productive life of a dairy cow is between four and five lactations, although this varies from one or two to over ten. Since the fifties artificial insemination (AI) has increasingly replaced natural service. Embryo transplants (ET) are also being developed.

A surrogate mother may have difficulty delivering, since the foreign calf is often larger than nature intended. When this happens, a rope or chain is attached to the calf's forelegs, and, with a winch-like device the newborn calf is pulled from the womb. Unnecessary suffering for both cow and calf.

Regular Calving

A high annual milk production per cow is dependent on regular calving at about 12- month intervals. Poor reproductive efficiency leads to longer calving indices and a reduction in milk sales per cow per annum. It also leads to an increase in the number of cows culled for failing to conceive. On average, in the UK, only about 85 cows calve per annum for every 100 cows available for breeding. Over thirty-five per cent of all cows culled are disposed of because of reproductive disorders.

Obtaining Semen

Photographer Henry Heap visited the Milk Marketing Board stud bull operation in Mid Wales. He provided Turning Point magazine with a complete set of photographs and commentary of his visit. "A large, handsome, tan bull stood stock still, tethered by the ring tethered through his nose to what looked like a giant circular clothes line or some grotesque merry-go-round. He was kept so motionless by a rubber hood that covered his eyes, which had a similar effect to that of hooding a falcon.

A worker commented that the bull would stay still all day if they left the blindfold on. However, this bull was not the source of the semen, but instead what is known as a 'teaser'. The stud bulls are led out by the nose one at a time to mount the motionless hooded bull. After they have mounted the teaser their penis is diverted into a leather vagina which is used to collect the semen.

As soon as the process is complete, another bull is led out. The hood keeps the teaser bull stationery for each successive beast to mount. These stud bulls never even see females. During embryo transfer an embryo can be flushed from the uterus of a genetically superior cow and implanted in a receiver cow of so-called 'poorer quality' for foetal development. Embryos are also taken from dead cows at the slaughterhouse. Hormone treatment is used to induce multiple ovulation and the cow is then inseminated. A few days later several embryos can be flushed from her. For both embryo collection and implanting the cow receives a local anaesthetic injection in the tail. Embryos of the desired breed can then be used. ET maximises the number of offspring from selected breeding females and maximises profit. Do-it-yourself embryo transfer is now taking place in the UK. Courses are run to train in direct embryo transfer techniques using the quick-thaw method for frozen embryos.

Although vets are not required to be present at the time of transfer, they must check each animal is suitable for implantation. Farmers genetically improve the herd by using techniques such as MOET (multiple ovulation embryo transfer) which involves inducing a genetically 'superior' cow to produce up to 20 embryos rather than one. For ease of management, farmers prefer cows to calf at the same time. They, therefore, need to detect when cows come into oestrous. The oestrous detection rate (ODR) is the number of cows observed in oestrus during a 3 week period, in relation to the number of cows expected to be observed.

The average ODR on farms is about 60 per cent. One alternative is to synchronise the oestrous period of several animals by injections of prostaglandin or administration of progesterone as an impregnated intraveginal coil implanted for 9-12 days.

Artificial Insemination

Artificial insemination (AI) now accounts for around 75 per cent of British cattle births. The cow is tied up and one hand of the inseminator manipulates the cervix through the rectum wall whilst the other discharges semen into the vagina and cervix using an inseminating gun. A regional manager with AI firm Genus the problems associated with the procedure. "Placing a quarter of a cc of diluted semen into an area as small as a penny coin, when the target area is inside 500kg of unco-operative cow, requires a high degree of expertise.

The more the cow is able to move about, the harder it becomes to hit the target. Poor handling facilities cause stress to you, the cow, the inseminator and both our bank balances." Natural service by bulls is still used, but the risk of slowing down genetic progress, and the danger of handling bulls means farmers mainly use AI. There are many invasive procedures carried out on farm animals that the general public rarely witness. One of many includes making sure the cow is in calf. This is carried out using an ultrasound device much like a small tv set with a long thick wire attachment.

A vet will insert their hand and arms length into the vagina of the cow. The animal is held in a holding device to keep her still. The screen shows the small calf in the middle of a dark shaped uterus. Herds usually calve at the same time (due to human intervention) so all the cows on the farm will usually be checked at the same time.

Jersey Cattle

A small, honey-brown breed of dairy cattle, the Jersey is renowned for the high butterfat content of its milk, as well as a genial disposition.

The Jersey cow is quite small, ranging from only 800 to 1200 pounds (360 to 540 kg). The main factor contributing to the popularity of the breed has been their greater economy of production, due to:

- The ability to carry a larger number of effective milking cows per unit area due to lower body weight, hence lower maintenance requirements
- High butterfat conditions, and to thrive on locally produced food.

Bulls are also small, ranging from 1200 to 1800 pounds (540 to 820 kg), and are notoriously very aggressive. Castrated males can be trained into fine oxen which, due to their small size and gentle nature make them popular with young teamsters. Jersey oxen are not as strong as larger breeds however and are generally out of favour among competitive teamsters.

Due to the small size, curious docile character and attractive features of the Jersey cow, small herds were imported into England by aristocratic landowners as props for aesthetic landscapes. As its name implies, the Jersey was bred on the British Channel Island of Jersey.

It apparently descended from cattle stock brought over from the nearby Norman mainland, and was first recorded as a separate breed around 1700. Frank Falle has speculated (on the basis of DNA evidence) that the oldest settlers in Jersey were Danish vikings who had been to Nantes with Hatain, where it is recorded that tribute was given by the King of France to Hatain of 500 cattle to leave that area, whereupon they settled in the Normandy area.

The marked resemblance between Jersey cattle and the Nantaise cow would seem to bear this out.

While the breed is isolated from outside influence today, this was not always the case. Before 1789 cows would be given as dowry for inter-island marriages between Jersey and Guernsey. This was, however, not widespread. Since 1789, imports of foreign cattle into Jersey have been forbidden by law to maintain the purity of the breed, although exports of cattle and semen have been an important economic resource for the island.

The restriction on the import of cattle was initially introduced in 1789 to prevent a collapse in the export price. The United Kingdom levied no import duty on cattle imported from Jersey. Cattle were being shipped from France to Jersey and then being shipped onward to England to circumvent the tariff on French cattle. The increase in the supply of cattle, sometimes of inferior quality, was bringing the price down and damaging the reputation of Jersey cattle. The import ban stabilised the price and enabled a more scientifically controlled programme of breeding to be undertaken.

Sir John Le Couteur studied selective breeding and became a Fellow of the Royal Society - his work led to the establishment of the Royal Jersey Agricultural and Horticultural Society in 1833. At that time, the breed displayed greater variation than it does today with white, dark brown and mulberry beasts.

SHEEP BREEDS

The domestic sheep is a multi-purpose animal, and the more than 200 breeds now in existence were created to serve these diverse purposes. Some sources give a count of a thousand or more breeds, but concrete documentation has not been produced for such a number. Almost all sheep are classified as being best suited to furnishing a certain product: wool, meat, milk, hides or a combination thereof (i.e. dual-purpose breeds). Other features that are important when classifying sheep include face colour (generally white or black), the length of the tail, the presence or lack of horns, and the topography for which the breed has been developed.

This last point is especially stressed in the United Kingdom, where breeds are described as either upland (hill or mountain) or lowland breeds. A sheep may also be of a fat-tailed breed, which is a type of dual-purpose sheep common in Africa and Asia with larger deposits of fat within its tail, or the meat from such a breed. Ovine breeds are also grouped based on how well they are suited to producing a certain type of breeding stock. Generally, sheep are thought to be either "ewe breeds" or "ram breeds". Ewe breeds are those that are hardy, and have good reproductive and mothering capabilities—they are for replacing breeding ewes in standing flocks.

Ram breeds are selected for rate of weight gain and carcass quality, and are mated with ewe breeds to produce meat lambs. Many breeds, especially rare or primitive ones, fall into neither category strictly. Lowland and upland breeds are also crossed in this fashion, with the hardy hill ewes crossed with larger, fast-growing lowland rams to produce a type of ewe called a mule, which can then be crossed with meat-type rams to produce prime market lambs.

Sheep breeds are categorized by the type of wool they exhibit, irrespective of whether they are favoured for meat or wool production. Fine wool breeds are those that have wool of great crimp and density, and are consequently preferred by the textile industry. Most of these breeds were derived from Merino sheep, and Merinos continue to dominate the world sheep industry.

The record for the most valuable sheep ever belongs to an Australian Merino ram which was sold for $16,000 AUD. Medium wool breeds have wool between the extremes, and are typically fast-growing meat and ram breeds with dark faces. Some major medium wool breeds, such as the Corriedale, are dual-purpose crosses of long and fine-wooled breeds created for high-production commercial flocks. Long wool breeds are the largest of sheep, with lengthy wool and a slow rate of maturation.

Long wool sheep are most valued by modern agriculture for crossbreeding to improve the attributes of other sheep types.

For example: the American Columbia breed was developed through crossing Lincoln rams—a long wool breed—with fine-wooled Rambouillet ewes. Coarse or carpet wool sheep are those with a medium to long length wool of characteristic coarseness. Breeds traditionally used for carpet wool show more variability in general traits than other types, the chief requirement being a wool that will not break down under heavy use (as would that of the finer breeds).

As the demand for carpet-quality wool declines, some breeders of this type of sheep are attempting to reposition a few of the traditional breeds to alternative purposes. Others have always been primarily meat-class sheep. A last group of sheep breeds is that of fur or hair sheep, which do not grow wool at all. Hair sheep are similar to the very early type of domesticated sheep kept before wooly breeds were developed, and are primarily raised for meat and pelts. Some modern breeds of hair sheep, such as the Dorper, are the result of crosses between wooled and hair breeds.

For producers uninterested in selling wool, hair sheep are highly advantageous, as those who keep hair sheep need not go to the expense of shearing. Hair sheep are also resistant to parasites and hot weather, and are consequently increasing in popularity among a subset of producers. With the modern rise of corporate agrobusiness and the decline of localized family farms, many breeds of sheep are in danger of extinction.

The Rare Breeds Survival Trust of the U.K. lists 25 native breeds as having only 3,000 registered animals, and the American Livestock Breeds Conservancy lists 14 as being at less than 10,000.

Preference for only those breeds with uniform characteristics and fast growth rates has pushed heritage (or heirloom) breeds to the margins of the sheep industry. Those that remain are maintained through the efforts of conservation organizations, breed registries, and individual farmers dedicated to their preservation.

7

Dairy Farming

Dairy farming is a class of agricultural, or an animal husbandry, enterprise, for long-term production of milk, usually from dairy cows but also from goats and sheep, which may be either processed on-site or transported to a dairy factory for processing and eventual retail sale.

Most dairy farms sell the male calves born by their cows, usually for veal production, or breeding depending on quality of the bull calf, rather than raising non-milk-producing stock. Many dairy farms also grow their own feed, typically including corn, alfalfa, and hay. This is fed directly to the cows, or is stored as silage for use during the winter season. Additional dietary supplements are often added to the feed to increase quality milk production.

BACKDROP OF DAIRY FARMING

Dairy farming has been part of agriculture for thousands of years. Historically it has been one part of small, diverse farms. In the last century or so larger farms doing only dairy production have emerged. Large scale dairy farming is only viable where either a large amount of milk is required for production of more durable dairy products such as cheese, or there is a substantial market of people with cash to buy milk, but no cows of their own.

HAND MILKING

Centralized dairy farming as we understand it primarily developed around villages and cities, where residents were unable to have cows of their own due to a lack of grazing land. Near the town, farmers could make some extra money on the side by having additional animals and selling the milk in town.

The dairy farmers would fill barrels with milk in the morning and bring it to market on a wagon. Until the late 1800s, the milking of the cow was done by hand. In the United States, several large dairy operations existed in some northeastern states and in the west, that involved as many as several hundred cows, but an individual milker could not be expected to milk more than a dozen cows a day. Smaller operations predominated. Milking took place indoors in a barn with the cattle tied by the neck with ropes or held in place by stanchions.

Feeding could occur simultaneously with milking in the barn, although most dairy cattle were pastured during the day between milkings. Such examples of this method of dairy farming are difficult to locate, but some are preserved as a historic site for a glimpse into the days gone by. One such instance that is open for this is at Point Reyes National Seashore.

VACUUM BUCKET MILKING

The first milking machines were an extension of the traditional milking pail. The early milker device fit on top of a regular milk pail and sat on the floor under the cow. Following each cow being milked, the bucket would be dumped into a holding tank. This developed into the Surge hanging milker.

Prior to milking a cow, a large wide leather strap called a surcingle was put around the cow, across the cow's lower back. The milker device and collection tank hung underneath the cow from the strap. This innovation allowed the cow to move around naturally during the milking process rather than having to stand perfectly still over a bucket on the floor.

With the availability of electric power and suction milking machines, the production levels that were possible in stanchion barns increased but the scale of the operations continued to be limited by the labour intensive nature of the milking process.

Attaching and removing milking machines involved repeated heavy lifting of the machinery and its contents several times per cow and the pouring of the milk into milk cans. As a result, it was rare to find single-farmer operations of more than 50 head of cattle.

STEP-SAVER MILK TRANSPORT

As herd size began to increase, the bucket milker system became laborious. A vacuum milk-transport system known as the Step-Saver was developed to transport milk to the storage tank. The system used a long vacuum hose coiled around a receiver cart, and connected to a vacuum-breaker device in the milkhouse, allowing farmers to milk many cows without the necessity of walking increasingly longer distances carrying heavy buckets of milk.

MILKING PIPELINE

The next innovation in automatic milking was the milk pipeline. This uses a permanent milk-return pipe and a second vacuum pipe that encircles the barn or milking parlor above the rows of cows, with quick-seal entry ports above each cow. By eliminating the need for the milk container, the milking device shrank in size and weight to the point where it could hang under the cow, held up only by the sucking force of the milker nipples on the cow's udder.

The milk is pulled up into the milk-return pipe by the vacuum system, and then flows by gravity to the milkhouse vacuum-breaker that puts the milk in the storage tank. The pipeline system greatly reduced the physical labour of milking since the farmer no longer needed to carry around huge heavy buckets of milk from each cow.

The pipeline allowed barn length to keep increasing and expanding, but after a point farmers started to milk the cows in large groups, filling the barn with one-half to one-third of the herd, milking the animals, and then emptying and refilling the barn. As herd sizes continued to increase, this evolved into the more efficient milking parlor.

MILKING PARLORS

Innovation in milking focused on mechanizing the milking parlor to maximize throughput of cows per operator which streamlined the milking process to permit cows to be milked as if on an assembly line, and to reduce physical stresses on the farmer by putting the cows on a platform slightly above the person milking the cows to eliminate having to constantly bend over. Many older and smaller farms still have tie-stall or stanchion barns, but worldwide a majority of commercial farms have parlors.

The milking parlor allowed a concentration of money into a small area, so that more technical monitoring and measuring equipment could be devoted to each milking station in the parlor. Rather than simply milking into a common pipeline for example, the parlor can be equipped with fixed measurement systems that monitor milk volume and record milking statistics for each animal. Tags on the animals allow the parlor system to automatically identify each animal as it enters the parlor.

RECESSED PARLORS

More modern farms use recessed parlors, where the milker stands in a recess such that his arms are at the level of the cow's udder.

Recessed parlors can be herringbone, where the cows stand in two angled rows either side of the recess and the milker accesses the udder from the side, parallel, where the cows stand side-by-side and the milker accesses the udder from the rear or, more recently, rotary, where the cows are on a raised circular platform, facing the centre of the circle, and the platform rotates while the milker stands in one place and accesses the udder from the rear. There are many other styles of milking parlors which are less common.

HERRINGBONE AND PARALLEL PARLORS

In herringbone and parallel parlors, the milker generally milks one row at a time. The milker will move a row of cows from the holding yard into the milking parlor, and milk each cow in that row. Once all or most of the milking machines have been removed from the milked row, the milker releases the cows to their feed.

A new group of cows is then loaded into the now vacant side and the process repeats until all cows are milked. Depending on the size of the milking parlor, which normally is the bottleneck, these rows of cows can range from four to sixty at a time.

ROTARY PARLORS

In rotary parlors, the cows are loaded one at a time onto the platform as it slowly rotates. The milker stands near the entry to the parlor and puts the cups on the cows as they move past. By the time the platform has completed almost a full rotation, another milker or a machine removes the cups and the cow steps backwards off the platform and then walks to its feed.

AUTOMATIC MILKER TAKE-OFF

It can be harmful to an animal for it to be over-milked past the point where the udder has stopped releasing milk. Consequently the milking process involves not just applying the milker, but also monitoring the process to determine when the animal has been milked out and the milker should be removed. While parlor operations allowed a farmer to milk many more animals much more quickly, it also increased the number of animals to be monitored simultaneously by the farmer. The automatic take-off system was developed to remove the milker from the cow when the milk flow reaches a preset level, relieving the farmer of the duties of carefully watching over 20 or more animals being milked at the same time.

FULLY AUTOMATED ROBOTIC MILKING

In the 1980s and 1990s, robotic milking systems were developed and introduced. Thousands of these systems are now in routine operation. In these systems the cow has a high degree of autonomy to choose her time of milking within pre-defined windows.

These systems are generally limited to intensively managed systems although research continues to match them to the requirements of grazing cattle and to develop sensors to detect animal health and fertility automatically.

DAIRY CATTLE

Dairy cattle (dairy cows) are cattle cows (adult females) bred for the ability to produce large quantities of milk, from which dairy products are made. Dairy cows generally are of the species Bos taurus.

Historically, there was little distinction between dairy cattle and beef cattle, with the same stock often being used for both meat and milk production. Today, dairy cows are specialized and most have been bred to produce large volumes of milk, with little or no regard for their production of meat. Between 1959 and 1990, US milk production doubled while the number of dairy cows declined 40 per cent.

COW

Dairy cows may be found either in herds on dairy farms where dairy farmers own, manage, care for, and collect milk from them, or on commercial farms. Dairy cow herds range in size from small farms of fewer than five cows to large herds of about 20,000.

The average dairy farmer in the United States manages about one hundred cows but this varies from an average of 800 cows in California to under 80 in the North East states. Herd sizes vary around the world depending on landholding culture and social structure. In many European countries the average herd size is well below 50. In the UK it is over 100 in New Zealand 350 and Australia 280.

To maintain high milk production, a dairy cow must be bred and produce calves. Depending on market conditions, the cow may be bred with a "dairy bull" or a "beef bull." Female calves (heifers) with dairy breeding may be kept as replacement cows for the dairy herd. If a replacement cow turns out to be a substandard producer of milk, she then goes to market and can be killed for beef.

Male calves can either be used later as a breeding bull or sold and used for veal or beef. Most dairy farmers begin breeding heifers at fifteen months of age. A cow's gestation period is approximately nine months, so most heifers give birth at around two years of age.

A Californian industry report, the natural lifespan of a dairy cow is approximately 15–25 years, however dairy cows are rarely kept longer than five years prior to slaughter. Herd life is strongly correlated with production levels. Approximately 17 per cent of the US beef supply comes from cull dairy cows: cows that can no longer be seen as a economic asset to the dairy farm. These animals may be sold due to common diseases of milk cows including mastitis, lameness, or other diseases.

In India, the Hindu majority holds the cow as sacred and a motherly figure due to her capacity to give milk. Cow slaughter is legally banned in India (except in the state of Kerala, West Bengal and the seven north eastern states). Thus, spent dairy cows don't go to slaughter, but are often seen as roaming on the city streets and die of old age or disease. Some pious Hindu organizations manage "old age homes" for old dairy cows.

CALF

Market calves are generally sold at two weeks of age and bull calves may fetch a premium over heifers due to their size, either current or potential. Calves may be sold for veal, or for one of several types of beef production, depending on available local crops and markets. Such bull calves may be castrated if turnout onto pastures is envisaged, in order to render the animals less aggressive. Purebred bulls from elite cows may be put into progeny testing schemes to find out whether they might become superior sires for breeding. Such animals may become extremely valuable.

Most dairy farms separate calves from their mothers within a day of birth to reduce transmission of disease and simplify management of milking cows. Studies have been done allowing calves to remain with their mothers for 1, 4, 7 or 14 days after birth. Cows whose calves were removed longer than one day after birth showed increased searching, sniffing and vocalizations. However, calves allowed to remain with their mothers for longer periods showed weight gains at three times the rate of early removals as well as more searching behaviour and better social relationships with other calves. After separation, most young dairy calves subsist on commercial milk replacer, a feed based on dried milk powder. Milk replacer is an economical alternative to feeding whole milk because it is cheaper, can be bought at varying fat and protein percentages. A day old calf consumes around 5 liters of milk per day.

BULL

A bull calf with high genetic potential may be reared for breeding purposes. It may be kept by a dairy farm as a herd bull, to provide natural breeding for the herd cows. A bull may service up to 50 or 60 cows during a breeding season. Any more and the sperm count will decline, leading to cows "returning to service" (to be bred again). A herd bull may only stay for one season since over two years old their temperament becomes too unpredictable. Bull calves intended for breeding commonly are bred on specialized dairy breeding farms, not production farms. These farms are the major source of stocks for artificial insemination (AI).

MILK PRODUCTION LEVELS

A cow will produce large amounts of milk over her lifetime. Certain breeds produce more milk than others; however, different breeds produce within a range of around 15,000 to 25,000 lbs of milk per lactation. The average for dairy cows in the US in 2005 was 19,576 pounds.

Production levels peak at around 40 to 60 days after calving. The cow is then bred. Production declines steadily afterwards, until, at about 305 days after calving, the cow is 'dried off', and milking ceases.

About sixty days later, one year after the birth of her previous calf, a cow will calve again. High production cows are more difficult to breed at a one year interval. Many farms take the view that 13 or even 14 month cycles are more appropriate for this type of cow. Dairy cows may continue to be economically productive for many lactations. Ten or more lactations are possible. The chances of problems arising which may lead to a cow being culled are however, high; the average herd life of US Holsteins is today fewer than 3 lactations. This requires more herd replacements to be reared or purchased.

Over 90% of all cows are culled for 4 main reasons:

1. *Infertility*: Failure to conceive and reduced milk production. Cows are at their most fertile between 60 and 80 days after calving. Cows remaining "open" (not with calf) after this period become increasingly difficult to breed, which may be due to poor health. Failure to expel the afterbirth from a previous pregnancy, luteal cysts, or metritis, an infection of the uterus, are common causes of infertility.
2. *Mastitis*: Persistent and potentially fatal mammary gland infection, leading to high somatic cell counts and loss of production. Mastitis is recognized by a reddening and swelling of the infected quarter of the udder and the presence of whitish clots or pus in the milk. Treatment is possible with long-acting antibiotics but milk from such cows is not marketable until drug residues have left the cow's system.
3. *Lameness*: Persistent foot infection or leg problems causing infertility and loss of production. High feed levels of highly digestible carbohydrate cause acidic conditions in the cow's rumen. This leads to laminitis and subsequent lameness, leaving the cow vulnerable to other foot infections and problems which may be exacerbated by standing in feces or water soaked areas.
4. *Production*: Some animals fail to produce economic levels of milk to justify their feed costs. Production below 12 to 15 liters of milk per day are not economically viable.

Herd life is strongly correlated with production levels. Lower production cows live longer than high production cows, but may be less profitable. Cows no longer wanted for milk production are sent to slaughter. Their meat is of relatively low value and is generally used for processed meat.

REPRODUCTION

Since the 1950s, artificial insemination (AI) is used at most dairy farms; these farms may keep no bull. Advantages of using AI include its low cost and ease compared to maintaining a bull, ability to select from a large number of bulls to match the anticipated market for the resulting calves, and predictable results. More recently, embryo transfer has been used to enable the multiplication of progeny from elite cows. Such cows are given hormone treatments to produce multiple embryos. These are then 'flushed' from the cow's uterus. 7-12 embryos are consequently removed from these donor cows and transferred into other cows who serve as surrogate mothers. The result will be between 3 and 6 calves instead of the normal single, or rarely, twins.

HORMONE USE

Hormone treatments are given to dairy cows to increase reproduction and to increase milk production. The hormones are used to produce multiple embryos have to be administered at specific times to dairy cattle to induce ovulation. Frequently, for economic considerations, these drugs are also used to synchronize a group of cows to ovulate simultaneously. The hormones Prostaglandin, Gonadotropin Releasing Hormone, and Progesterone are used for this purpose and sold under the brand names Lutalyse, Cystorelin, Estrumate, Factrel, Prostamate, Fertagyl. Insynch, and Ovacyst. They may be administered by injection, insertion or mixed with feed.

About 17% of dairy cows in the United States are injected with Bovine somatotropin, also called recombinant bovine somatotropin (rBST), recombinant bovine growth hormone (rBGH), or artificial growth hormone. The use of this hormone increases milk production from 11%-25%, but also increases the likelihood of cattle developing mastitis, reduction in fertility and lameness. The U.S. Food and Drug Administration (FDA) has ruled that rBST is harmless to people, although critics point out increased levels of insulin-like growth factor 1 (IGF-1) in milk produced using this hormone. The use of rBST is banned in Canada, parts of the European Union, Australia and New Zealand.

NUTRITION

Nutrition plays an important role in keeping cattle healthy and strong. Implementing an adequate nutrition programme can also improve milk production and reproductive performance. Nutrient requirements may not be the same depending on the animal's age and stage of production.

Forages, which refer especially to hay or straw, are the most common type of feed used. Cereal grains, as the main contributors of starch to diets, are important in meeting the energy needs of dairy cattle. Barley is one example of grain that is extensively used around the world. Barley is grown in temperate to sub-artic climates, and it transported to those areas lacking the necessary amounts of grain. Although variations may occur, in general, barley is an excellent source of balanced amounts of protein, energy, and fibre.

Ensuring adequate body fat reserves is essential for cattle to produce milk and also to keep reproductive efficiency. However, if cattle get excessively fat or too thin, they run the risk of developing metabolic problems. Scientists have found that a variety of fat supplements can benefit conception rates of lactating dairy cows. Some of these different fats include oleic acids, found in canola oil, animal tallow, and yellow grease; palmitic acid found in granular fats and dry fats; and linolenic acids which are found in cottonseed, safflower, sunflower, and soybean. It is also important to note that proper levels of fat also improve cattle longevity. Using by-products is one way of reducing the normally high feed costs. However, lack of knowledge of their nutritional and economic value limits their use. Although the reduction of costs may be significant, they have to be used carefully because animal may have negative reactions to radical changes in feeds. Such a change must then be made slowly and with the proper follow up.

PESTICIDE USE

A survey of the primary dairy producing areas in the US indicated that 13 per cent of lactating animals were treated with insecticides permethrin, pyrethrin, coumaphos, and dichlorvos primarily by daily or every-other-day coat sprays. Workers, particularly in stanchion barns, may be exposed to higher than recommended amounts of these pesticides.

BREEDS

In the United States, dairy cattle are divided into six major breeds. These are the: Holstein-Friesian, Brown Swiss, Guernsey, Ayrshire, Jersey, and Milking Shorthorn. In Rajasthan, an indigenous breed called Tharparkar exists, named from the Tharparkar District, now in Sindh Pakistan. Another type of dairy cow known as Nagauri from Nagaur District, the bull of which is renowned for its ability to plow fields and run. Traditionally, they used to pull covered wagons, known as rath, and in marriages to transport the newlywed couple. They are now a crutch for thriving agricultural and livestock rearing societies of the Thar Desert. Many other breeds are used nearly exclusively for beef, or for both dairy and beef purposes.

ANIMAL WASTE FROM LARGE DAIRIES

As measured in phosphorus, the waste output of 5,000 cows roughly equals a municipality of 70,000 people. In the U.S., dairy operations with more than 1,000 cows meet the EPA definition of a CAFO, and are subject to EPA regulations. For example, in the San Joaquin Valley of California a number of dairies have been established on a very large scale. Each dairy consists of several modern milking parlor set-ups operated as a single enterprise. Each milking parlor is surrounded by a set of 3 or 4 loafing barns housing 1,500 or 2,000 cattle. Some of the larger dairies have planned 10 or more series of loafing barns and milking parlors in this arrangement, so that the total operation may include as many as 15,000 or 20,000 cows. The milking process for these dairies is similar to a smaller dairy with a single milking parlor but repeated several times.

The size and concentration of cattle creates major environmental issues associated with manure handling and disposal, which requires substantial areas of cropland for manure spreading and dispersion, or several-acre methane digesters.

Air pollution from methane gas associated with manure management also is a major concern. As a result, proposals to develop dairies of this size can be controversial and provoke substantial opposition from environmentalists including the Sierra Club and local activists.

The potential impact of large dairies was demonstrated when a massive manure spill occurred on a 5,000-cow dairy in Upstate New York, contaminating a 20-mile stretch of the Black River, and killing 375,000 fish. On Aug. 10, 2005, a manure storage lagoon collapsed releasing several million gallons of manure into the Black River. Subsequently the New York Department of Environmental Conservation mandated a settlement package of $2.2 million against the dairy.

USE OF HORMONES

It is possible to maintain higher milk production by injecting cows with growth hormones known as recombinant BST or rBGH, but this is controversial due to its effects on animal and possibly human health. The European Union, Japan, Australia, New Zealand and Canada have banned its use due to these concerns.

In the US however, no such prohibition exists, and approximately 17.2% of dairy cows are treated in this way. The U.S. Food and Drug Administration states that no "significant difference" has been found between milk from treated and non-treated cows but based on consumer concerns several milk purchasers and resellers have elected not to purchase milk produced with rBST.

MANAGEMENT OF THE HERD

Modern dairy farmers use milking machines and sophisticated plumbing systems to harvest and store the milk from the cows, which are usually milked two or three times daily. During the warm months, in the northern hemisphere, cows may be allowed to graze in their pastures, both day and night, and are brought into the barn only to be milked. Many barns also incorporate tunnel ventilation into the architecture of the barn structure.

This ventilation system is highly efficient and involves opening both ends of the structure allowing cool air to blow through the building. Farmers with this type of structure keep cows inside during the summer months to prevent sunburn and damage to udders. During the winter months, especially in northern climates, the cows may spend the majority of their time inside the barn, which is warmed by their collective body heat. Even in winter, the heat produced by the cattle requires the barns to be ventilated for cooling purposes. Many modern facilities, and particularly those in tropical areas, keep all animals inside at all times to facilitate herd management.

Housing the cow can be either loose housed or stalls. There is little research available on dimensions required for cow stalls, and much housing can be out of date, however increasingly companies are making farmers aware of the benefits, in terms of animal welfare, health and milk production. In the southern hemisphere cows spend most of their lives outside on pasture, although they may receive supplementation during periods of low pasture availabliity. The production of milk requires that the cow be in lactation, which is a result of the cow having given birth to a calf. The cycle of insemination, pregnancy, parturition, and lactation, followed by a "dry" period of a few weeks before calving which allows udder tissue to regenerate. Dairy operations therefore include both the production of milk and the production of calves. Bull calves are either castrated and raised as steers for beef production or veal.

HEALTH AND WELL-BEING

Common ailments affecting dairy cows include infectious disease, metabolic disease and injuries caused by their environment.Lameness is commonly considered one of the most significant animal welfare issues for dairy cattle. It can be caused by a number of sources, including infections of the hoof tissue and physical damage causing bruising or lesions. While housing and management features common in modern conventional dairy farms have been identified as contributing risk factors, small farms in developing countries can also demonstrate high rates.

DAIRY PRODUCTS

Dairy products are generally defined as foods produced from cow's or domestic buffalo's milk. They are usually high-energy-yielding food products. A production plant for such processing is called a dairy or a dairy factory. Raw milk for processing mostly comes from cows and to a lesser amount from domestic buffalos, but occasionally from other mammals such as goats, sheep, yaks, or horses. Dairy products are commonly found in European, Middle Eastern and Indian cuisine, whereas they are almost unknown in East Asian cuisine.

TYPES OF DAIRY PRODUCTS

- Milk after optional homogenization, pasteurization, in several grades after standardization of the fat level, and possible addition of bacteria Streptococcus lactis and Leuconostoc citrovorum
  - Crème fraîche, slightly fermented cream
    a. Smetana, Central and Eastern European variety of sour cream
    b. Clotted cream, thick spoonable cream made by heating
  - Cultured buttermilk, fermented concentrated (water removed) milk using the same bacteria as sour cream
  - Kefir, fermented milk resembling buttermilk but based on different yeast and bacteria culture
  - Kumis/Airag, slightly fermented mares' milk popular in Central Asia
  - Milk powder (or powdered milk), produced by removing the water from milk
    a. Whole milk products
    b. Buttermilk products
    c. Skim milk
    d. Whey products

e. Ice Cream
f. High milk-fat and nutritional products (for infant formulas)
g. Cultured and confectionery products

- Condensed milk, milk which has been concentrated by evaporation, often with sugar added for longer life in an opened can
- Evaporated milk, (less concentrated than condensed) milk without added sugar
- Ricotta cheese, milk heated and reduced in volume, known in Indian cuisine as Khoa
- Infant formula, dried milk powder with specific additives for feeding human infants
- Baked milk, a variety of boiled milk that has been particularly popular in Russia

- Butter, mostly milk fat, produced by churning cream
  - Buttermilk, the liquid left over after producing butter from cream, often dried as livestock food
  - Ghee, clarified butter, by gentle heating of butter and removal of the solid matter
  - Smen, a fermented clarified butter used in Moroccan cooking.
  - Anhydrous milkfat
- Cheese, produced by coagulating milk, separating from whey and letting it ripen, generally with bacteria and sometimes also with certain molds
  - Curds, the soft curdled part of milk (or skim milk) used to make cheese (or casein)
  - Whey, the liquid drained from curds and used for further processing or as a livestock food
  - Cottage cheese
  - Quark
  - Cream cheese, produced by the addition of cream to milk and then curdled to form a rich curd or cheese made from skim milk with cream added to the curd
  - Fromage frais
- Casein
  - Caseinates
  - Milk protein concentrates and isonates
  - Whey protein concentrates and isonates
  - Hydrolysates
  - Mineral concentrates
- Yogurt, milk fermented by Streptococcus salivarius ssp. thermophilus and Lactobacillus delbrueckii ssp. bulgaricus sometimes with additional bacteria, such as Lactobacillus acidophilus

  - Ayran
  - Lassi
- Clabber (food), milk naturally fermented to a yogurt-like state
- Gelato, slowly frozen milk and water, lesser fat than ice cream
- Ice cream, slowly frozen cream and emulsifying additives
  - Ice milk
  - Frozen custard
  - Frozen yogurt, yogurt with emulsifiers that is frozen
- Other
  - Viili
  - Kajmak
  - Filmjölk
  - Piimä
  - Vla
  - Dulce de leche

HEALTH RISKS OF CONSUMING DAIRY PRODUCTS

Most dairy products contain large amounts of saturated fat. Some dairy products may cause health issues for individuals who have a lactose intolerance and milk allergies. Some dairy products such as blue cheese may become contaminated with the fungus Aspergillus fumigatus during ripening, which can trigger asthma and other respiratory problems in susceptible individuals.

Vegans and some vegetarians avoid dairy products due to a variety of ethical, dietary, environmental, political, and religious concerns.

PASTURE USE IN ORGANIC MILK PRODUCTION

The most common technology used on conventional dairy operations confines milk cows in large barns and limits access to pasture. This means that forage from grazing comprises very little of the forage fed to dairy cows on most conventional dairies. Access to pasture is a requirement for the organic certification of dairy operations. Forage from grazing is also an important element of the feeding programme on many organic dairies.

Milk producers may find it easier and less expensive to maintain organic pastures for grazing dairy cows than to either purchase or produce and harvest organic crops and forage for dairy feed. Also, grazing systems may be less stressful for dairy cattle and contribute to lower veterinary expenses. Pasture-based production may also have environmental benefits, such as improved soil quality and reduced soil erosion. Pasture use for dairy feed on organic dairies, relative to conventional dairies.

More than 60 per cent of organic milk producers reported that at least half of their total forage ration came from pasture during the grazing months (an average of 6.5 months per year), compared with 18 per cent of conventional dairies. Nearly 90 per cent of organic dairies sourced at least 25 per cent of their total forage ration from pasture. In contrast, 70 per cent of conventional dairies reported that less than 25 per cent of their total forage ration fed to dairy cows came from pasture, while 40 per cent obtained none from pasture.

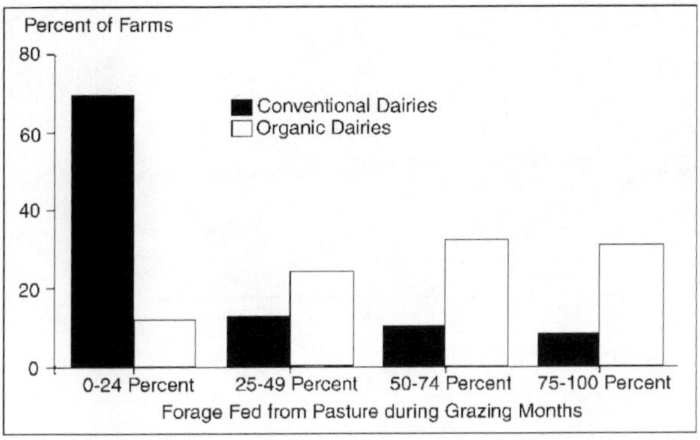

Fig. Producer Type, by Share of Dairies Forage Fed from Pasture

Notes: More than 60 per cent of organic dairies obtained at least 50 per cent of the forage fed to dairy cows from pasture, while nearly 70 per cent of conventional dairies obtained less than 25 per cent

To evaluate the role of pasture grazing in organic milk production, surveyed producers were divided into groups that indicate the level of pasture grazing as a source of dairy cow forage during the grazing months:

- 0-24 per cent;
- 25-49 per cent;
- 50-74 per cent; and
- 75-100 per cent.

CHARACTERISTICS AND PRACTICES BY LEVEL OF PASTURE USE

Almost two-thirds of organic milk producers reported that at least 50 per cent of the forage fed to dairy cattle during the grazing months came from pasture, and a third reported that 75 per cent or more came from pasture. Only 12 per cent of organic milk producers reported that less than 25 per cent of forage fed came from pasture. Differences between organic dairies where pasture use was highest (75-100 per cent of forage fed) and lowest (0-24 per cent of forage fed) were significant and are emphasized in this section.

Farm characteristics among organic dairies in the middle groups (25-49 per cent and 50-74 per cent of forage fed from pasture) showed little difference. Operations that relied on pasture use the most were the smallest in size and produced the least milk per cow. Organic dairies with 75-100 per cent of forage fed from pasture had an average of 64 milk cows that averaged 11,289 pounds of milk per cow.

Table. Characteristics and Production Practices, by Level of Pasture use on Organic Dairy Operations, 2005

	Forage Fed from Pasture During Grazing Months			
Item	**0-24 Per cent**	**25-49 Per cent**	**50-74 Per cent**	**75-100 Per cent**
Per cent of farms	12	24	32	31
Per cent of milk cows	21	23	31	25
Per cent of milk production	25	23	31	21
Milk cows *(number per farm)*	135	79	78	64
Milk production *(pounds per cow)*	16,560	13,594	13,319	11,289
Region:			*Per cent of farms*	
Northeast (ME, NY, PA, VT)	21	42	43	55
Upper Midwest (MI, MN, WI)	58	44	40	35
Corn Belt (IL, IN, IA, MO, OH)	13	8	5	9
West (CA, ID, OR, WA)	8	6	12	2
Region:			*Per cent of cows*	
Northeast (ME, NY, PA, VT)	9	29	28	45
Upper Midwest (MI, MN, WI)	30	46	28	29
Corn Belt (IL, IN, IA, MO, OH)	10	9	4	8
West (CA, ID, OR, WA)	51	16	40	18
In dairy business:			*Per cent of farms*	
Less than 10 years	11	18	19	27
10 years or more	89	82	81 73	
Pasture rotation:				
At least once a day	39	48	58	72
Every 2-7 days	30	33	38	26
Less than weekly	4	1	id	1
Never	24	18	3	1
Pasture not used for feed	4	0	0	0

Notes: id=Insufficient data for disclosure.

In contrast, dairies with 0-24 per cent of forage from pasture averaged more than twice the number of milk cows (135) and more than 5,000 pounds more milk production per cow (16,560 pounds). Thus, operations that relied less on pasture use (12 per cent) had a disproportionately large share of milk cows (21 per cent) and milk production (25 per cent), while a third of operations that relied most on pasture use accounted for only 25 per cent of milk cows and 21 per cent of production.

Organic dairies with the highest pasture use for feed were located in the Northeast. More than half of these farms were in the Northeast and operated with 45 per cent of the organic milk cows in this group. Operations in the Upper Midwest accounted for the majority of farms using the least pasture (0-24 per cent) and 35 per cent of farms using the most (75-100 per cent). Organic dairies in the West accounted for only 8 per cent of the farms using the least pasture, but these large operations held 51 per cent of the milk cows in this group.

Large operations relied less on pasture as a forage source because of the significant land requirements necessary to supply pasture for large herds. However, some large operations in the West used pasture as an important dairy feed source. Among the farms with the highest pasture use, the West (at 2 per cent) included 18 per cent of the milk cows. Organic dairies that relied the most on pasture use were in business relatively less time than other organic dairies.

Twenty-seven per cent of dairies that relied on pasture for 75-100 per cent of their forage feed had been in business less than 10 years, compared with only 11 per cent that relied on pasture for 0-24 per cent of their forage feed. Also, farm operators who relied the most on pasture forage were younger than other farm operators.

Sixteen per cent of these farm operators were younger than 35 and only 4 per cent were older than 65, compared with 6 and 10 per cent, respectively, of the operators who used the least pasture. Younger farm operators on relatively new dairies were more likely to use pasture-based feeding for organic milk production. Pasture as a source of forage on organic dairies was closely related to the intensity of pasture management.

Rotating pastures is a management strategy that can increase the total volume and quality of dry matter produced on pastures by allowing cows to graze for specified periods before moving them to new pastures. Seventy-two per cent of organic dairies that reported the highest use of pasture forage rotated pastures at least once a day, compared with less than 40 per cent of dairies that used pasture forage the least. Among dairies that used pasture the least, 24 per cent reported that pastures were never rotated and 4 per cent indicated that pasture was not a source of dairy feed.

PRODUCTION COSTS AND NET RETURNS BY LEVEL OF PASTURE USE

Substituting pasture for more expensive feed sources may appear to be an efficient way to lower dairy feed costs. Average feed costs per cwt of milk, however, were lowest on the organic operations using the least pasture for dairy forage (0-24 per cent), between $1 and $2 per cwt less than on the operations using more pasture. This relationship changes little even if charges for pasture resources are excluded from the feed costs. Average feed costs per cow declined as pasture use for dairy forage increased. Total feed costs were $1,902 per cow on organic dairies that relied on pasture for 0-24 per cent of forage fed, compared with only $1,409 on organic dairies that relied on pasture for 75-100 per cent of forage fed, a savings of about 25 per cent. Breaking down the distribution of milk production per cow for each group provides an indication of how milk yields declined as pasture was substituted for higher energy feed in the dairy ration. Nearly half of organic dairies feeding 0-24 per cent of forage from pasture had annual milk yields at 15,000 pounds per cow or more, compared with just 10 per cent of dairies feeding 75-100 per cent.

Table. Production Costs and Net Returns on Organic Dairy Operations, by Level of Pasture use, 2005

Forage Fed from Pasture During Grazing Months				
Item	**0-24 Per cent**	**25-49 Per cent**	**50-74 Per cent**	**75-100 Per cent**
		Dollars per cwt sold		
Feed:	11.61	13.14	13.42	12.90
Purchased feed	7.43	5.15	7.19	6.09
Homegrown harvested feed	3.89	7.51	5.41	5.69
Grazed feed	0.29	0.48	0.82	1.13
Hired labour	2.44	1.85	2.17	2.27
Capital costs	4.11	5.36	4.48	6.70
Opportunity cost of unpaid labour	2.82	6.47	6.24	8.36
Cost summary				
Operating costs	17.74	18.55	18.99	19.44
Operating and capital costs	21.85	23.91	23.47	26.14
Total economic costs	25.33	31.53	30.64	35.99
Gross value of production	24.10	24.49	23.59	25.63
Returns above total economic costs	-1.22	-7.04	-7.05	-10.36

Notes: Measuring Milk Production Costs. The gross value of production includes milk sales, cull and breeding animal sales, revenue from leasing dairy animals or space to other operations, co-op patronage dividends, dairy assessment rebates or refunds, and the value of manure. The value of milk sales was determined from prices reported by the survey respondents.

In contrast, more than 40 per cent of dairies using the most pasture had an annual milk yield of less than 10,000 pounds per cow, compared with 19 per cent of those using the least pasture. This suggests that improving pasture quality to achieve higher milk production can contribute to the success of organic dairies using pasture-based feeding. The relationship between pasture use and total production costs follows much the same pattern as seen for size and production costs on organic dairies. That is, the operations using less pasture were much larger and had lower average costs per cwt than the smaller operations using more pasture. Capital costs were more than $2 per cwt higher and the charge for unpaid labour was more than $5 per cwt higher on the operations where pasture forage accounted for 75-100 per cent of forage fed than on those where pasture forage accounted for 0-24 per cent of forage fed.

Average returns net of total economic costs were negative for dairies in each of the pasture use groups, but were significantly higher among the dairies using the least pasture. This disparity suggests that organic operations using conventional dairy feeding methods, such as confining cows and feeding higher energy feed, were more likely to generate higher returns to capital and labour resources than those relying more on pasture-based feeding. Thus, the technologies and production methods used on organic dairies could become more like those used on conventional dairies. Changing technologies and production methods may depend on the extent to which pasture requirements for organic dairies must include pasture as part of the dairy feeding programme.

LABOUR USE BY LEVEL OF PASTURE USE

The biggest difference in organic dairy costs between those using the most and the least pasture for dairy forage was the unpaid labour charge. The labour charge on organic dairies feeding 75-100 per cent of forage from pasture was more than \$8 per cwt, compared with less than \$3 per cwt on dairies feeding 0-24 per cent of forage from pasture. The labour charge contributed about half the difference in total costs that were more than \$10 per cwt higher for dairies using the most pasture. Table includes labour use per cow and per cwt for each level of pasture feeding. Unpaid labour use per cwt on organic dairies feeding 75-100 per cent of forage from pasture was about three times that for those feeding 0-24 per cent of forage from pasture. Part of this difference came from fixed amounts of labour being spread over more output on the larger operations that used the least amount of pasture. Some of the labour difference, however, may be due to the pasture feeding system's being more labour intensive than handling dairy cows in confinement. Among farms with fewer than 100 cows, the unpaid labour use on dairies feeding the least forage from pasture was still about a third less than on those feeding the most forage from pasture (0.45 hour versus 0.65 hour). In addition, total labour use among these small dairies increased steadily as more pasture was used for dairy feed. This increase suggests that pasture-based feeding systems are inherently more labour intensive than the conventional practice of handling dairy cows in confinement, possibly due to the labour required for moving cows to, from, and between pastures.

Table. Organic Dairy Labour Use, by Level of Pasture Use, 2005

Forage Fed from Pasture During Grazing Months				
Item	**0-24 Per cent**	**25-49 Per cent**	**50-74 Per cent**	**75-100 Per cent**
Per cow				
Hired labour hours	30	20	22	17
Unpaid labour hours	27	52	48	56
Total labour hours	57	72	70	73
Per cwt				
Hired labour hours	0.18	0.15	0.17	0.15
Unpaid labour hours	0.16	0.39	0.36	0.49
Total labour hours	0.34	0.53	0.52	0.65
Per cwt—dairies with fewer than 100 cows				
Hired labour hours	0.15	0.18	0.17	0.16
Unpaid labour hours	0.45	0.52	0.59	0.65
Total labour hours	0.60	0.70	0.76	0.81

Note: Totals may not equal the sum of items due to rounding

ECONOMIES OF SIZE IN ORGANIC MILK PRODUCTION

The emergence of large conventional dairy farms and the continued shift of production towards such farms suggest that economies of size play a major role in the structure of dairy farming. Economies of size and the role it takes in the structural changes observed in conventional milk production were documented by MacDonald *et al.* This section reports on the relationship between operation size and production costs in organic milk production, including what size and costs might mean for the structure of the industry.

To evaluate the relationship between organic milk production costs and operation size, surveyed producers were divided into groups by size, and differences in farm characteristics and practices, production costs and net returns, and labour use were compared among the groups.

The size groups were defined by the largest number of cows milked on the operation during 2005:

- Fewer than 50 cows;
- 50-99 cows;
- 100-199 cows; and
- 200 cows or more.

CHARACTERISTICS AND PRACTICES BY SIZE

Data for 2005 indicated that about 45 per cent of organic dairies milked fewer than 50 cows, while 87 per cent had fewer than 100 cows. These small operations accounted for a disproportionately small share of organic milk cows and production. In contrast, about 13 per cent of organic dairies had 100 cows or more, but they accounted for 44 per cent of milk cows and nearly half of production.

The largest organic dairies—those with 200 cows or more—had an average herd size of almost 500 cows per farm and included nearly a third of organic milk cows. These large dairies had a higher average milk yield per cow than smaller dairies and produced 37 per cent of organic milk.

The most striking difference among organic dairies in each size group was their location. Northeast and Upper Midwest farms accounted for 96 per cent of dairies with fewer than 50 cows. Farms in the West accounted for 80 per cent of dairies with 200 cows or more. Within each region, 97 per cent of organic dairies in the Northeast, 88 per cent of organic dairies in the Upper Midwest, and 85 per cent of organic dairies in the Corn Belt had fewer than 100 cows. In contrast, 83 per cent of organic dairies in the West had 100 cows or more and 63 per cent had 200 cows or more.

The use of several milk production technologies and practices tended to increase with size of organic operation, but most of the differences were between the smallest and other organic farms. For example, use of a milking parlor increased from 22 per cent for the smallest dairies to nearly all dairies with

100 cows or more. Thirty-eight per cent of the smallest farms and over 50 per cent of other farms participated in the Dairy Herd Improvement programme.

Table. Dairy Farm Characteristics and Production Practices, by Size of Organic Operation, 2005

	Size of Organic Dairy			
Item	**Fewer than 50 Cows**	**50-99 Cows**	**100-199 Cows**	**200 Cows or More**
Per cent of farms	45	42	8	5
Per cent of milk cows	21	35	13	31
Per cent of milk production	18	33	12	37
Milk cows (number per farm)	37	68	132	490
Milk production (pounds per cow)	11,884	12,796	12,008	16,133
Per cent of farms				
Region:				
Northeast (ME, NY, PA, VT)	51	46	17	0
Upper Midwest (MI, MN, WI)	45	39	52	20
Corn Belt (IL, IN, IA, MO, OH)	4	12	15	0
West (CA, ID, OR, WA)	0	3	16	80
Production practices:				
DHI programme participation[1]	38	51	53	60
Pasture-based feeding[2]	70	60	49	54
Artificial insemination	70	76	71	87
Embryo transplants or sexed semen	3	3	6	2
Controlled breeding/calving season	38	35	36	20
Regular veterinary services	28	41	72	55
Nutritionist services	30	53	66	70
Computerized milking system	1	4	4	7
Computerized feeding system	id	3	5	31
Milking parlor	22	45	88	100
Kept individual cow records	59	64	62	70
Johne's disease programme participation	18	28	44	45
Onfarm computer records	13	25	34	41
Dairy information from Internet	30	46	56	64
Forward-purchased inputs	4	10	22	16
Negotiate input price discounts	18	21	33	36

Notes: id=Insufficient data for disclosure.

[1]Dairy Herd Improvement.

[2]Pasture-based feeding is defined as providing at least half of the forage fed to milk cows during the grazing months from pasture. Organic dairies reported an average grazing period for milk cows of 6. 5 months in 2005.

Use of regular veterinary and nutritionist services was also much lower on the smallest farms. Likewise, the use of business management tools, such as computer records, the Internet, and forward-purchasing inputs, was less. Pasture-based feeding was used by 70 per cent of the smallest dairies and was also used by more than half of the largest dairies.

PRODUCTION COSTS AND NET RETURNS BY SIZE

Average feed costs differed little among the dairies in each size group, but average operating costs for organic dairies increased from the smallest to largest size group due to the additional costs of hired labour on the largest farms. Total operating costs were about $2 per cwt higher on the largest organic dairies than on the smallest. Once capital costs were added, cost differences among the size groups were negligible, as lower capital costs on the larger operations offset the additional hired labour costs.

The smallest farms, however, used significantly more unpaid labour, mainly from farm operators and other family members, than did larger farms. After adding an opportunity cost for this unpaid labour, total economic costs for organic dairies declined with size and were nearly $14 per cwt less on the largest farms than on the smallest.

Table. Production Costs and Returns, by Size of Organic Dairy Operation, 2005

	Size of Organic Dairy			
Item	**Fewer than 50 Cows**	**50-99 Cows**	**100-199 Cows**	**200 Cows or More**
Dollars per cwt sold				
Feed:	12.96	12.67	11.93	13.06
Purchased feed	5.16	4.94	5.29	8.99
Homegrown harvested feed	6.96	7.00	6.06	3.51
Grazed feed	0.83	0.73	0.58	0.56
Hired labour	0.79	1.75	2.25	3.21
Capital costs	6.02	6.20	5.04	3.58
Opportunity cost of unpaid labour	13.42	7.45	4.84	1.15
Cost summary:				
Operating costs	17.65	18.25	18.10	19.66
Operating and capital costs	23.67	24.44	23.13	23.25
Total economic costs	38.50	33.36	29.13	24.84
Gross value of production	25.57	24.88	23.46	23.59
Returns above total economic costs	-12.93	-8.48	-5.67	-1.26

Notes: Costs are defined in Appendix A: Measuring Milk Production Costs. The gross value of production includes milk sales, cull and breeding animal sales, revenue from leasing dairy animals or space to other operations, co-op patronage dividends, dairy assessment rebates or refunds, and the value of manure. The value of milk sales was determined from prices reported by the survey respondents.

Average production costs for conventional milk production declined from about $30 per cwt on operations with fewer than 50 cows to about $18 per cwt on operations with 200-499 cows. Average costs dropped to less than $14 per cwt on conventional operations with more than 1,000 cows. Lower costs among the largest conventional producers suggest that similar economies of size may be available to organic producers if they increase the size of their operation above the industry norm.

However, the additional costs of complying with pasture requirements and securing organic inputs in large volume may limit the cost advantages for larger organic operations. Net returns to the organic dairy enterprise were computed as the difference between the gross value of production and total economic costs. Average net returns in 2005 were negative for all organic dairy size groups (from-$12.93 per cwt among dairies with fewer than 50 cows to-$1.26 per cwt among dairies with 200 cows or more).

Farmers are unlikely to start or expand a dairy operation if they are unable to recover their investment in capital and labour. Farmers who have already made substantial capital investments in dairy facilities and equipment, however, with little or no opportunity for use outside the dairy operation. These costs, therefore, have no bearing on the decision to operate in the short term, only becoming important when these capital assets must be replaced.

To better measure costs relevant to shortterm operating decisions, the share of farms with gross returns exceeding all costs except for capital costs was computed. About a fourth of organic dairies with fewer than 50 cows covered these costs, compared with almost threefourths of those with 200 cows or more. Some small organic dairy farms appeared to be earning enough to keep operating, but on average, farms in the smaller size groups were not covering the opportunity costs of their investments in capital and the operator's time.

In contrast, larger organic dairies were much more likely to generate returns above capital and labour costs. These conditions suggest that organic milk production may migrate towards larger operations, as has conventional production. This production change will likely occur over an extended period as existing smaller operations use up their production facilities and equipment, ultimately facing decisions about replacing capital assets or exiting the industry.

Also, in the short term, some small operators may continue producing organic milk as a lifestyle choice despite returns that fail to cover the opportunity cost of their time.

LABOUR USE BY SIZE

Small organic dairies are at a cost disadvantage relative to larger dairies due to the opportunity cost assigned to the unpaid labour provided by operators, partners, and family members. Unpaid labour costs for organic dairies with fewer than 50 cows were $13.42 per cwt, compared with only $1.15 per cwt on dairies with 200 cows or more. If this opportunity cost is ignored, the net returns to organic milk production were more similar among the size groups. The largest dairies used substantially more hired labour than the smallest (11,530 hours versus 557 hours), but the hired labour hours per unit of production differed little among the size groups because of much higher production on the largest dairies. The difference in unpaid labour use among the farm size groups was much less. Organic dairy operations with 200 cows or more used about 700 more hours of unpaid labour, about 20 per cent more than those with fewer than 50 cows. The largest dairies, however, produced about 1,700 per cent more milk as a result of having more cows and an average productivity per cow that was 36 per cent higher than the smallest dairies. Enhanced productivity for these larger dairies means that unpaid labour hours per unit of production fell dramatically across the size groups from 0.82 hour per cwt on the smallest organic dairies to only 0.05 hour on the largest.

Table. Labour Use, by Size of Organic Dairy Operation, 2005

	Size of Organic Dairy			
Item	**Fewer than 50 Cows**	**50-99 Cows**	**100-199 Cows**	**200 Cows or More**
Per farm				
Hired labour hours	557	1,648	3,051	11,530
Unpaid labour hours	3,612	3,759	4,265	4,322
Total labour hours	4,169	5,407	7,316	15,852
Milk production (cwt)	4,367	8,748	15,825	78,977
Per cwt				
Hired labour hours	0.13	0.19	0.19	0.15
Unpaid labour hours	0.82	0.43	0.27	0.05
Total labour hours	0.95	0.62	0.46	0.20

Unpaid labour is an opportunity cost defined as the time spent working on a dairy operation charged at a wage rate that represents what unpaid workers could earn in off-farm employment. Opportunity costs of labour may vary significantly among producers, and some producers may be willing to accept returns lower than they could earn in nonfarm employment because of lifestyle preferences and costs of switching occupations, among other reasons. The incentive to earn a competitive return for the time spent on the dairy operation, however, may be a motivating force for increasingly larger organic dairies.

CURRENT DAIRY POLICIES TO ASSIST PRODUCERS

Indian dairy policy has been developed over the last seven decades.

The early policies addressed three main problems:

1. Producers lacked bargaining power with milk buyers;
2. Producers suffered from volatile or low prices; and
3. Market participants encountered severe shortages/gluts resulting from marketing a highly perishable commodity.

The policy response resulted in the development of two major government activities that still function today: federal milk marketing orders and the Dairy Product Price Support Programme. While both FMMOs and the DPPSP have their roots in the 1930s and 1940s, the programmes have changed modestly over the years as the industry structure and markets changed. Two other components of Indian dairy policy are relatively new programmes. First, the

1985 farm bill established the Dairy Export Incentive Programme to counter foreign competitor subsidies.

Second, the Milk Income Loss Contract programme was established in the 2002 farm bill as a government payment for dairy farmers in times of low milk prices. Like Indian crop programmes, the MILC programme pays dairy producers when prices decline below a specified level. The following sections describe each of these four components and how they relate to the current market situation. Lower milk and dairy product prices since late 2008 have generated new programme activity. Purchasing dairy products last fall under the DPPSP; MILC payments were triggered beginning in February.

MILK INCOME LOSS CONTRACT PROGRAMME

The Milk Income Loss Contract (MILC) programme pays dairy farmers when farm milk prices fall below an established target price. Section 1506 of the 2008 farm bill extends authority for the MILC programme until September 30, 2012. This programme is similar to long-time subsidy programmes for crops that pay farmers when farm prices drop below certain levels. Farm Service Agency implements the MILC programme. Under MILC, participating dairy farmers nationwide are eligible for a federal payment whenever the minimum monthly market price for farm milk used for fluid consumption in Boston falls below $16.94 per cwt.

Eligible farmers then receive a payment equal to 45% of the difference between the $16.94 target price and the lower monthly market price. The payment quantity is limited to 2.985 million pounds of annual production. Since the inception of the MILC programme, large dairy farm operators have expressed concern that the payment limit has negatively affected their income. For larger farm operations, their annual production is well in excess of the limit, and any production in excess of that receives no federal payments.

To address the issue of rising feed costs, the 2008 farm bill includes a provision that adjusts upward the $16.94 target price in any month when feed prices are above a certain threshold. The law requires calculating monthly a National Average Dairy Feed Ration Cost based on a formula that currently uses to calculate feed costs. In any month that the average feed cost is above $7.35 per cwt., the $16.94 target price will be increased by 45% of the difference between the monthly feed cost and $7.35. For the latter half of 2007 and all of 2008, farm milk prices remained well above the MILC trigger price, precluding the need for any MILC payments.

However, milk prices have since declined below the trigger for MILC payments. The Class I Boston farm milk price for February 2009 was $13.97 per cwt. With the adjustment for feed costs raising the trigger to $17.33 per cwt., MILC payments were activated for the first time in two years at a payment rate of $1.51. The payment rate rose to $2.01 per cwt. in March. Given current prospects in the futures markets for milk, corn, and soybeans, payments are expected to continue during 2009, but at smaller rates.

Individual producers must select which month to begin receiving payments, based on their projection of potential payment rates and the possibility of hitting the production payment limit. As of October 26, 2009, total MILC payments distributed to date were $775 million. The timing of the payments has caused some concern for producers this spring. While milk price data become available during the payment month, data needed for the feed cost adjustor are not available until publishes monthly average feed prices in *Agricultural Prices* at the end of the next month. Consequently, MILC payments for a particular month are not processed until two months later.

DAIRY PRODUCT PRICE SUPPORT PROGRAMME (DPPSP)

The Agricultural Act of 1949 first established a dairy price support programme by permanently requiring supporting the farm price of milk. Since 1949, Congress has regularly amended the programme, usually in the context of multiyear omnibus farm acts and budget reconciliation acts. Historically, the supported farm price for milk is intended to protect farmers from price declines that might force them out of business and to protect consumers from seasonal imbalances of supply and demand. Commodity Credit Corporation (CCC) supports milk prices by its standing offer to purchase surplus nonfat dry milk, cheese, and butter from dairy processors. Whenever market prices fall to product support levels, processors generally make the business decision of selling surplus product to the government rather than to the marketplace. Consequently, the government purchase prices usually serve as a floor for the market price, which in turn indirectly supports the farm price of milk for all dairy farmers.

The effectiveness of the dairy price supports depends on removal of products from the market and placement into government storage. The Dairy Product Price Support Programme (DPPSP) as authorized by the 2008 farm bill requires purchasing products at the following minimum prices: block cheese, $1.13/lb.; barrel cheese, $1.10/lb.; butter, $1.05/lb.; and nonfat dry milk, $0.80/lb. Under previous law, the support price for farm milk was statutorily set at $9.90 per cwt., and given the administrative authority to establish a combination of dairy product purchase prices that indirectly supported the farm price of milk at $9.90. Although the 2008 law does not specifically state that the overall support price is $9.90 per cwt, each of the mandated product prices in the law is equivalent to the existing product purchase prices, so farm milk prices effectively continue to be supported at $9.90.

In late 2008 and 2009, after several years of relative inactivity, the price support programme resumed purchases when dairy product prices approached support levels. As of September 11, 2009, estimated that it purchased 111 million pounds of nonfat dry milk under the programme in 2008 and expects to purchase 379 million pounds in 2009, along with small amounts of butter and cheese (including amounts exported under the Dairy Export Incentive Programme).

Total expenditures on the DPPSP were $223 million from October 1, 2008, through September 10, 2009. With an expected rise in milk and product prices next year, Forecasts only a small amount of butter to be purchased in 2010. Following heightened industry and congressional interest in taking action to boost milk prices for farmers, July 31, 2009, a temporary increase in price support for cheese and nonfat dry milk from August 2009 through October 2009.

Subsequently, the Senate approved an amendment to the Senate-passed FY2010 agriculture appropriations bill to increase Farm Service Agency funding by $350 million, ostensibly for an additional increase in dairy product price support levels. However, the conference agreement for the FY2010 Agriculture appropriations bill, which was enacted on October 21, 2009, provides for a different use of the funds.

MILK MARKETING ORDERS

Federal milk marketing orders (FMMOs) mandate minimum prices that processors must pay producers for milk depending on its end use. This compares with the MILC programme, which provides direct payments to producers, and the DPPSP, which buys surplus dairy products at specified minimum prices. The DPPSP serves as a price floor for products and under girds FMMO minimum milk prices. The farm price of approximately two-thirds of the nation's fluid milk is regulated under FMMOs.

Federal orders, which are administered Agricultural Marketing Service, were instituted in the 1930s to promote orderly marketing conditions by, among other things, applying a uniform system of classified pricing throughout the market. Some states, California for example, have their own state milk marketing regulations instead of federal rules. FMMOs also address how market proceeds are distributed among producers delivering milk to federal marketing order areas. Producers are affected by two fundamental marketing order provisions: the classified pricing of milk according to its end use, and the pooling of receipts to pay all farmers a blend price. Federal orders regulate dairy handlers (processors) who sell milk or milk products within a defined marketing area by requiring them to pay not less than established minimum class prices for the Grade A milk they purchase from dairy producers, depending on how the milk is used.

This classified pricing system requires handlers to pay a higher price for milk used for fluid consumption than for milk used in manufactured dairy products such as yogurt, ice cream, and sour cream, cheese and butter and dry milk products. These differences between classes reflect the different market values for the products. Blend pricing allows all dairy farmers who ship to the market to pool their milk receipts and then be paid a single price for all milk based on order-wide usage (a weighted average of the four usage classes).

Paying all farmers a single blend price is seen as an equitable way of sharing revenues for identical raw milk directed to both the higher-valued fluid market and the lowervalued manufacturing market. Manufactured class prices are the same in all orders nationwide and are calculated monthly based on current market conditions for manufactured dairy products. The Class I price for milk used for fluid consumption varies from area to area. Class I prices are determined by adding, to a monthly base price, a "Class I differential" that generally rises with the geographical distance from milk surplus regions in the Upper Midwest, the Southwest, and the West. Class I differential pricing is a mechanism designed to ensure adequate supplies of milk for fluid use at consumption centers. The supply of milk may come from local supplies or distant supplies, whichever is more efficient. However, local dairy farmers are protected by the minimum price rule against lower-priced milk that might otherwise be hauled into their region. Over the years, dairy farmers have supported minimum prices afforded by FMMOs because they help balance marketing power traditionally held by processors. In contrast, dairy processors generally oppose them.

Mandated minimum prices, they say, do not allow for timely adjustments in a rapidly changing market and can leave product manufacturers in unprofitable situations. Also, they contend that the FMMO system distorts markets, saying fixed differentials contributed to high fluid milk prices last year.

DAIRY EXPORT INCENTIVE PROGRAMME (DEIP)

First authorized in 1985, the Dairy Export Incentive Programme (DEIP) provides cash bonus payments to Indian dairy exporters. The programme was initially intended to counter foreign—mostly European Union—dairy subsidies (while removing surplus dairy products from the market), but subsequent farm bill reauthorizations have added market development to the role of DEIP. Payments since the program's inception have totaled \$1.1 billion. The programme was active throughout the 1990s, peaking in 1993 with \$162 million in bonuses. DEIP funding is a mandatory account provided through the Commodity Credit Corporation (CCC) borrowing authority from the Indian Treasury, rather than through annual appropriations bills. The programme had not been used since FY2004 until announced its reactivation on May 22, 2009. Indian dairy product exports made with DEIP bonuses are subject to annual limitations under the Uruguay Round Agreement of the World Trade Organization (WTO). The limits are 68,201 metric tons of skim milk powder, 21,097 tons of butterfat, 3,030 tons of various cheeses, and 34 tons of other dairy products (quantity limits are on a July-June year). Total expenditures under WTO commitments are now capped at \$117 million per year (value limits on a October-September year).

8

Food and Beverage Packaging in Dairy Farming

Food and beverage packaging comprises 55% to 65% of the $130 billion value of packaging in the United States. Food processing and packaging industries spend an estimated 15% of the total variable costs on packaging materials. Industrial processing of food, reduced consumption of animal protein, importation of raw materials and ingredients to be converted in the United States, and scarcity of time to select/prepare food from fresh ingredients have enhanced innovation in food and beverage packaging.

The continued quest for innovation in food and beverage packaging is mostly driven by consumer needs and demands influenced by changing global trends, such as increased life expectancy, fewer organizations investing in food production and distribution, and regionally abundant and diverse food supply. The use of food packaging is a socioeconomic indicator of increased spending ability of the population or the gross domestic product as well as regional food availability.

This Scientific Status Summary provides an overview of the latest innovations in food packaging. It begins with a brief history of food and beverage packaging, covering the more prominent packaging developments from the past, and proceeds to more modern advances in the packaging industry.

The article then delves into current and emerging innovations in active and intelligent packaging (such as oxygen scavengers and moisture control agents), packaging mechanisms that control volatile flavours and aromas (such as flavour and odour absorbers), and cutting-edge advances in food packaging distribution (such as radio frequency identification and electronic product codes). Finally, the nano-sized components that have the potential to transform the food packaging industry.

HISTORY OF FOOD AND BEVERAGE PACKAGING

Modern food packaging is believed to have begun in the 19th century with the invention of canning by Nicholas Appert. After the inauguration of food microbiology by Louis Pasteur and colleagues in the 19th century, Samuel C. Prescott and William L. Underwood worked to establish the fundamental principles of bacteriology as applied to canning processes. These endeavors to preserve and package food were paralleled by several other packaging-related inventions such as cutting dies for paperboard cartons by Robert Gair and mechanical production of glass bottles by Michael Owens.

In the beginning of the 20th century, 3-piece tin-plated steel cans, glass bottles, and wooden crates were used for food and beverage distribution. Some food packaging innovations stemmed from unexpected sources. For example, Jacques E. Brandenberger's failed attempts at transparent tablecloths resulted in the invention of cellophane. In addition, wax and related petroleum-based materials used to protect ammunition during World War II became packaging materials for dry cereals and biscuits.

Many packaging innovations occurred during the period between World War I and World War II; these include aluminum foil, electrically powered packaging machinery, plastics such as polyethylene and polyvinylidene chloride, aseptic packaging, metal beer cans, flexographic printing, and flexible packaging. Most of these developments helped immeasurably in World War II by protecting military goods and foods from extreme conditions in war zones.

Tin-plated soldered side-seam steel progressed to welded sideseam tin-free steel for cans, and 2-piece aluminum with easy open pop tops were invented for beverage cans, spearheading the exponential growth of canned carbonated beverages and beer during the 1960s and 1970s. The development of polypropylene, polyester, and ethylene vinyl alcohol polymers led the incredible move away from metal, glass, and paperboard packaging to plastic and flexible packaging.

Later 20th century innovations include active packaging (oxygen controllers, antimicrobials, respiration mediators, and odour/aroma controllers) and intelligent or smart packaging. Distribution packaging is already influenced by the potential role of radio frequency identification for tracking purposes. Products such as retort pouches and trays, stand-up flexible pouches, zipper closures on flexible pouches, coextrusion for films and bottles, and an inexorable drive by injection stretch blowmolded polyester bottles and jars for carbonated beverages and water have emerged as rigid and semi-rigid packaging.

Multilayer barrier plastic cans, microwave susceptors, dispensing closures, gas barrier bags for prime cuts of meat, modified atmosphere packaging, rotogravure printed full-panel shrink film labels, and dual oven able trays are examples of innovations for the convenience attributes that have propelled food and beverage packaging into the 21st century.

Moreover, some 21st century innovations are related to nanotechnology whose future may lie in improving barrier and structural/mechanical properties of packaging materials and development of sensing technologies. The principal drivers for most of these innovations have been consumer and food service needs and demands for global and fast transport of food. These packaging innovations are derived largely from industry research and development programmes.

ACTIVE AND INTELLIGENT FOOD PACKAGING

Traditional food packages are passive barriers designed to delay the adverse effects of the environment on the food product. Active packaging, however, allows packages to interact with food and the environment and play a dynamic role in food preservation. Developments in active packaging have led to advances in many areas, including delayed oxidation and controlled respiration rate, microbial growth, and moisture migration. Other active packaging technologies include carbon dioxide absorbers/emitters, odour absorbers, ethylene removers, and aroma emitters. While purge and moisture control and oxygen removal have been prominent in active packaging, purge control is the most successful commercially.

An example is the drip-absorbing pad used in the poultry industry. In addition, active packaging technology can manipulate permselectivity, which is the selective permeation of package materials to various gases. Through coating, microperforation, lamination, coextrusion, or polymer blending, permselectivity can be manipulated to modify the atmospheric concentration of gaseous compounds inside a package, relative to the oxidation or respiration kinetics of foods.

Certain nanocomposite materials can also serve as active packaging by actively preventing oxygen, carbon dioxide, and moisture from reaching food. Intelligent or smart packaging is designed to monitor and communicate information about food quality. Examples include time-temperature indicators (TTIs), ripeness indicators, biosensors, and radio frequency identification. These smart devices may be incorporated in package materials or attached to the inside or outside of a package.

As of summer 2008, the commercial application of these technologies to food packaging has been small. However, the U.S. Food and Drug Administration (FDA) recognizes TTIs in the 3rd edition of the *Fish and Fisheries Products Hazards and Control Guidance*, so their importance may increase in the seafood industry. Moreover, Wal-Mart, Home Depot and other retail outlets use radio frequency identification, so it is likely to become very prominent as a mechanism for tracking and tracing produce and others perishable commodities.

OXYGEN SCAVENGERS

The presence of oxygen in a package can trigger or accelerate oxidative reactions that result in food deterioration: Oxygen facilitates the growth of aerobic microbes and molds. Oxidative reactions result in adverse qualities such as off-odours, off-flavours, undesirable colour changes, and reduced nutritional quality. Oxygen scavengers remove oxygen (residual and/or entering), thereby retarding oxidative reactions, and they come in various forms: sachets in headspace, labels, or direct incorporation into package material and/or closures.

Oxygen scavenging compounds are mostly agents that react with oxygen to reduce its concentration. Ferrous oxide is the most commonly used scavenger. Others include ascorbic acid, sulfites, catechol, some nylons, photosensitive dyes, unsaturated hydrocarbons, ligands, and enzymes such as glucose oxidase. To prevent scavengers from acting prematurely, specialized mechanisms can trigger the scavenging reaction. For example, photosensitive dyes irradiated with ultraviolet light activate oxygen removal. Oxygen scavenging technologies have been successfully used in the meat industry.

CARBON DIOXIDE ABSORBERS AND EMITTERS

Carbon dioxide may be added for beneficial effects, for example, to suppress microbial growth in certain products such as fresh meat, poultry, cheese, and baked goods. Carbon dioxide is also used to reduce the respiration rate of fresh produce and to overcome package collapse or partial vacuum caused by oxygen scavengers. Carbon dioxide is available in various forms, such as moisture-activated bicarbonate chemicals in sachets and absorbent pads. Conversely, high levels of carbon dioxide resulting from food deterioration or oxidative reactions could cause adverse quality effects in food products. Excess carbon dioxide can be removed by using highly permeable plastics whose permeability increases with higher temperatures.

MOISTURE CONTROL AGENTS

For moisture-sensitive foods, excess moisture in packages can have detrimental results: for example, caking in powdered products, softening of crispy products such as crackers, and moistening of hygroscopic products such as sweets and candy. Conversely, too much moisture loss from food may result in product desiccation.

Moisture control agents help control water activity, thus reducing microbial growth; remove melting water from frozen products and blood or fluids from meat products; prevent condensation from fresh produce; and keep the rate of lipid oxidation in check. Desiccants such as silica gels, natural clays and calcium oxide are used with dry foods while internal humidity controllers are used for high moisture foods (for example, meat, poultry, fruits, and vegetables).

Desiccants usually take the form of internal porous sachets or perforated water-vapour barrier plastic cartridges containing desiccants. They can also be incorporated in packaging material. Humidity controllers help maintain optimum in-package relative humidity (about 85% for cut fruits and vegetables), reduce moisture loss, and retard excess moisture in headspace and interstices where microorganisms can grow. Purge absorbers remove liquid squeezed or leaking from fresh products and can be enhanced by other active additives such as oxygen scavengers, antimicrobials, pH reducers, and carbon dioxide generators.

ANTIMICROBIALS

Antimicrobials in food packaging are used to enhance quality and safety by reducing surface contamination of processed food; they are not a substitute for good sanitation practices. Antimicrobials reduce the growth rate and maximum population of microorganisms (spoilage and pathogenic) by extending the lag phase of microbes or inactivating them. Antimicrobial agents may be incorporated directly into packaging materials for slow release to the food surface or may be used in vapour form.

Research is underway on the antimicrobial properties of the following agents:

- *Silver Ions*: Silver salts function on direct contact, but they migrate slowly and react preferentially with organics. Research on the use of silver nanoparticles as antimicrobials in food packaging is ongoing, but at least 1 product has already emerged: FresherLongerTM storage containers allegedly contain silver nanoparticles infused into polypropylene base material for inhibition of growth of microorganisms.
- *Ethyl Alcohol*: Ethyl alcohol adsorbed on silica or zeolite is emitted by evaporation and is somewhat effective but leaves a secondary odour.
- *Chlorine Dioxide*: Chlorine dioxide is a gas that permeates through the packaged product. It is broadly effective against microorganisms but has adverse secondary effects such as darkening meat colour and bleaching green vegetables.
- *Nisin*: Nisin has been found to be most effective against lactic acid and Gram-positive bacteria. It acts by incorporating itself in the cytoplasmic membrane of target cells and works best in acidic conditions.
- *Organic Acids*: Organic acids such as acetic, benzoic, lactic, tartaric, and propionic are used as preservative agents.
- *Allyl Isothiocyanate*: Allyl isothiocyanate, an active component in wasabi, mustard, and horseradish, is an effective broad spectrum antimicrobial and antimycotic. However, it has strong adverse secondary odour effects in food.
- *Spice-based Essential Oils*: Spice-based essential oils have been studied for antimicrobial effects: for example, oregano oil in meat, mustard oil in bread, oregano, basil, clove, carvacol, thymol, and cinnamon.
- *Metal Oxides*: Nanoscale levels of metal oxides such as magnesium oxide and zinc oxide are being explored as antimicrobial materials for use in food packaging.

ETHYLENE ABSORBERS AND ADSORBERS

Ethylene is a natural plant hormone produced by ripening produce. It accelerates produce respiration, resulting in maturity and senescence. Removing ethylene from a package environment helps extend the shelf life of fresh produce. The most common agent of ethylene removal is potassium permanganate, which oxidizes ethylene to acetate and ethanol. Ethylene may also be removed by physical adsorption on active surfaces such as activated carbon or zeolite. Potassium permanganate is mostly supplied in sachets while other adsorbent or absorbent chemicals may be distributed as sachets or incorporated in the packaging materials.

TEMPERATURE CONTROL: SELF-HEATING AND COOLING

Self-heating packaging employs calcium or magnesium oxide and water to generate an exothermic reaction. It has been used for plastic coffee cans, military rations, and on-the-go meal platters. The heating device occupies a significant amount of volume (almost half) within the package. Self-cooling packaging involves the evaporation of an external compound that removes heat from contents (usually water that is evaporated and adsorbed on surfaces).

SUSTAINABLE FOOD PACKAGING

Three concepts populate the key areas of change within food packaging. The first is the trend towards more sustainable packaging. While there are multiple definitions of sustainable packaging, the Sustainable Packaging Coalition, an international consortium of more than 200 industry members, offers the most accepted definition.

Sustainable packaging is characterized by the following criteria:

- It is beneficial, safe, and healthy for individuals and communities throughout its life cycle.
- It meets market criteria for performance and cost.
- It is sourced, manufactured, transported, and recycled using renewable energy.
- It maximizes the use of renewable or recycled source materials.
- It is manufactured using clean production technologies and best practices.
- It is made from materials healthy in all probable end-of-life scenarios.
- It is designed to optimize materials and energy.
- It is recovered effectively and used in biological and/or industrial cradle-to-cradle cycles.

Sustainability initiatives led by global legislation, retailers, and corporations guide package material choices, design, and food packaging sales for food packaging professionals. The revised 1997 European Commission's Packaging and Packaging Waste Directive, the 2007 REACH (Registration, Evaluation, and Authorization of Chemicals), and the BS EN 13432 standard (which defines compost-ability, degradability, and biodegradability) are examples of effective global legislative guides in a majority of the Group of 8 countries (Canada, France, Germany, Italy, Japan, Russia, United Kingdom, United States, and the European Union), BRIC (Brazil, Russia, India, and China), and the developing world. In retail, IKEA and United Kingdom retailers have been long-term promoters of sustainable packaging and have launched impressive initiatives such as Plan A byMarks and Spencer, which defines food packaging material use. Also, Wal-Mart, the world's largest retailer, entered the sustainable packaging arena in 2006. Introduced in 2007, the Wal-Mart scorecard ranks packages compared with category competitors, based on environmental scores.

Finally, because a packaging material's source has been shown to be a defining factor in the final package's sustainability, corporations and environmental coalitions are working together to reduce the effect of packaging materials on global sources. International paper companies, the World Wildlife Fund, the Sustainable Forestry Initiative, and the Forest Stewardship Council verify the sources of wood for use in packaging.

Groups such as the Confederation of European Paper Industries (CEPI) in the European Union are addressing solutions to the remaining high environmental impact of paper, which is heavily dependent on water, gas, and energy. Other material-based working groups are taking similar actions.

USE OF PACKAGING SUPPLIER RELATIONSHIPS FOR COMPETITIVE ADVANTAGE

The second key area of change in food packaging is the use of packaging-development value chain relationships for competitive advantage. Suppliers to the food packaging industry are adding value in relationships within the traditional package development value chain. This chain extends from raw material generation, to conversion, to production and distribution, to retail, to consumer use and disposal.

But the formerly linear nature of this chain has become an integrated sphere that enables cross-fertilization of ideas fromvarious supply-chain functions directly to the food manufacturer. Investing in packaging supply chain relationships offers opportunities for focus, innovation, and technology transfer for a competitive advantage. For example, Starbucks engaged their packaging value chain—Mississippi River Corp., MeadWestvaco, and the Solo Cup Co.—to create an FDA-approved 10% recycled-fibre coffee cup.

In addition, Green Mountain Coffee Roasters used its supply chain effectively when it introduced a compost able cup (the Ecotainer) with members of their supplier chain: DaniMer Scientific, NatureWorks, and International Paper. Also, Naturipe has used the value chain optimization to enable consistent packaging to stores from over 50 global locations.

EVOLUTION OF FOOD SERVICE PACKAGING

The third and final key area of change in food packaging is the evolution of food service packaging. Food service has grown to become a major part of consumer spending. As this trend increases, packaging plays a key role in ensuring food safety and providing convenience to consumers. For instance, proper package labeling allows food preparers to know the source of a food, its proper holding temperature, and the adequate cooking needed.

Ease of package opening allows for fewer utensils needed to open packages, which lessens contamination. While tracking and shelf-life extension technologies are employed in the food service industry to reduce the risk of food borne illness, proper heating and heat retention continue to be challenges. Consequently, packaging's role in ensuring proper heating and heat retention continues to increase. For example, CuliDish is a new product that uses varying levels of aluminum within a tray to package foods that require heating together with those that do not require heating.

The tray allows foods requiring high heat to be heated in the microwave at the same time with foods that do not require heat (such as salads). The food can be served in the same tray, thus reducing handling. More innovations in the areas of heat and heat retention are expected to assist in reducing food safety risks associated with improper cooking.

Two major convenience trends—meals eaten in transit and multi-component meals—have also advanced the food service packaging industry. The popularity of meals eaten in transit is evident since only 60% of meals are prepared and eaten at home and 20% of consumers eat on their way to another location instead of in a fixed position. On-the-go food consumption has resulted in packaging that contains a greater variety of foods.

Technologies that support this trend include edible films to enrobe food particles and edible wraps that peel off, allowing consumers to eat numerous foods while in transit. Food package design innovations such as modular folding cartons with flip-off lids, pouches that are easily opened or have a seatbelt flap to hold food close to consumers, and reusable packaging have contributed to the increase of eating while on the go. The growth of multi-component meals in food service stems from multiple types of foods being ordered at quick-service restaurants. This innovation allows for ease of food service preparation and less waste. Multi-component packaging also provides consumers with a presentation platform that makes multi-component foods easy to consume. Examples include reusable trays that clip together into 1 larger serving tray such as that used in the Les Petits Grande line of products; KFC's triple dip strip cartons, which allow consumers to select from 1 of 3 dips when dipping chicken strips; and the folding tray that allows consumers to carry 2 cups of coffee with 1 hand.

9

Fish Farming and Pond Culture

Fish culture is the rational cultivation of fish in a confined water area where the practices of both agriculture and animal husbandry are applicable. The soil and water management aspect of fish culture practice involving application of organic manures and inorganic fertilizers for the production of microscopic plants, the phytoplankton, is basically similar to agriculture while husbandry of fish such as feeding, breeding and health care is more or less similar to a livestock farming system.

This farming system is also unique in that the farmed animal is cold-blooded or poikilothermic and lives in a water medium. Although this fish farming is approximately 2000 years old, the importance of it has been realized only recently in the face of mounting pressure on land resources and scarcity of animal protein for the ever increasing human population.

FISH FARMING WITH PALM TREES

Though fish farming with cultivation of palm trees comes under the broader subject of fish farming cum agriculture it was thought that to have a separate booklet on this topic itself because of the nature of the palm trees and all the possible designs one can make in fish farming cum palm tree cultivation. In our country large tracts of coastal areas which are usually low lying and occasionally flooded are cultivated under coconut palms. Rows of coconut palm trees with intermittent broad and narrow ditches, trenches and irregularly shaped and sized pools and waterways is a common sight in many parts of the coastal regions of Karnataka, Kerala, Tamil Nadu, Andhra Pradesh, Orissa and West Bengal. No doubt fish grow naturally in those trenches, ditches and pools and the people are harvesting them once a year. However these ditches, trenches and pools, if they are properly shaped and arranged along with the coconut trees, can be an immense source of quality fish production.

Areas away from the coastal region wherever water table is high or good irrigation facilities are available areca nut palms are cultivated. Rows of areca nut palms with intermittent narrow water channels are frequently seen in most of the above mentioned states. These channels, if made little more wider, deeper and kept clean from naturally growing vegetation is another potential area for fish farming. In the same way the oil palm cultivated areas also can be incorporated with fish farming provided the water table is high or there is sufficient water available from some natural source.

Resetting the coastal and low lying land into fish cum coconut gardens or the interior areas of areca nut and oil palm growing areas into fish cum areca nut gardens or oil palm growing farms, though difficult due to problems of ownership by various people and restructuring, is a worthwhile attempt from the point of view of long term profitability 'and environmental management. Certainly, some policy level changes and decision making should precede such reorganisation of these areas. However, interested farmers, if they are wanting, with minimum expenses can modify and incorporate fish rearing into their coconut and areca nut gardens.

Areas cultivated with palm trees and having high water table are specially suitable for fish farming because of the following reasons.

- The palms are without branches and grow straight up allowing a lot of sunlight to reach the ground which can be used for fish growth.
- The distance between the palm trees can be adjusted and the planting geometry can be arranged so that a number of fish ponds can be fitted in between the rows of palm trees. The distance between the coconut and oil palms ranges from 25 to 33 feet and broad strips of fish ponds can be fitted in between the rows of coconuts and oil palms. The areca nuts are planted at 9-10 ft distance and the fish ponds of 20 to 30 ft width and any convenient length can be fitted between block plantations (several lines of areca nut trees of areca nuts.
- Palms in general require higher amount of water than most other crops and the intermittent fish ponds become a source of steady supply of moisture to them.
- The fish waste is a very good manure for the palm trees. Similarly the oil cakes from the coconuts and oil palms.are good feed for the fish. Hence there exists a symbiotic or complementary relationship between the fish and the palms in a palm cum fish rearing garden or agro-eco-system.

The purpose of this booklet is to promote the idea of fish rearing along with the cultivation of palm trees wherever the water table is high and it provides the basics of the planning and management of fish rearing cum palm tree cultivation to those who are interested and have the facilities to implement the same.

DESIGNS OF THE FARM

Depending on the availability of the water, the topography of the land, the type of palm trees planted, the type and extent of intercropping being planned, the needs and priorities of the farmer there can be many designs that could be followed in a fish farming cum palm tree cultivation. Some of these designs are complex and the implementation requires meticulous planning and good engineering skill. It is beyond the scope of this booklet to explain and elaborate all the intricacies of the possible designs; only some basic ideas are explained here as the number of designs vary depending on many factors.

Factors Affecting the Designs

The design for fish pond construction in the palm garden depends on the distance which is maintained between palms in the same row and between rows. Therefore the possible designs for different palms are discussed here.

The desires and the ideas given here are by no means claims to be exhaustive. Each farmer can and should modify and develop better ways and designs in their own garden. However some common factors that influence the type of design adopted can be listed as follows.

Climate

Among all the factors the main deciding factor in the choice of the palm for cultivation is climate though it is not so much of a deciding factor for the fish cultivation. The choice of palm and the type of palm itself will determine the design of the arrangement of fish ponds and the land area allotted for palm trees.

Quantity of Water Available

Depending on the quantity of water and the drainage facilities available the area under the fish ponds may be more or less compared to the area under oil palm cultivation. More the water and lesser the drainage facilities the farmer will be compelled to allot greater area under the fish ponds and lesser area to palm trees and vice versa. However, even if the water availability is more but the drainage facility is also good, then the farmer is tree in allotting whatever amount of area to fish ponds depending on other factors affecting both the cultivation of palm trees and fish rearing. Accordingly the designs also will change.

Market

Depending on the demand for the palm and the fish there can be equal or more importance to both or either of them. Hence the possible designs also will be different.

Intercropping

The design also depends on the amount of area the farmer may be wanting to have for intercropping. If he intends to have more area under intercropping more land will be allotted to the palm trees and less will be allotted to the fish rearing.

Easy Movement in the Farm

The fish ponds and the strips of palm tree plantation should be so arranged that it should help in the easy movement of the people, vehicles, trolleys, etc. Such arrangement will help in the smooth management of the entire farm.

Farming System

Under the broad farming system of palm tree cum fish culture we can have several sub-farming systems such as coconut based farming system in which coconut will be the main crop and all others (inter crops and fish rearing) will be subsidiary to it. Similarly we can have a fish farm centred farming system in which the main crop is fish and all others will be subsidiary to it or we can have arecanut centered farming system. We can also include animal husbandry and ladder cultivation along with the system of fish cum palm tree cultivation. In all these circumstances the design will change.

Others

Other factors like availability of the fry and fingerlings, transport and communication, availability of fish feed, facilities of processing of palm products, the cultural factors of the people are all taken into consideration in deciding the design for the fish cum palm trees cultivation.

Designs for Coconut Garden

In coconut plantations, a spacing of 25–33 ft. is maintained between the palms and rows as well. Between the rows of coconuts the fish ponds can be fitted in a number of ways as explained here. The design also is affected by the quantity of water available in the garden either in the form of high water table or from a perennial river, canal, lake or any other source. Accordingly the area covered under water will be more or less; if the water available is more the area covered will be more and vice versa. Hence the design also can be changed.

Coconuts are planted normally keeping a distance of 33 feet distance between plants and rows with fish ponds having a width of 15 to 20 feet and any convenient length between rows. This again can be two types (a) fish ponds in parallel rows and (b) fish ponds in criss-cross fashion.

Ponds in Parallel Rows

Between two parallel rows of coconuts a long fish pond is fitted. Depending on the water availability and the desire of the farmer the garden can be covered with this design fully or partially. If the farmer needs some area for intercropping he can leave some area free. Or if there is not enough water to cover the whole area then he will be forced to leave some area free which he can cultivate under some intercrops. The width of the pond will vary from 15 to 20 it. and any convenient length can be adopted.

A modification of this design is accomplished by planting two rows of coconut trees at a distance of 20 to 25 feet and between every two such sets of double rows leave a distance of 33 to 50 feet in which we can fit in fish pond having a width varying between 20 to 38 feet and any convenient length.

Similarly one can have coconuts in triple lines planted at 25 to 30 feet distance and between two sets of triple lines leave a distance of 33 to 50 ft in which we can fit a pond of 20 to 38 ft width and of any length. Thus, depending on the convenience, availability of water, needs and priorities of the farmer the parallel arrangement of coconut trees and fish ponds can be implemented. It may be better to connect these parallel ponds with a small open or underground channel for easy management of the water flow. But these channels should have facilities to fit wire mesh fixed frames so that fish is prevented from going from one pond to another.

Ponds in Criss-Cross Fashion

If the water is in abundance and more area can be set apart for the fish ponds the arrangements of the coconut palms and the ponds is done in criss-cross fashion. Coconuts are planted at least 33 ft distance (row to row and plant to plant) and fish ponds of 15 to 20 ft are constructed between the rows in both ways resulting in having each coconut tree standing on an island of 13 to 18 sq. ft. land. In the actual situation in a water logged area islands of about 13 to 18 square ft. are created on which the coconuts are planted.

In this design the area under water will be more compared to the area under coconuts. Hence, the area available for intercropping between the coconuts also will be much less. Variations of this design can be done having bigger islands with four coconuts each and the rest of the area under fish ponds. This means criss-crossing is made after every two rows of coconuts.

A number of crops are suitable for intercropping between the coconut trees. If the coconut trees are planted at 33 ft distance any crop can be cultivated as intercrop. We can even go for multilayered multiple cropping in the coconut garden if the distance is maintained at 33 feet. Black pepper can be grown as climber to the coconut tree itself. We can go for extensive fodder cultivation in between the coconuts. The fodder can be fed to the animal in our own dairy farm or sold to some body else who is having the dairy farm.

It should be mentioned here that coconut cum fish farm cum dairy cum fodder cultivation is excellent farming system. All these four have a symbiotic relationship with each other. The fodder cultivated in the coconut garden is fed to the cattle in the dairy farm, the dung and the urine generated in the dairy farm will be fed to the coconuts, fodder crops and to the fish ponds. The oil cake from the coconut will be feed for the cattle and fish. The fish waste converted into fish meal is excellent feed supplement to the cattle and manure to the coconuts and fodder crops. The usages of by-products from the coconut are many leaves for thatching, leaf fibs for broom making, the leaf, petiole, inflorescence sheath and the stem for fuel wood, the fibre for ropes and brush making, trunk as building material and making furnitures, the tender inflorescence for toady tapping, the shells for handicrafts, etc., are some of the usages.

Designs for Areca Nut Garden

The plant to plant and row to row distance in areca nut is maintained at 9 to 10 feet. Obviously fish ponds cannot be fitted in between the rows of areca nut trees. What is possible is to have block plantations consisting of several lines of areca nuts alternated with fish ponds of 20 to 30 ft wide and any convenient length. These blocks of areca nut plantations and long fish ponds are arranged in parallel designs. Perhaps not as many variations are possible in areca nut cum fish pond farming as in the coconut palm plantations. Areca nuts do not provide any supplementary feed to the fish though the fish pond can provide some supplementary nutrition to the areca nuts.

A number of crops fit in as intercrops in the areca nut plantations. Prominent among them are: black pepper, dwarf varieties of coffee, yams both creeping and herbaceous, fodder grasses, betel leaves, ginger, turmeric, medicinal plants, etc. Depending on the need and priority of the cropping the intercropped area can be increased or decreased in the areca nut plantation.

Design for Oil Palms

The oil palms are normally planted at 30 ft distance since they grow as big as a coconut tree in their canopy size. But in the case of oil palm cum fish cultivation they should be planted at the same distance as the coconut trees and all the possible designs under the coconut cum fish cultivation can be followed.

As already mentioned the variety of designs depends on the amount of water available, the priority of the farmer is whether to go for fish farming or oil palm cultivation, higher profitability from either of them, amount of land needed for intercropping, etc.

Planning of the Farm

The success of the fish cum palm tree farming system depends very much on the planning. No doubt a well planned farm will save a lot in the management of the daily operations in the entire farm. Before setting up the farm one should have well defined aims, objectives and strategies.

In planning the fish cum palm tree farming we must, check on the following points.

- First of all, on the economic viability of the whole farming system. Nothing can exist in the long run unless it is economically viable.

Make a detailed study of the economic viability taking into consideration all the costs including the opportunity costs.

- Secondly, we must ensure the technical feasibility or the establishment and management of this farming system in a particular place. The required technical know-how for the maintenance of the system should be locally available.
- Thirdly, the system should be environmentally suitable. It should fit in with over all environmental system of the place avoiding pollution or draining out the resources too much from the place.
- Fourthly, it should be socially acceptable to the people of the localilty. Though in the beginning there may not be any difficulty, in the long run social acceptance is an important point to be considered in planning this farming system. Involvement of the local people in the management of the farm and making them feel that they too benefit proportionately from the farming system is also a part of the social acceptance besides the possible need for cultural and religious acceptance.
- Fifthly, it should be educative. The farming system introduced into a locality should provide enough technical and manegerial skills to the local people so that more and more people take up to this type of farming system and thus create an association of the fish cum palm tree growers. Such unions and associations are necessary to build up economic and political strength among the owners of this farming system.
- Sixthly, there should be a possibility for product diversification and value addition by which the producers retain maximum control over their products. There should be provision for processing of raw materials. This will generate maximum employment opportunities to the local people and the owners of the farms will get maximum support from them. While planning it is very much required to make a blue print of the farm plan showing all the details of the components, location of each component, size, acreage occupied, etc., even it should show even the number and type of trees planted and the breed of fish reared. In short the blue print should be as detailed and meticulous as possible; nothing should be left to chance.

Before making the final blue print consult as many people as possible and collect sufficient data on the availability of all the externally needed inputs both for establishment and maintenance.

While planning we should pay attention to the flow of organic matter and water through the various components in the farm. We must ensure that all the organic matter that is generated at the farm is recycled and used as manure and feed. Similarly all the water that comes into the farm in the form of rain, over ground or underground should be stored and recycled as far as possible. In planning the sustainability of the farm should be given top priority. As far as possible try to supply all the inputs internally or follow the principle of low external inputs. Heavy dependence on external inputs in the long run is detrimental to the sustainability of the farm. The agro-ecosystem we create in any given piece of land should be balanced from the point of view of inputs and outputs.

- Seventhly prepare a detailed and feasible time planning. In the time planning what activities and operations will be carried out when and in how much time etc should be planned out. The time planning consists of chronological or synchronological order of the various activities and the duration of the time taken for the completion of each activity or unit.

After preparing the time plan arrange the labour force required for completing the operations and the finance required for each unit and activity is arranged. The reader may feel that it is beyond his ability to do such meticulous planning. For ordinary people it may not be possible. But for the success of this type of farming system such planning is necessary. What the people should do is to form an association and get things done through it. An association can get things done which an individual cannot do. Re- member these booklets are not written only for guiding individuals to do their day to farming operations but to promote long term sustainable farming systems that should last for several generations.

Establishment of the Farm

In the establishment of the farm the best way is to follow the time bound operational planning. However if the finance and labour is available then different units can be completed at the same time. Before the actual start of the work arrange for all the materials required including the planting material. Seedlings of the coconuts, arecanuts and oil palms are arranged sufficiently early enough so that we can get quality seedlings at the time of planting. The seedlings should be minimum two to three years old in the case of palms. Such arrangements are done outside the farm and not directly related to the work inside the farm in establishing the farm.

Among the works done inside the farm in establishing it, first of all locate and demarcate various units on the land in detail according to the blue print prepared already. Secondly structure the farm by laying out the roads and paths so that movement of material to the farm and from the farm during the construction can be done easily. Thirdly, construct the ponds and prepare the planting beds for the palm trees. The soil that is excavated during the construction of the fish pond will have to be adjusted on the planting beds or can be sold out.

The first stage consists in completing all the structures within the farm mainly the fish ponds and the planting beds for the palm trees. After the structures are complete the fish ponds can be operationalised at the first instance itself because in the first year itself we can obtain considerable amount of income from the farm. Mean time demarcate the spots of the pits and prepare pits for the plantation of the palm trees at the already decided spots according to the design planned. The readers are advised to refer the booklet on "Planting and Management of Trees" (Booklet No. 21) for more details about planting and caring of palms. Similarly whatever perennial inter crops are cultivated prepare the pits and planting materials in advance. Pits should be dug at least one year ahead and leave them exposed to at least one rainy and summer seasons for seasoning of the soil in the pits.

When all the perennial trees are planted and the fish ponds are put into operation the farm is said to be established. Normally it will take ten years for the palm trees to come to full production and to yield substantial economic return. But from the fish ponds we can get substantial return from the first year itself. By the end of second year in every pit where a palm tree or a perennial intercrop is planted should have a surviving and healthy plant. The same should be the fact even at the end of five years and at the time of flowering and fruiting. This is an important point because every tree missing from the spot for whatever reason is a huge loss as each tree can yield for many years and the total, income amount to be surprising.

For example, the coconut is expected to provide about 300 nuts per year; assuming each nut fetches five rupees the expected income from the nuts per year will be 1500 rupees.

A coconut tree can yield up to 40 to. 50 years. The total ! expected income from one coconut tree will amount to be I 60,000 to 75,000 rupees. And the total income from 100 coconuts (planted at 33 ft distance in two and half acres of planted area) is estimated to be 60,00,000 rupees at price level of 40 to 50 years before. Without calculating the possible interest on the early years of income. This is only the income from the nuts; income from other products of the coconut trees is not yet to be estimated. We may level that income with the maintenance expenditure.

FISH AS FOOD

Malnutrition and starvation are the two serious problems being faced by millions of rural poor in most of the developing countries. The problem of malnutrition is in fact more serious and of a bigger dimension than the starvation problem and is caused mainly due to animal protein-deficient diets. Animal protein is essential for proper growth, repair and maintenance of body organs and tissues.Fish contain about 16–20 per cent protein compared to about 12 per cent in egg, 3.5 per cent in milk and 6–8 per cent in rice and wheat.

Moreover, it is wholesome, tasty, highly nutritive and an excellent source of essential minerals, vitamins and essential amino acids.

At present about 31 per cent of the total animal protein supply in the Asian region is in the form of fish protein. For the poorest segments of the population, fish is not only the most important animal protein source, but often the only one.

MINIMAL WATER REQUIREMENT

The reproductive potential of fish compared to any other farmed animal is also very high. A kilogram of female cultivable carp species yields on an average about 0.1 million eggs, each of which has the potential to become 1 kg fish in about a year.

No livestock animal possesses this magnitude of fecundity. Although fish needs water as a medium to survive and grow, it consumes minimal quantity of water compared with any livestock or agricultural crop. Fish also enriches the water with its voided metabolites thus making the water more productive for agriculture.

Low Energy Requirement for Protein

Fish culture systems require a relatively less amount of energy for protein production than any other farming system. Carp culture, depending upon culture practices, requires energy at the rate of 22–468 KJ/g of protein production while a land animal farming system needs over 550 to 3 400 KJ/g.Fish are cold blooded or poikilothermic animals. In other words they cannot maintain a constant and high body temperature like other livestock animals. Instead, their body temperature fluctuates according to the surrounding temperature. In warmer climates, their metabolism accelerates and they grow faster, while in colder climates, the metabolic rate slows down, resulting in a reduced rate of growth. In this way they save energy by not spending it for maintaining a higher and constant body temperature.

AQUACULTURE PRODUCTION POTENTIAL

Although the world's total marine production now stands at more than 80 million tons per year, all trends indicate that a saturation point is quickly approaching. During the past decade the growth rate hovered around 2 per cent, much lower than earlier decades. On the other hand, tremendous potential exists in aquaculture.

Aquaculture presently produces over 8 million tonnes of fish and shellfish annually.It is estimated that Asian aquaculture production could be raised to 20–30 million tonnes a year by the end of the century. Aquaculture production has increased at an annual growth rate of nearly 7 per cent between 1975–84.

BIOLOGY OF MAJOR CULTIVATED FISHES

China has a vast area of waster, with a distribution of inland rivers, lakes, reserviors and ponds throughout its territory, and hence, it is rich in fishery resources. There is an ample variety of fishes, of which more than 50 per cent are carps. Up until now, there have been about 20 species cultured in ponds in integrated fish farms. The introduction here focuses on the four main well-known carps: Silver carp, Bighead, Grass carp and Black carp, and other good stocks------Common carp, Crucian carp, Chinese bream (Wuchang fish), Mud carp and Tilapia, etc.

Understanding the habits of fish, mastering the laws of their growth, development, propagation and feeding and satisfying their ecological requirements accordingly will be of great practical significance to the development of fishery production, to the application of farming techniques and to the increase of fish yield.

MORPHOLOGY

Silver carp (Hypophthalmichthys molitrix)

Silver carp belong in taxonomy to Family Cyprinidae, Subfamily Hypophthalmichthyinae. Body: compressed Scales: small, Mouth: in front, with lower jaw slightly slanting upward. Eyes: comparatively small, situated below the horizontal axis of body. Gill membrane: unconnected to isthmus. Gill rakers: dense, interlaced and connected, covered with a sponge-like membraneous sieve. Abdominal keel: extending from the base of pectoral fins to the anus. Pectoral fin: its terminal tip does not exceed the base of ventral fin. Pharyngeal teeth: one row in 4/4, with fine lines and tiny grooves on surface. Intestinal length: 6---10 times that of body length. Colour of body: silvery white while alive; colour of dorsal, very dark brown. The largest body so far discovered is about 20 kg.

Fig. Hypophthalmichthys Molitrix

Big Head (Aristichthys Nobilis)

Bighead are similar to Silver carp in shape. They belong to the same genus Hypophthalmichthys. Head: bigger. Snout: short and blunt. Eyes: small, situated below the horizontal axis of body. Gill membrane: unconnected to isthmus. Gill rakers, dense and separated; without spongelike sieve.

Abdominal keel: between the bases of ventral fins and the anus. Pectoral fin: the terminal tip reaches 1/3---2/5 of the base of ventral fin. Pharyngeal teeth: one row in 4/4, surface flat. Intestinal length: about 5 times that of body length. Colour of body: while alive, dorsal and upper sides, light black, scattered with irregular yellowish black spots; ventral surface, silvery white. The largest body so far found is about 40 kg.

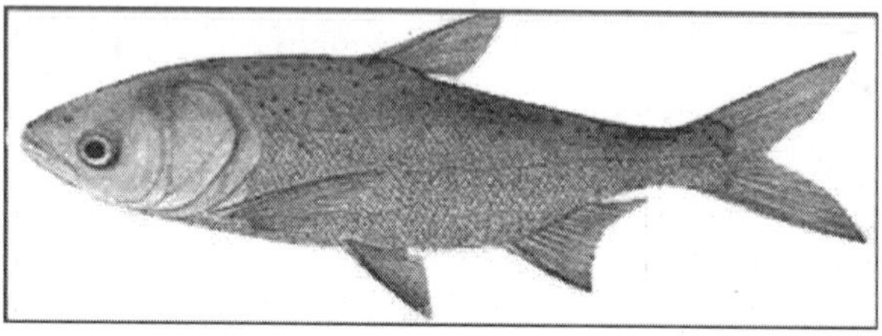

Fig. Aristichthys Nobilis

Grass Carp (Ctenopharyngoden Idellus)

Grass carp are larger-size fish in sub-family Leuciscinae of family Cyprinidae. Body shape: almost cylindric, with flat head and round abdomen. Scales: big. Mouth: in front; lower jaw, shorter. Gill membrane: connected to isthmus. Gill raker: small and short, in scattered arrangement. Pharyngeal teeth: 2 rows in 2,5/4,2 compressed like combs. Intestinal length: 2.3---3.3 times that of body length. Colour of body: while alive, dorsal, grey; abdomen, light grey; sides, greenish yellow; fins, a lighter colour. The largest body so far found is about 35 kg.

Fig. Ctenopharyngodon Idellus

Black Carp (Mylopharyngodon Piceus)

Black carp are more close to Grass carp among Cyprinidae. Body shape: like Grass carp with a pointed head. Scales: big and circular. Mouth: arc-shaped in front. Eyes: medium size, situated in the middle part of head sides. Gill raker: short. Gill membrane: connected to the isthmus. Pharyngeal teeth: one row in 5/4, big, short and molar-like; surface smooth. Intestinal length: 1.2---2 times that of body length. Colour of body: while alive, greenish black; dorsal darker; abdomen light grey; Fins, black. The largest body so far discovered is about 70 kg.

Fig. Mylopharyngodon Piceus

Common Carp (Cyprinus Carpio)

Body: compressed. Dorsal: projected in arch-shape. Abdomen: round. Mouth: slightly downwards with a long blunt snout and with two pairs of barbels on upper jaw; lower pair a little longer. Dorsal fin: long. Scales: big and thick. Pharyngeal teeth: 3 rows in 1,1,3/3,1,1; teeth on inner sides, molar like. Intestinal length: 1.5---2 times that of body length. Colour of body: while alive, varying with different living conditions, usually dark grey or yellowish brown on dorsal; sides golden yellow; lower part of caudal fin, red. The largest body so far discovered is about 40 kg. China has a long history of culturing Common carp, which have a wide distribution and strong adaptability. There are a lot of morphological variations through artificial breeding and natural selections, such as Scale carp, Mirror carp, Wu Yuan red purse carp, Xing Guo red carp, etc.

Fig. Cyprinus Carpio

Crucian Carp (Carassius Auratus)

Crucian carp are very close to Common carp. Body: compressed and relatively thick. Abdomen: round. Head: small and short. Snout: blunt. Mouth: arc-shaped in front. Lip: thick without barbels. Pharyngeal teeth: compressed, one row in 4/4. Intestinal length: 2.7--3.2 times that of body length, some even reaching 5 times. Colour of body: silvery grey when alive; darker on dorsal and lighter on abdomen. The largest body found is about 1.5 kg.They have a wide distribution and a strong adaptability. They can live in different water body such as rivers, lakes, ponds and ditches with some variations and differentiations in characteristics.

Carassius auratus gibelio are sub-species of Crucian carp. Body: higher than usual Crucian carp. Abdomen: round. Colour of body: bluish grey with silvery white on abdomen. The largest body is about 3 kg. Japanese Crucian carp originated from Lake Pipa in Japan and was introduced into China. They grow faster than Carassius auratus gibelio. A large variety of golden fish are bred out through long period of artificial breeding.

fig. Carassius Auratus

Chinese bream, or Wuchang fish

(Megalobrama amblycephala) Chinese bream belong to subfamily Abramidinae. Body: high, compressed and lozenge-shaped. Head: small and short Mouth: slanting. Abdominal keel: extending from the base of pelvic fin to the anus. Pharyngeal teeth: 3 rows arranged in 2,4,5/4,4,2. Intestinal length: 2.7 times that of body length. Colour of body: dark grey while alive, darker on dorsal; scale, dark grey in the middle and lighter on its edge. The largest body so far caught, is about 3 kg.

Fig. Megalobrama Amblycephala

Mud Carp (Cirrhina Molitorella)

Mud carp belong to family Cyprinidae, subfamily Barbinae, genus Cirrhina. Body: long and compressed. Abdomen: round and slightly flat. Snout: short, round and blunt.

Mouth: inferior and transverse, with two pairs of barbels; snout barbels, strong and thick; jaw barbels, small and short. Caudal fin: deeply separated with upper part a little longer than the lower one. Lateral line: from the upper part of the pectoral fin around the lateral line, there are 8--12 scales with dark dots at their bases which form lozenge-shaped spots. Fins: dark-grey. Pharyngeal teeth: 3 rows in 2,4,5/5,4,2. Intestinal length: about 14 times that of body length. The largest body caught is about 4 kg.

Fig. Cirrhina Molitorella

Tilapia

Tilapia belong to order Perciformes, family Cichlidae, genus Tilapia. This genus comprises more than 100 species including subspecies. At present, 15 of them are taken as cultured fishes all over-the world. In China, mainly Tilapia mossambica and *T*. nilotica, etc., are reared.

Body of *T*. mossambica: short and compressed. Dorsal: a little higher. Shape: similar to Crucian carp. Mouth: bigger. Lip: thick, with lower jaw a little long than upper jaw. Scales: circular. Lateral line: disjointed. The intestinal length about 7 times that of body length. Colour of body: dark grey while alive; during spawning stage, body colour of male fish, dark green; edges of dorsal, anal and caudal fins, obvious red; female, greyish yellow. The largest body weight so far discovered is about 0.5 kg.

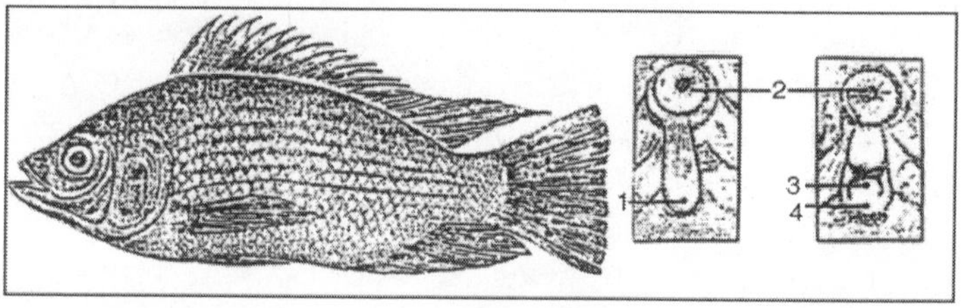

Fig. Tilapia Mossambica **Fig.** Male **Fig.** B Female

Colour of *T*. nilotica: changing with external conditions, light black; abdomen white, with 9 longitudinal black stripes on body surface, of which, 7 below dors; fins and 2 on the peduncle; caudal fin with 10 clear vertical black stripes for life. Scales: ctenoid. The largest body weight is 2.5 kg.

Fig. Tilapia Nilotica

Nuptial colour can be found on the male and the female of Tilapia at reproductive period. But their external genital organs are different in appearance. The male has two pores; anus in front and cloaca in the rear. The female, 3 pores, *i*.e,. anus in front, genital pore in the middle and excretory pore in the rear

FISH SPECIES FOR CULTURE

The choice of fish species is very important in maximizing production, both in terms of quantity and quality.Since considerable amount of energy is lost in successive trophic levels of the food chain, efficient fish culture always aims at making the chain as short as possible. Because of this, herbivorous fishes are always preferred to carnivorous fishes, the latter being mostly excluded because of their longer food chains. Mixed species farming or polyculture yields a higher production than single species farming.

It is obvious that any single species cannot utilize all the available food in a pond because of its specific feeding habit and hence a combination of compatible species with complementary feeding habits are usually stocked to make better use of the natural food available in the pond. Selection of the species should be based on the productivity of a pond, availability of artificial food resources, availability of seed and the marketing prospects. The principal considerations in species combination are that they have complementary feeding habits, they occupy different ecological niches, they attain marketable size at more or less the same time, they tolerate each other and they be non-predatory in nature.

A combination of plankton and macrophyte feeders is most usual. Ungrazed phytoplankton is fed upon by zooplankton and to utilize them, the zooplankton feeders are included in the combination. The combination of the phytoplankton-feeding silver carp (Hypophthalmichthys molitrix), the zooplankton-feeding bighead carp (Aristichthys nobilis) and the weed-eating grass carp (Ctenopharyngodon idella) is well known in China and Southeast Asia.

In India, under composite fish culture, six species of fish viz. catla (Catla catla), rohu (Labeo rohita) and mrigal (Cirrhinus mrigala) along with three Chinese carps such as grass carp, silver carp and common carp (Cyprinus carpio) are stocked together so as to utilize most of the fish food organisms present in the pond Other similar combinations may work just as well, but the most important aspect is to try to establish a balance between the species based on the food spectrum of the pond Normally fish production increases with the increase in the number of fish stocked per unit area upto a level and then starts decreasing. Higher stocking density results in increased total production, as there is better utilization of the available food, but in such cases the individual weight and size is reduced. On the other hand, lower stocking density yields larger individual fish.Proper stocking rate for a pond is that optimum level which results in a given time, usually a year, in a production which is highest in quantity and quality of fish and most profitable. In ponds where no artificial feed is used, the total crop becomes dependent on the primary production and in such cases simply by increasing the stocking density, the increase in the total production is not possible.

Even with supplementary feeding the scope of increasing stocking density and fish yield is limited; it increases to an optimum level and then starts decreasing.Under crowded conditions fish compete for food and space and are stressed due to aggressive interaction. Fish under stress exhibit decreased feed consumption and slow growth and are predisposed to many parasitic and microbial infections. Increase in stocking density simultaneously increases the total oxygen demand with obvious dangers. In undrainable ponds, accumulation of excretory products of the fish population also suppresses their growth rate. With efficient removal of such metabolites by aerating the pond water, the stocking rate can be increased further, thereby enhancing production.

CHARACTERISTICS OF DRAINABLE PONDS

In India ponds are relatively small and shallow bodies of impounded water with limited wind action. They may be called perennial if they retain water the year round or temporary/seasonal, if they do so seasonally. They may be further classified as drainable ponds and undrainable ponds, depending upon the drainage facility by gravity. Drainability imparts a very desirable feature to a pond and some authors prefer to call only the drainable type of ponds as fish ponds. However, Indian experience has shown that experimental fish production to the tune of over 10 tonnes/ha/yr can be achieved even from such undrainable ponds through a proper understanding of the biotic and abiotic components of the ecosystem and adoption of suitable culture technologies. Hence, before adopting any culture technology, it is imperative to have an idea about the basic biology of the pond types in terms of environmental factors, community structure and community metabolism.

UNDRAINABLE PONDS

The periods of 'plenty rain' and 'no rain' usually prevail in regions having undrainable ponds. With the onset of monsoon, torrential downpours sweep across the land and the amount and frequency of rain decrease towards the end of the monsoon. Severe floods may occur, whereas a late monsoon or early monsoon of short duration may result in serious drought. Both flood/ rain and drought influence the ecosystem of the ponds on such lands. Small, shallow and seasonal ponds get filled or dry, whereas deeper perennial ponds exhibit considerable fluctuations in water levels accordingly. Though these ponds are basically constructed for storing water in such areas for multiple uses, ranging from supplying drinking water for human population, live stock, etc., to supplying water for agriculture, recent trend is to utilize them for fish culture. The description of the undrainable ponds is based upon the studies conducted at the Central Institute of Freshwater Aquaculture, Dhauli, Bhubaneswar, India, under an extensive environmental monitoring programme of rural undrainable ponds.

General Morphometry

Undrainable ponds in general are relatively small, perennial or seasonal water bodies constructed or excavated for multiple uses. Some of these ponds have proper embankments. They greatly vary in their dimensions ranging from 0.02 ha to over 2.5 ha in water surface area and 50 cm to 250 cm in depth.

Larger ponds are relatively deeper while smaller and seasonal ponds are shallower.Unlike shallow seasonal ponds, the bottom of the perennial ponds is never exposed to sunlight and therefore the whole ecosystem of such ponds is quite different from those of shallow and seasonal ponds. Use of these ponds by villagers for multipurpose provides the source of organic enrichment. Usually the only source of water for these undrainable ponds is the heavy rainfall during the monsoons.

However, in some cases, the pond bottom is cut below the water table so that ground water enters the ponds. As soon as the monsoon ceases, the water level starts decreasing gradually and shortage of water is quite common during the pre-monsoon season. Water is lost from the pond through evaporation, seepage and transpiration by aquatic macrophytes and the trees and shrubs planted along the pond sides. Macrophytes tend to appear in both perennial and seasonal types of ponds but with increased intensity in shallower ponds.Presence of thick sediment layers in the bottom is the most characteristic feature of these ponds which gradually get accumulated during the course of time and vary between a few centimeters to over a meter and half in thickness.

The quality and quantity of sediment deposition depend mainly upon the original soil, method of construction, nature of embankments, macrophyte cover, pond productivity, organic and inorganic additions, species cultured, etc. Although sedimentation is relatively faster in smaller ponds, there is a positive correlation between the age of a pond and its sediment thickness.

Physico-chemical Environment

The water depth and total volume of water available for individual fish are crucial in fish culture systems. Adequate water depth is needed not only for optimum growth, but also to provide enough space and oxygen for fish life. Water levels in these ponds are mainly dependent upon monsoon rains.After the monsoon season, the water level starts decreasing gradually and shortage of water is quite common during the summer season which is the most crucial time for fish culture since the fish growth rate is faster in this period. In fact, during the time of lowest water level the ponds contain the maximum biomass.

In shallow and seasonal ponds, sufficient phytoplankton population fails to appear and the soft sediment layer is vigorously stirred up by fish, making the water more turbid, thereby reducing the photosynthetic process by limiting light penetration. Eventually, the total amount of available dissolved oxygen may not be, at times, sufficient to meet the demand for total community respiration and the chemical oxygen demand of the sediment, resulting sometimes in mass fish kill and planktonic collapse. On the other hand, in deeper perennial ponds where the water column is more than 3 m, fish life is again adversely affected. In such ponds the photosynthetic or oxygen producing zone is less in comparison with the oxygen consuming layer.In addition, the sediment proper and the sediment community also consume a considerable amount of oxygen. All such conditions lead to a negative oxygen balance. The water in most of these ponds remains slightly alkaline (pH 7.0–9.0).

The NH_4-N (ammonia-nitrogen) content of the waters remain below 0.02 mg/l with even lesser quantities of No_3-N (nitrate-nitrogen). The PO_4^-P (phosphate) concentrations remain low and these chemical features of such ponds suggest that these waters are highly nutrient-deficient, particularly in nitrogen. On the contrary, the pond sediment is rich in organic and inorganic nutrients.The organic carbon ranges between 3 (in newly excavated ponds) and 50 mg/g dry sediment weight (in older ponds).

The nutrient status of the sediment differs completely from that found in the overlaying water column In general, all the basic nutrients in the pond sediment are about thousand times higher than in their respective water column.

Carbon dioxide and oxygen are the most important gases affecting the pond community including fish. During the photosynthetic activity, carbon dioxide is usually at zero level while during the darker period its concentration increases. At higher concentrations it may be toxic to fish life. Carbon dioxide toxicity increases with decreasing level of dissolved oxygen. Carbon dioxide concentration can be tolerated upto 20–30 ppm in these ponds provided oxygen is near saturation.

Community Structure and Function

Bacterioplankton and phytoplankton constitute the basic food for the fine filter feeder fish species and also for the zooplankton which form the main food of the rough filter-feeder species. The bacterioplankton population is always higher in those ponds which are associated with the activities of larger human and livestock populations.Most of the relatively older ponds with frequent appearance of Microcystis bloom have higher levels of bacterioplankton population (3–10 million/ml).

On the contrary, the newly constructed and recently desilted ponds have less dense bacterioplanktonic community, around 1–2 million/ml. Macrophytic infestation also significantly limits the bacterioplankton production.

The planktonic detritus originates mainly from decomposing fragments of the phytoplankton and zooplankton and has generally a concentration range of 2,000 – 20,000 number/l. The main groups of phytoplanktonic population are Myxophyceae, Chlorophyceae, Eugleno-phyceae and Bacillariophyceae; whereas copepods, cladocerans and rotifers constitute the majority of zooplanktonic population. Some of the very old ponds having excessively thick sediment layer face Microcystis blooming. In such ponds the benthic animal fauna is represented by a very small number. In the majority of the ponds the benthic animal communities are dominated by red chironomids and oligochaetes indicating the general oxygen deficiency in the sediment layer.The bacterial decomposition and nutrient recycling in ponds are greatly influenced by the anaerobic nature of the sediment. At the initial stage of Microcystis bloom in older ponds, the oxygen production has been found to be the highest (1) (over 15 g O_2/m^2), whereas the total community respiration remains considerably low (2) (below 10 g O_2/m^2). However, during the active decomposition stage of Microcystis (plantonic collapse stage) the total oxygen production level goes lower (5–6g O_2/m^2) than that of the community oxygen consumption (4) (6–7 g O_2/m^2).

In older ponds, especially those having thick anaerobic sediment, the biochemical oxygen demand ranged between 70 per cent and 90 per cent of the total oxygen production, ultimately causing anoxic condition leading to fish kills.The majority of rural undrainable ponds are characterized by anaerobic benthic sediments. The dead and decaying organic matter settles down to the pond bottom (sedimentation) where it is subjected to further decomposition and mineralisation.The upper layer of the sediment remains aerobic while the deeper layers are deficient in oxygen and thus anaerobic. These perennial undrainable ponds in tropical monsoon lands with yearround warm water under plenty of light offer an excellent possibility for fish culture.

Most of the species cultured greatly depend upon natural fish food resources and with a limited dependence upon artificial supplementary feed.However, without proper environmental management, the water remains infertile due to the overall nutrient deficiency with a very pronounced nitrogen limitation, although they possess a very high production potential. On the contrary, the pond sediments have extremely high level of organic nutrients in almost locked-up conditions which remain unutilized due to the anaerobic nature of the pond bottom. However, though regular raking up of the pond sediment, either by manual or biological means, the organic nutrients could be released for making the pond water more productive.

Proper management methods can optimise fish production in perennial ponds at most economical rates while seasonal ponds can suitably be utilized for fish seed rearing and also for short-term fish production depending upon the duration of water retention. The general feature of the properly managed and ill-managed undrainable fish p and 8b.

DRAINABLE PONDS

Ponds which can be supplied with water and drained of its water according to the requirements of the fish farming operation are known as drainable ponds. These ponds require suitable ground with proper embankments, inlet and outlet structures and adequate supply of water on regular basis.

CARP CULTURE

The most successful system of pond fish culture is the polyculture of three Indian major carp species-catla, rohu and mrigal along with three Chinese carps viz. silver carp, grass carp and common carp. In India this is commonly known as composite fish culture.

The best results in terms of fish production in this system results not only through a judicious combination of species, but also due to appropriate management techniques including pond fertilization, supplementary feeding and health care.On the basis of growth performance of different species, modifications are often made in stocking density, species ratio, fertilization schedule and supplementary feeding programme in different agroclimatic conditions.

Fig. Carp culture

High rates of fish production to the tune of over 5 500 kg/ha/6 months, 7 200 kg/ha/8 months and over 10 tonnes/ha/yr have been achieved in composite fish culture trials conducted in different agroclimatic conditions of India.The carp culture system as a whole is operated as a three-tier culture system where the practices are adopted for rearing fish during their different stages till they are harvested. Spawn (post larvae) are reared upto fry (2–3 cm) stage in nursery ponds, fry to fingerlings (8–12 cm) in rearing ponds and finally fingerlings to table-size fish in composite fish culture ponds or stocking ponds. Relatively smaller, seasonal ponds are mainly used for rearing spawn to fry stage and harvested after 2–3 weeks.

Several crops of fry are usually taken during the season. Pond fertilization by cattle manure and feeding with 1:1 mixture of oil cakes and rice bran is the usual practice. Fry raised in nurseries are reared upto fingerlings in slightly bigger ponds (0.05 – 0.1 ha) of seasonal or perennial in nature. Fingerlings are removed after 3 months and stocked in composite fish culture ponds.

INTEGRATED CARP FARMING

An integrated approach of composite fish culture together with compatible combination(s) with poultry, duckery, pig rearing and cattle raising is now being adopted.

Table. Annual production through integrated carp livestock farming

Integration	Fish production	Animal production (live weight)
Fish + Pig farming	6 – 7 ton/ha	4 000–5 000 kg pig meat
Fish + Duck farming	3 – 4 ton/ha	500 kg duck meat + 17 000–20 000 eggs
Fish + Poultry farming	4–5 ton/ha	60 000–70 000 eggs + 1 500– 2 000 kg meat

Under this system of farming small livestock and farm yard animals, viz. pigs, poultry, ducks, etc., are integrated with composite fish culture by siting animal housing units on the pond embankments in such a way that the animal wastes and washings are diverted into fish ponds for recycling. The fish not only utilize spilled animal feed but also directly feed on fresh animal excreta which is partially digested and is rich in nutrients. Surplus excreta supports the rich growth of planktonic fauna. Fertilizers and supplementary feed are not used, resulting in drastic cost reduction.

Integrated Fish-pig Farming

Pigstyes are constructed either on the pond embankment or near the pond to facilitate easy drainage of waste directly into the pond which acts as pond fertilizer and supports dense growth of natural fish food organisms. Besides, fish also feed directly on the pig excreta. No other feed or fertilizer is applied to the pond. A pond is prepared by following the usual pond preparation techniques and stocked with fingerlings of all the six species of carps cultured under composite fish culture at higher of 8 000–9 000/ha with surface, column, bottom feeders and grass carp in the ratio of 40:20:30:10. Marketable size fish are sold by partial harvesting while final harvesting is done only after 12 months of farming.

About 2 months-old weaned piglets are fattened for six months when they attain slaughter size (60–70 kg) and similarly a second crop is raised within the next six months. About 30–40 pigs should be kept for proper fertilization of the pond. Pigs are fed on mash at an average rate of 1 kg/day. Green grasses or animal fodder is also provided. Grass with interlocked soil in root system (sod) are provided once a week to avoid mineral deficiency. Grass carp is fed with aquatic weeds or green animal fodder. Fish yields ranging from 6 000–7 000 kg/ha/yr are generally obtained.

Integrated Fish-duck Farming

This is also an efficient integrated system based on the principle of waste recycling. Pond preparation technique is basically the same. A duck house is normally constructed on the pond embankment or on the pond water on a floating platform When given free range, ducks feed on aquatic organisms such as insect larvae, tadpoles, molluscs, weeds, etc. The duck droppings like pig excreta act as fertilizer.

Ponds are prepared and stocked with fingerlings of all the six carp species at 6 000 ha with surface, column, bottom feeder and grass carp in the ratio of 40:20:30:10. Fingerlings of over 10 cm are preferred for stocking. About 200–400 ducks are sufficient to adequately fertilize a 1 ha pond.

Normally 2–3 months old ducklings are preferred. Although ducks are able to feed upon natural food from the pond, they are also provided with duck feed at the rate of 100 g/bird/day. Ducks start laying after 5–6 months and continue for 2 years. Fish yields ranging from 3 000–5 000 kg/ha/yr are generally obtained.

Integrated Fish-Poultry Farming

Under this system of integration the poultry birds are raised in cages under a shed normally constructed over the pond embankments or in the vicinity of the pond. The space requirement in such a system of poultry raising is about 1 sq.ft. per bird. The droppings of the birds fall on the floor from where these are collected and applied to the pond. The chicken house can also be built directly over the pond water so that the excreta may fall in the pond water underneath.

Usually, 400–600 chickens/ha of pond water surface are used. No feed or fertilizer is applied in the pond, except aquatic vegetation for the grass carp. Fish production at the rate of 4–5 t/ha is possible using this system. In India, this system of freshwater fish culture has assumed greater significance in view of its potential role in recycling of organic wastes and in integrated rural develop.

AIR-BREATHING FISH CULTURE

Besides freshwater ponds, there are many low-lying areas which become waterlogged during the rainy season. In course of time these areas get infested with dense aquatic vegetation and turn into swamps. Swamps are also formed along the irrigation canals due to profuse seepage. These areas are best suited for culturing airbreathing fishes such as Koi Singhi, Magur and Murrels without getting involved in costly processes of their reclamation essentially needed for carp culture.

There are three levels of culture practices. low cost culture, semi-intensive culture and intensive culture depending on inputs and level of management Fingerlings (6–10 g) and feed are the two material inputs used in the culture system. Fertilizer is not used. However, replenishment of water becomes an essential input in case of intensive system of culture in ponds where very high stocking rate and intensive feeding is practised to obtain very high yields Monoculture as well as polyculture of these fishes are commonly undertaken, yielding production to the tune of 3 000 to 7 000 kg/ha/yr.

Production at the rate of over 3 000 kg/ha/8 months is possible through monoculture of Channa marulius in swampy ponds. In intensive system of monoculture of magur and singhi with frequent change of water, yields of over 15 t/ha/yr have been obtained. Presence of naturally occurring weeds in airbreathing fish culture ponds not only provide protection against poachers, but also encourage the growth of insects which are consumed by the fish.

SEWAGE-FED FISH CULTURE

The wastes, including sewage and waste water produced by the human community hold high potential for fish production. In India itself there are about 150 sewage-fed fish farms covering an area of about 12 000 ha. Very high production in the order of 7–10 tonnes/ha/yr has been obtained from ponds fed with sewage which invariably contains high percentage of N,P,Ca,K,etc.

An average production of about 7 t/ha/yr is easily obtained using a mix culture of 5 carp species The sewage fed ponds are generally dewatered completely during summer so as to remove all the carnivorous fishes.

The pond is initially fertilized by introducing partially treated sewage effluent upto about 75–80 cm and then clean water is pumped in to raise the pond water level to 1.5m. Within a month the pond stabilises with respect to dissolved oxygen and becomes suitable for stocking with fish seed. During raising of marketable size fish, additional fertilization with sewage effluent is carried out in small doses every month and the pond is netted frequently to help oxygenate the water and in course of netting the marketable size fish are also harvested.

FEEDING HABITS

Although Silver carp, Bighead, Grass carp, Black carp, Common carp and Mud carp, belong to Cyprinidae, they have formed their respective ways of food in-take and their respective food chains at different stages of the development, It is due to their long-term adaptations to ecological conditions, the restrictions of water body environment and the perfection of body structure as well as the development of foodintake and digestive organs along with their growth.

- The main natural food Silver carp and Bighead feed on are plankton. Silver carp mainly feed on zooplankton at larval stage, whereas their feeding turns to phytoplankton after grown-up. Bighead feed on zooplankton mainly all their life. The differentiation of feeding habits are caused by the structure and the density of their filtering organs------ gill rakers.

The gill rakers of Silver carp and Bighead are situated in the operculum with 4 pairs on either side, and the 5th pair of gill arch which was specialised into inferior pharyngobranchial. The gill arch bone is the base to which gill rakers and gill filaments attach.

The gill rakers of Bighead are delicate and sabre-shaped, each consisting of a neck, a stem and a base. The neck is narrow and short. The stem is the principal part of the gill raker. The dorsal part of it is thicker, while the ventral part thinner. On either side, there is a row of wartlike protuberance, so to speak, lateral protuberance. The lateral protuberance of a gill-raker is interwoven with the next one in serration. Occasionally there are opposite protuberances, their triangle-shaped bases cling to the gill arch bones. Besides, gill rakers can be divided into two kinds, broad and narrow ones. There exists one broad gill raker every 3---6 narrow gill rakers.

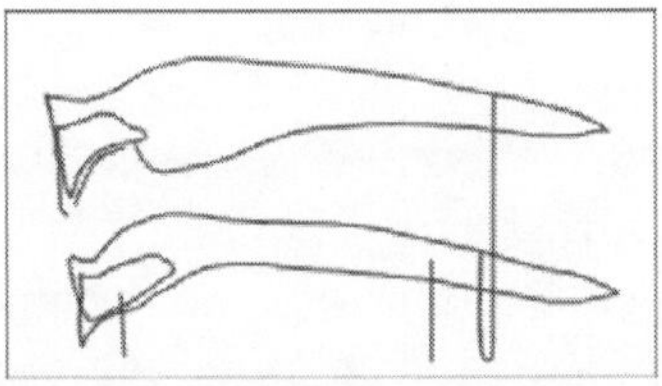

The gill rakers of Silver carp are different from that of Bighead. The gill rakers are inter-connected by minute bony bridges, which are covered with spongy sieve membranes. The density of gill rakers is higher than that of Bighead's gill raker.

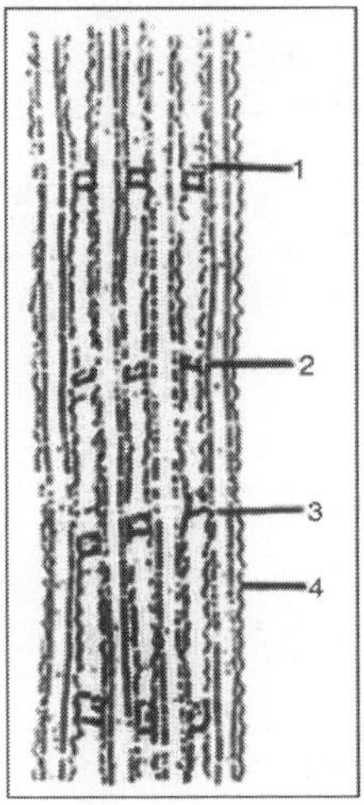

- Inner gill raker
- Bony bridge
- Gill raker
- Lateral protuber

The newly hatched Silver carp and Bighead fry nourish themselves with egg yolk and begin to take in plankton in the water when their body length reaches 7---9 *mm*, 3---4 days after hatching.

The food-filtering organs are imperfect with short and sparse gill rakers before the body length reaches 15 *mm*. At that time, both Silver carp and Bighead eat the same food. The major groups of zooplankton that Silver carp and Bighead eat are rotifera, nauplius of copepods and tiny cladocera.

The shape and structure of filtering organs of fry about 20–30 *mm* in body length, are generally the same as those of adult fish's. The fry of Silver carp will take in phytoplankton as its main foods, since, it has nearly 200 gill rakers, 1 mm in length each, with minute bony bridges between them Sieve membranes covering the bridges form a fine. The gill rakers of Bighead fry, short and sparse, are separated with larger spaces and difficult to detain tiny phytoplankton. As a result, their main food changes from tiny zooplankton to all sorts of zooplankton.

Beside the filtering organs------gill rakers, Silver carp and Bighead have accessary organs------palatine folds, that is, inter-gill raker folds. Palatine folds, located at the apex of mouth cavity, consist of 9 vertical ridges of mucus membrane, four on either side and one in between. The middle ridge is very short like an inverted letter *Y* with the branched end backwards. Palatine folds act in coordination with gill rakers, to get the food through filtering.Under culture conditions, Silver carp and Bighead can also feed on commercial feeds such as cakes, brans, dregs and wastes.

FEEDING HABITS OF GRASS CARP AND BLACK CARP

The fry of both fishes under 15 mm in body length have the same feeding habits, mainly feeding on zooplankton. But the feeding habits of the fry above 20–30 mm in body length will differentiate. The fry of Grass carp start to take in tender aquatic plants, while the fry of Black carp eat Benthos such as snails and corbicula spp., etc. Grass carp are typical herbivorous species taking in all sorts of aquatic and land grasses, hence, the pharyngeal teeth are well developed, tough and strong (tooth formula: 2,5/4.2). The teeth are shaped like choppers with saw-toothed edges. Pharyngeal teeth at both sides are interlaced with each other. They are against callous pad of the basioccipital, grinding food into pieces for digestion in intestines. Grass carp are voracious eaters, feeding amount usually could reach 40 per cent. But the plant cellulose could not be digested and utilised.

Black carp are carnivorous fish, and usually feed on molluscs like snails, clams and corbicula spp. By using pharyngeal teeth and callous pad, they crush hard shells, then swallow the meat. Their pharyngeal teeth are strong and tough and molar-shaped (tooth formula: 4/5).

Under culture conditions, both fishes tend to be more or less omnivorous, feeding on oil seed cakes, brans, dregs and animal feeds like silkworm pupae, earthworm, animal entrails.

FEEDING HABITS OF COMMON CARP AND CRUCIAN CARP

At the larval stage, feeding habits of the two are basically similar. They chiefly feed on rotifer, cladoceran, copepoda, chironomid larvae and other insects' larvae. Common carp and Crucian carp about 50 *mm* in body length tend to be omnivorous. Common carp are inclined to be more carnivorous, whereas Crucian carp more herbivorous. Both have feeding habits of phagotrophy. The pharyngeal teeth of Common carp are relatively developed, molarshaped in 3 rows, with transverse grooves on the rest of the inner row except the first tooth, which is smooth. Common carp display a wide adaptation to foods. Their common natural foods are benthos such as snails, young clams, corbicula spp. cladocera, copepod, chironomid larvae, shrimps and insects' larvae. Also they take certain amount of detritus of higher aquatic plants and plant seeds.The nasal bone of Common carp is well developed so that their premaxilla and mandible could be projected out like a tube to dig the mud for organic detritus.

Foods for Crucian carp are chiefly large amounts of detritus, diatom, filament algae, aquatic grasses and plant seeds, including a certain amount of cladocera, copepod, chironomid larvae and water earthworm, etc. Under culture conditions, Common carp and Crucian carp also like to take commercial feeds such as oil seed cakes, brans, crops and silkworm pupae, etc.

FEEDING HABITS OF CHINESE BREAM WUCHANG FISH

At larval stage, Chinese bream feed mainly on zooplankton such as cladocera and copepod, whereas, at adult stage, mainly on aquatic grass such as Vallisneria spiralis, Hydrilla verticillata, and Potamogeton malainus and secondarily on Potamogeton crispus, Myriophyllum spicatum, Spirogyra and plant detritus. With a mouth small in size, and the pharyngeal teeth and callous pad small and weak, the ability and intensity of food intake of Wuchang fish fail to come up with Grass carp.

FEEDING HABITS OF MUD CARP

With a small mouth, transverse in inferior position, under the natural conditions, mud carp use the bony edges of upper and lower jaws to scrape diatom, green algae chlamydomonas and filament algae, etc., attached to stones and something else.

Not seldom, they take detritus of higher plants, bottom humus and a little zooplankton as feeds. Under culture conditions, they like to eat commercial feeds such as seedcakes, dregs, brans, and animal manures.

FEEDING HABITS OF TILAPIA

Tilapia are omnivorous fish, with a tendancy to be herbivorous. Tilapia feed mainly on zooplankton at larval stage. The scope of food enlarges with the growth. The usual common foods are all kinds of planktonic, benthic and epiphytic algae, tender higher aquatic plants, all organic detritus and a little animal feeds such as earthworms, small shrimps and aquatic insects.Tilapia nilotica, possessing denser gill rakers, are more likely to feed on phytoplankton and could utilise the green algae Chlorophyta and blue algae Cyanophyceae which can not be digested by other fishes, but detritus occupies greater proportion among the natural feeds for Tilapia mossambica because they only have 14--- 19 gill rakers.

Under culture conditions they can utilise all kinds of vegetable leaves tender grass, animal manures brans, seed cake and pelleted feeds if applied. At a proper temperature, they ability and intensity of their food intake are much higher.

SELECTION OFSUITABLE FISH SPECIES

- Adaptability to undrainable pond environment
- Faster growth rate
- Efficient utilizers of natural food resources of the pond
- Efficient converter of artificial feed
- Hardy and not easily susceptible to disease
- Easy to breed and rear the seed
- Prolonged breeding period or multiple breeding frequency
- Non-predaceous, planktophagous and preferably herbivorous and detritus feeder
- Compatability with other cultivable species of fish
- Palatable with high nutritive value
- High market demand and high price.

To find all these qualities in one fish species would be very unlikely. Therefore, the species having maximum required traits are considered to be desirable for cultivation in undrainable ponds. Carps fit well to these criteria and hence the most widely cultivated food fishes in South Asia are the quick-growing, non-predatory carps.

There are two major systems of carp culture in Asia: the Chinese polyculture system where Chinese carps are cultured together and the Indian composite fish culture system where the Indian major carps and Chinese carps are combined.

CATLA

Catla is the fastest growing Indian major carp species and widely distributed throughout India, Nepal, Pakistan, Burma and Bangladesh. It inhabits the surface layer of water and feeds upon plankton. Adult stages are predominantly zooplankton feeder, occasionally taking in decaying macrovegetation, phytoplankton and smaller molluscs. It attains maturity in the second year of life and carry over 70 000 eggs per kg body weight.

It naturally breeds in rivers during monsoon season and under control conditions in bundhs as well. It does not breed in ponds.However, it responds well to hypophysation techniques. Seeds are easily reared in undrainable ponds of relatively smaller size. Under composite fish culture in ponds it usually grows to over 1 kg in one year.

Rohu

Rohu is the natural inhabitant of river systems of India, Nepal, Pakistan, Bangladesh and Burma. In recent years it has been transplanted to many countries of the world including Sri Lanka, Mauritius, USSR, Japan, Philippines, Laos, Malaysia and Thailand. Normally it occupies the column region of the aquatic ecosystem and feeds mostly on vegetable matter including higher plants, detritus, etc. Like catla it naturally breeds in rivers and under special conditions in bundhs.Except by hypophysation to which it responds quickly, it never breeds in ponds. It attains sexual maturity during the second year. However, certain percentages of pond-reared specimens mature within one year. Fecundity varies from 226 000 to about 2 800 000 depending upon the size. Rohu spawns during the monsoon.Seeds collected from rivers or produced by bundh breeding or induced breeding are reared with ease in seasonal or perennial undrainable ponds. Under pond culture conditions it grows upto 900 g within one year.

Fig. Rohu

Mrigal

Mrigal inhabits all the major river systems of India, Pakistan, Bangladesh and Burma. The adult fish feeds upon filamentous green algae, diatoms, pieces of higher plants, decayed vegetable, mud and detritus. It is basically a bottom feeder and hence suitable for cultivation with column and surface feeder carps in ponds. Mrigal usually attains maturity within 1 or 2 years depending upon the agroclimatic conditions of the location.

Fecundity ranges between 124 000 to over 1 900 000 depending upon size. Spawning season is linked with the onset and duration of the southwest monsoon. It does not breed in ponds, but can be easily bred in bundhs and by hypophysation. It is now being induced to breed twice within the same spawning season. Rearing of seed is usually undertaken in seasonal or perennial undrainable ponds. Under pond culture conditions it grows to over 1 kg in one year.

Silver Carp

Silver carp is basically inhabitant of major river systems of South and Central China and in the Amur Basin of USSR from where it has been transplanted throughout the Indo-Pacific region including India. It is a surface dweller feeding mainly upon zooplankton during its early stages and gradually becomes predominantly a phytoplankton feeder.

Its relatively longer branchiospines provide a fine filter capable of retaining planktonic organisms. It readily accepts supplementary feed like oil cakes and rice bran mixture in pond culture systems. It does not breed in pond condition. However, through the technique of hypophysation they are induced to breed in ponds during the monsoon season. Fecundity varies greatly with the size and agroclimatic condition. A fecundity range of 145 000 to 2 044 000 has been found from silver carp.

It takes about 2–6 years to mature in China, whereas in India it matures very early, within 2 years. Males mature earlier than the females. In composite fish culture ponds it usually attains over 1.5 kg within one year of rearing. Seed rearing is done in smaller seasonal or perennial undrainable ponds with a high rate of survival.

Grass Carp

Grass carp is a native of the river systems of South-Central and North China and the Amur river of USSR. Its suitability in aquaculture and biological control of aquatic weed infestation has resulted in wide-scale transplantation throughout the world.

In early life it feeds on planktonic organisms and gradually switches over to macrophytes. They are voracious eaters and show distinct preference for vegetable food materials such as grass, leaves, weeds, etc.

However, they also accept supplementary artificial feed materials. Usually only a portion of ingested food is digested and the rest is voided in semidigested or undigested form which, in turn, becomes choice food for the bottom dweller common carp In China it takes about 3–4 years to achieve maturity whereas in India it usually takes 2 years.

The total number of eggs range between 308 800 and 618 100 from the fishes weighing between 4.7 kg to 7.0 kg. The fish does not breed under pond condition and hence seed production is achieved through hypophysation. Growth mainly depends on the rate of feeding. Under optimum feeding rate it can grow over 5 kg in one year Usually it grows to over 1.5 kg in composite fish culture ponds.

Common Carp

Originally a native of temperate region of Asia, especially China, the common carp is now the most domesticated and cultivated carp species throughout the world. It is an omnivorous bottom dweller subsisting mainly on benthic fauna and decaying vegetable matter. It frequently burrows the pond bottom in search of food.

This habit of digging the pond bottom helps in maintaining the productivity of undrainable ponds and hence culture of common carp with other carp species is of great advantage. Moreover, it also feeds directly on the undigested excreta of grass carp.

Growth mainly depends upon the bottom biota, stocking density and the rate of supplementary feed.In composite fish culture ponds it grows to about 1 kg within one year. In a tropical climate it spawns throughout the year in the pond environment with two peak periods, one from January to March and the other during July and August. Eggs are small and adhesive in nature. In tropical conditions it attains maturity within 12 months.

POND ECOSYSTEM

Like agriculture, fish culture is also based on a series of processes involving reception and transformation of solar energy. In the pond ecosystem solar energy is utilized for primary production by chlorophyll-bearing plants such as planktonic algae and macrophytes. This conversion of solar energy into chemical energy (food) is guided by the photosynthetic and chemosynthetic activities going on in the aquatic plant community and the rate at which this is carried out is called primary productivity of that ecosystem.A part of the primary production is cycled through different trophic levels resulting in fish production.

Here comes the community of consumers that comprise microscopic as well as large animals, which are unable to synthesize their own food and feed upon primary producers. Different forms of pond life are linked together through predator-prey relationship.

This chain of food production, which follows a general pattern, primary producers-herbivores-carnivores-appears too simple and straight. But, in fact, it is a complex food web with various cross linkages.Fish populations may be classified into several trophic levels, depending upon their position in this food chain. Phytophagous fish such as grass carp and silver carp belong to the second trophic level as they feed upon the first trophic level organisms. Likewise, zooplankton, feeding upon phytoplankton, also belong to the same category.

Carnivorous fish communities thriving upon zooplankton or herbivorous fishes occupy the third trophic level while other predatory fishes preying upon carnivorous fishes belong to the fourth trophic level A relatively simple food chain operates in fish ponds, but a complex one occurs in lakes and other larger aquatic ecosystems.

The picture becomes even more complicated in large water bodies such as rivers and seas where complex food chains are referred to as food webs which in fact represent several interconnected food chains. There are some fishes which occupy mixed positions, between different trophic levels. They consume both plants and animals and as such, cannot be naturally categorised into any one particular trophic level.

A properly managed pond presents an example of a simple food chain under simple conditions. Here the number of food chains is reduced by encouraging the growth of phytoplankton. The macrophytes such as rooted green plants, floating plants, etc., are not allowed to grow. Phytoplankton is consumed by the zooplankton in the water column, whereas its detritus is utilized by benthic invertebrates.

Phytoplankton, zooplankton, detritus and benthic organisms serve as food for the stocked fishes such as the desired carp species. Thus, as much of the available solar energy as possible is utilized for fish production by proper pond management.

Primary productivity is dependent on light, carbon dioxide, temperature and essential nutrients, each of which can be a limiting factor.

Of these factors affecting primary production in ponds, the one that can be manipulated easily is the quantity of nutrient elements through the application of nitrogenous, phosphatic and potassic fertilizers, as in agriculture. In ponds, only the top 2 to 5 cm of soil is concerned with nutriention exchange and the soil below plays a negligible role in the production cycle.Undrainable ponds receive dissolved nutrients and sedimentary particles carried by rain water from the catchment area.

Besides, production and decomposition of minute plant and animal organisms in ponds also modify the properties of the pond bottom to a great extent. The nature and quantity of fertilizers determines the species composition to be used in a culture system. At low phosphate concentration, diatoms are common, but with increasing concentrations green algae become more frequent, eventually giving way to blue-green algae. In addition, excessive phosphate gives rise to phytoplankton blooms which check the light penetration and thus lower the pond productivity through 'autoshading'.

Light energy is one of the major inputs in primary production and hence the success of fish culture depends largely on the efficient utilization of incident light. When incident light strikes the water surface, it is partially reflected and partially transmitted into the water where part of it is utilized in the process of photosynthesis and the rest is scattered or absorbed by suspended particles. In turbid waters, more light is scattered or absorbed, thus allowing the light penetration only to shallow depths.

The rapid disappearance of light in such waters affects adversely the growth of diatoms. The bottom layer of water, being devoid of photosynthetic plants and also being in close contact with the decaying organic matter, suffers from oxygen depletion causing critical stress conditions for the fish.

Thus, it is important that primary producers must provide oxygen to support the total biological respiration during darkness and also during the less favourable (warmer, overcast or rainy) days apart from providing food for the second and third trophic-level fish.

Table. Solubility of oxygen under different temperatures at 760 mm of Hg pressure

Temperature (°C)	*Solubility of oxygen (mg/l)*	*Temperature (°C)*	*Solubility of oxygen (mg/l)*
15	9.76	26	7.99
16	9.56	27	7.86
17	9.37	28	7.75
18	9.18	29	7.64
19	9.01	30	7.53
20	8.84	31	7.42
21	8.68	32	7.32
22	8.53	33	7.22
23	8.38	34	7.13
24	8.25	35	7.04
25	8.11		

The concentration of dissolved oxygen in the water, which depends on the temperature, is an essential component of the aquatic environment to govern the carrying capacity of a pond. Variations in concentration of dissolved oxygen may occur due to the following three important factors:

- The saturation level of oxygen in water decreases as the temperature rises;
- Supersaturation is an unstable state and
- Plants not only photosynthesize to produce oxygen, they also respire and consume oxygen.

The saturation value for dissolved oxygen available for fish life at 20°C water temperature is more than that at 30°C at a particular atmospheric pressure Dissolved oxygen (Do) concentration is always high at lower temperatures and gradually decreases with increase in temperature.In natural waters, including undrainable fish ponds, DO values are constantly changing because of biological, physical and chemical processes.

The air above the pond water surface may be considered to have a more or less constant percentage of oxygen. However, the partial pressure of oxygen in the air may vary slightly at a given location because of differences in atmospheric pressure. Transfer of oxygen from air to water will occur when water is undersaturated with DO and oxygen will diffuse from water to air when water is supersaturated with oxygen. However, the diffusion of oxygen into the pond water is very slow, except under conditions of strong turbulence, hence the most important source of oxygen is that generated during photosynthesis.Sometimes, phytoplankton blooms or algal scums limit light penetration causing reduction in photosynthetic rates, even in waters with adequate nutrient concentrations. Oxygen production by phytoplankton is greatest near the surface and decreases with the increase in depth because of self-shading. When heavy infestation of aquatic weeds and dense bloom of plankton occur, the situation becomes much more complex. On the other hand, these are additional sources of oxygen at daytime; but on the other hand, they also respire and consume oxygen throughout day and night. At times the pondwater is supersaturated with oxygen during the day, which is a highly unstable state, while during the night, a greater proportion of oxygen is used up for their respiration, thereby reducing the availability of oxygen to fish. Thus, it creates a wide fluctuation in the level of dissolved oxygen, adversely affecting fish life.

In fact, under such situations oxygen production increases to its maximum during the daytime leaving surplus for the fish even after consuming for their own respiration, but at night this surplus level drops down to critical level.Under conditions of heavy algal blooms and weed infestation, the phytoplankton and aquatic weeds actually consume more available oxygen during day and night than they produce during the whole day During cloudy days, when the incident light is inadequate for phytosynthesis, the situation in terms of availability of DO becomes worse.Aerobic decomposition of organic matter by bacteria is also an important drain on the oxygen supply in ponds.

Aerobic decomposition requires a continuous supply of oxygen and proceeds more rapidly when DO concentrations are near saturation. However, decomposition also occurs under anaerobic conditions, but the rate of degradation of organic matter is not as rapid and complete as under aerobic conditions. Under aerobic condition, the end product of decomposition is primarily carbon dioxide.

At times high rate of bacterial decomposition of dead organisms and other organic bottom deposits lead to a condition favouring the increase of the level of carbon dioxide and other abnoxious gases, with a simultaneous depletion of DO, resulting in fish kills and planktonic collapses Therefore, it is important that the pond water should provide adequate oxygen to support the total biological respiration during the hours of darkness.

EMPLOYMENT POTENTIAL

Aquaculture is also considered to be a potential source of employment for poor farmers and displaced capture fishermen. Rapid development of aquaculture has already generated considerable employment through culture of marketable fish, fish seed production and marketing of fish and fish seed.The National Agriculture Commission of India while estimating the employment potential of fish culture has indicated that every tonne of fish produced provides employment to 2.5 persons.

The other important advantages of fish farming are that the production is carried out within easy reach of consumers and also the harvesting can be adjusted to demand, thus minimising distribution problems and spoilage.Rural ponds in Asia, hitherto producing at subsistence level, have succeeded in increasing production per unit area through improved culture practices involving higher stocking densities, polyculture combinations, pond manuring and feeding.

The switch over from monoculture practice to polyculture has significantly contributed toward higher production and the prospect of polyculture appears very bright as the fish seed of desired species is becoming easily available due to the establishment of a large number of hatcheries.During recent years, advances have also been made in traditional aquaculture systems practised in rural India by the development of composite fish culture, a system of polyculture of a group of complementary and supplementary freshwater species of fast growing carps in undrainable ponds.

With the successful demonstration in different agroclimatic zones of India, gradual improvements in technology have been made and it is now possible to obtain a production rate of over 10 t/ha/yr in experimental ponds and up to about 5 t/ha/yr in farmers' ponds against the traditional average rate of production of 600 kg/ha/yr.

To meet the increasing demand of seed of culturable carps, hypophysation techniques have also been developed for both Indian and Chinese carps and as a result they are now being bred in captivity even by fish farmers in remote villages. The emergence of this culture technology suitable for undrainable ponds and the simultaneous development of hypophysation techniques for fish seed production has completely revolutionized fish farm productivity. This manual intends to provide the basic concept and practical guidelines of fish culture in undrainable ponds. Since it is prepared especially for extension agents and field workers, certain important things have been repeated and at times experimental results have been simplified with a view to making it more practical, simple and illustrative. It outlines the practices of procurement and propagation of fish seed, rearing of spawn to fry and fingerling stage and production of table-size fish following simple sequential steps. Like other farming systems this culture system is also prone to certain unexpected hazards for which one has to be prepared and properly equipped.Such hazards are disease outbreaks, oxygen depletion, pollution, flood, drought, poaching, etc.

The content of this manual is a synthesis of the author's personal field experience, the information gathered from published literature and the observations of other workers in India.Based upon Indian experience, water resources in other developing countries with similar agroclimatic conditions may be utilized for the development of fish culture.It is hoped that this manual will serve as a practical guide to extension workers in popularizing freshwater fish culture in undrainable ponds.

POND FERTILITY

Organic matter and mineral constituents of the pond soil supply the required nutrients for chemical and biochemical production processes. The pond bottom also provides a suitable environment for the decomposers like bacteria and fungi to mineralise organic components of the pond sediment and release soluble nutrients. Sometimes such nutrients are not available in sufficient quantity in the pond and hence they are added from outside in the form of fertilizers.

Since plankton production is often limited by inadequate quantity of phosphorus which is essential for the assimilation of nitrogen into cellular matter, phosphatic fertilizers are widely used in fish culture. Unlike phosphorus, availability of nitrogen does not depend on the inherent status of soil, since it is brought to the soil by different processes. Nitrogen fixation by azotobacteria, blue-green algae, atmospheric electric discharges and photochemical fixation are some of the potential sources of pond nitrogen. In a tropical climate, the fixation of atmospheric nitrogen by blue-green algae is of considerable importance.

However, when nitrogen is added from outside, its form, ammoniacal or nitrate is also very important. It is advisable to use the ammoniacal form of nitrogen in acid and neutral soils and the nitrate form in alkaline soil.Though there is a considerable loss of nitrogen from the ammoniacal form in alkaline soil, the use of ammonium sulphate in low doses is usually recommended, keeping in view the role of sulphate in reducing the soil alkalinity. Potassium is the other essential nutrient for plant growth. In ponds it is easily available both in soil and water and does not form insoluble salts and is rarely deficient except in acid peaty soil. Yet, a little potassium when added to the pond, stimulates the production of plankton.Organic matter of the pond sediment is also an essential factor regulating the bacterial activity. In this context, the ratio of organic carbon to total nitrogen (C/N ratio) is important. Periodic application of organic manures ensures to a certain extent replenishment of nutrients and also provides an energy base for bacterial activities. Apart from this, the organic matter and the bacterial flora are also directly consumed by zooplankton and some fish species.

DISEASES AND THEIR CONTROL

Various intensification approaches such as increased stocking rates, increased feeding, fertilization programmes, etc., sometimes result in nutrient accumulation, frequent appearance of algal blooms, dissolved oxygen deficiency and other water quality problems in undrainable ponds. As a result of such water quality and environmental problems, the infectious diseases and their control assume importance.A fish farming system is unique in that the farmed animal is poikilothermic and lives in water where respiratory oxygen level compared to air is limited and becomes critical at times.

Further, metabolic waste products, left-out feed materials and organic load of the pond bottom can affect certain exposed vital organs and tissues of fish.

All such factors affect fish health and contribute to the risk of disease outbreaks.The above basic facts need careful consideration while planning for freshwater pond fish culture. The habitat of an undrainable pond is very varied and dynamic, but can be monitored and managed for increasing fish production.

LIVING SPACE

It has been observed that under identical conditions of management levels and stocking density fish grow bigger in larger ponds. In Malaysia, grass carp, Puntius sp. and monosex Tilapia mossambica grew bigger in ponds of a larger area indicating the living space phenomenon In other words, the rate of production in a 0.2 ha pond will be more than double that of a 0.1 ha pond, despite the fact that the stocking rate per unit area is the same and all other management components including the genotype of the stocking materials and ecological conditions remain the same.

The ponds having larger surface area are subjected more often to wind action resulting in greater rate of diffusion of atmospheric oxygen into the water.Larger ponds have other advantages also, viz. better cooling action by wind. In smaller ponds, water tends to stagnate and in hot weather tends to heat up quickly. Though it is preferable to have ponds of a large size, there is a physical limitation. Large ponds are difficult to fill and even more difficult to harvest. There must be an optimum size and shape of the pond to balance size with practicability of management, i.e. large enough to allow proper growth of fish, but at the same time small enough to be manageable.

Supplementary Feeding

With the increase in carrying capacity of the pond either by aeration or circulation of water, fish growth can be increased further by supplementing the natural food with some artificial feed. This is the single most important management component for increasing production. In intensive and semi-intensive culture of fishes, supplementary feeding is indispensable. The quantity of feed and the form in which it is offered affect the rate of consumption. Temperature, dissolved oxygen level, crowding and health condition, etc., affect the rate of food consumption.

RENOVATION OF EXISTING PONDS

The majority of freshwater fish ponds in the Indian subcontinent are the dugout ponds of an undrainable nature which at times lack proper embankments.During the course of culture operations, such ponds receive huge amounts of feed, fertilizers and manures as critical inputs and sediment particles carried down by rain water from the catchment area.

A portion of the organic production in the pond also undergoes death and decay and gradually adds to the pond bottom sediment. Thus, with the advancement of time, a thick sediment layer is formed reducing the depth of the pond.They are quite rich in organic and inorganic nutrients, but due to slow bacterial action under prevailing anaerobic conditions the nutrients are almost locked up in the sediment and are not available for primary production. Further, the anaerobic decomposition of the organic matter accumulated in the sediment releases harmful gases and depletes dissolved oxygen level in the water.

Thus, it becomes necessary to renovate the existing ponds periodically every 4–6 years by removing sediment from the pond bottom, redressing and repairing the dykes, etc., in order to make the ponds more suitable and to regain their fertility. For this, the following practical measures are recommended.As soon as the water table of the area surrounding the pond goes down the renovation work can be initiated.

Summer is the most suitable period for this Purpose as complete drying of the water body is possible. In this period pond renovation can be carried out efficiently and economically. Removal of slushy silt from the partially dried pond bottom is difficult, laborious and expensive.

DEWEEDING

It has been observed that most of the rural ponds are not properly managed and become weed-infested in course of time. Before dewatering the pond, large floating weeds such as water hyacinth should be eradicated by pulling them out manually or mechanically. Otherwise, collection and removal of such weeds will require more labour and time. Other rooted emergent or submerged weeds can be taken care of only after draining the pond.

DEWATERING AND DRYING

Dewatering of the existing pond is possible either by draining the water after cutting a portion of the embankment or by pumping out. If the water table in the surrounding area is high, there is considerable inflow of water from the pond bottom. This phenomenon of sub-surface secretion is called percolation.In case the rate of percolation is high, several furrows or ditches may be made towards the lowest contour point where a pit may be dug out to drain all the percolated wall.

Periodical pumping of water from the pit facilitates keeping the bed dry.Where the bed is found to be uneven, contouring is necessary to estimate the amount of silt to be removed. It is done by taking the level measurements at certain spots on the pond bed. It will also help in redesigning the pond taking into consideration the highest flood level and maximum rain water level.

DESILTING

After complete dewatering, the pond bed is allowed to dry and develop cracks in the silt mass. The texture of silt is different from that of the bottom hard soil and cracks quickly. At this stage dried silt is cut and removed manually or mechanically and heaped at a suitable place for its utilization in agricultural fields. Where complete drying is not possible due to high rate of percolation, walking platforms made up of bamboos or wooden planks may be put on the slushy bed to facilitate desilting work.In some larger ponds it becomes difficult to dry the central portion of the pond bottom as it is nearer to underground water table. In such cases the slushy and loose silt should be scrapped and spread to the sides with the help of wooden planks tied with ropes for pulling. This helps in drying the silt and easy removal thereafter.

RECLAMATION OF WATER BODIES

Derelict waters in millions of hectares, lying unutilized, are common sights in most of the South Asian countries. Such untapped water bodies with potential for aquacultural production may be reclaimed and made suitable for fish culture by adopting more or less similar procedures.In case of larger water areas, it would be better if they are connected temporarily to nearby natural or man-made drainage systems having relatively lower bed level for complete dewatering by gravity and making the entire area completely dry.

However, if such topographic facilities are not available, heavy duty water pumps may be put into use for quicker dewatering.In extensively large areas dewatering by draining or by pumping is not feasible. Moreover, the dry period of the year also may not last long enough to permit the work to be completed. It has been experienced that such areas can also be successfully reclaimed and renovated by partitioning into smaller units by raising cross bundhs, farm roads, etc. Each newly formed unit then can be dewatered, dried and desilted.

MAINTENANCE OF DYKES

In general, rural ponds lack proper embankments. During high rainfall or peak irrigation periods in canal-irrigated areas such ponds get inundated with water from the neighbouring agricultural fields causing stocked fish to escape, predators and unwanted species to enter and at times results in mass fish kills due to pesticide pollution. Hence, provision of proper dykes is a must.

Existing pond dykes should be repaired every year after the monsoon.Rats and crabs cause great harm to pond dykes by making holes. Such holes allow serious leakage and if not checked immediately, may endanger the stability of the dykes. Periodically and especially at the time of renovation, such spots should be properly repaired by stuffing binding clay, claylime mixture or any other locally cheap cementing material.

Due to poor consolidation, erosion from the top of the dyke during heavy rains usually results in grooving out of small channels. These areas should be covered with earth, levelled, thoroughly rammed and grass turfed. In relatively larger ponds, wave action due to wind also causes large-scale dyke erosion.

By putting large floating aquatic plants such as water hyacinth along the sides of the dykes exposed to wave action during the windy season such erosion can be checked. Frequent erosion in steep dykes during heavy rain or wind can be avoided by strengthening the inner sides of the dykes with poles or bamboos or corrugated cement planks.

Most of the traditional pond dykes are below the required height; as a result, overflow of water occurs during heavy rains or flood. These dykes should be properly raised and the height may be kept at a minimum of one meter above the maximum water level recorded in that area. While raising the dykes, the top width may be kept at a minimum of 1.5 m with 2:1 slope (horizontal: vertical).

Cutting the dyke to allow water into the pond from the surrounding area without any secured screening is a normal practice, which however creates many management problems. The silt mass is very rich in organic and inorganic nutrients making it most suitable for application in agriculture and horticulture. Being non-cohesive and unstable, it is unsuitable for making dykes as it may be washed back in the pond.

CONSTRUCTION OF NEW PONDS AND FARMS

Village ponds, homestead or backyard kitchen ponds, garden or farm ponds, irrigation ponds and occasional ponds such as brick mine pits and quarries, etc., occupy enormous freshwater areas in the tropics and are used for fish culture with minor improvements. However, ponds designed and constructed for fish culture are easier to manage and are expected to give higher production.

Although certain well-defined guidelines do exist for the construction of fish ponds, it is mainly the topography of the site which determines the basic design of the pond/farm. There are, however, certain basic principles to be considered when choosing a site and deciding the method of pond construction.

SITE SELECTION

Selection of suitable sites for fish farm construction is very important. The following three essential conditions guide the proper site selection:

- Topography
- Source of water and its quality
- Soil type

Topography

It is economical and convenient to construct ponds in waterlogged areas, irrigation command areas or in marginal lands. In such areas construction cost is relatively low mainly due to limited earth cutting.For example, a pond of 100 m × 40 m (0.4 ha) of water area requires only 3 234 m^3 of earth to construct around a dyke of 2 m high above ground level (GL) with side slope ratio of 2:1 and top width of 1.5 m.

This quantity of earth may be obtained only from 1.1 m depth of cutting. This limited depth of cutting reduces the construction cost considerably. However, full consideration should also be given to the possible effects of flood.

The surface features of the area proposed for the pond or the farm is also equally important. A saucer-shaped area may be an ideal site for a large dug-out pond, because it may hold appreciable quantity of water with a small amount of earthwork.For smaller and flat areas eye estimation is enough, but for a big area proposed for farm construction with a number of ponds for different purposes and of different sizes, it is essential to conduct contour survey for determining the topography and land configuration. The site should be easily approachable so that there may not be any difficulty in the transportation of input materials and in the marketing of the produce. The labour and materials required for construction and operation should also be locally available as far as possible.From an efficient management point of view the pond site should, if possible, be within the sight of the farmer's house. It also reduces the risk of poaching. Siting fish ponds near the farmer's other agricultural or livestock farming activities makes it easier to integrate all the farming activities.

Dyke

The dyke should be properly designed so that it can hold maximum water in the pond and withstand the hydraulic pressure. The slope of the dyke usually depends on the type of soil. Suitable side slopes for different soil 7.

Table. Suitable Slopes for Different Soils

Soil type	*Soil (horizontal:vertical)*
Clay	1:1 to 2:1
Clay loam	1.5:1 to 2:1
Sandy loam	2:1 to 2.5:1
Sandy	3:1

Provision for a berm of sufficient width may also be provided for stabilizing the slopes. A wider berm also helps in operating the net in the pond. The berm should be 1 m or more in width. The top width of the dyke should be decided taking into account its usage. Usually the minimum top width of the dyke should be 1.5 m.

The wider crest requires not only a larger area for dykes, but also an increased amount of earth material involving heavy expenditure. It is always wise to design the dyke as per the quantity of earth expected to be available from excavation work.

A soil-type containing approximately 25 per cent silt, 35 per cent sand and 40 per cent clay is most suitable for dykes. However, if excavated soil quality is not up to the above standard, provision may be made for a clay core to make the dyke watertight. While designing, about 10–12 per cent allowance may be given for settling of earthwork.

Source of Water

A dependable source of water supply must be available within or near the site, even for undrainable ponds. However, unlike drainable ponds, undrainable ponds require just sufficient water to fill the ponds and to compensate the water loss through seepage and surface evaporation thereafter.

Equally important is the need for avoiding excess water and hence there must be arrangement for the excess water to escape through a bypass channel or a spillway.The water supply to the pond should as far as possible be natural, preferably rain water.

However, alternative arrangements of water supply should be made for dry season either from a deep tube well or irrigation canal or from perennial sources like spring, stream, river, etc. Ponds should be on the lower lands to allow accumulation of surface runoff from a larger catchment area.

However, care should be taken to provide proper bypass or spillway to avoid flooding. A higher subsoil water table due to irrigation in surrounding fields and percolation from artificial or natural channels, in addition to absorption from rain water, also helps in maintaining water level in undrainable ponds.

The quality of the available water is also equally important for fish culture.Pond fish production is influenced by the physical and chemical properties of the water. Water should be clear as far as possible. Turbid waters which carry suspended solids cut the light penetration, thus reducing primary productivity of the pond.

Excess of suspended solids also adhere closely to the gill filaments and cause breathing problems.Water temperature also significantly influences the feeding and growth of fish. Prevailing water temperature, ranging between 15°C and 35°C in tropical areas, is most suitable for carps. The chemical quality of water depends on its content of dissolved salts. Rain water does not carry any dissolved salts.

However, it collects nutrient salts from the ground surface of the catchment area. The water should be neither too acid nor too alkaline; neutral or slightly alkaline waters are most suitable for fish culture and hence acid water should be limed to make it neutral.Waters with p^H values below 5.5 or over 8.5 are not proper for fish culture. The farmer will need huge quantity of lime to neutralize it while highly alkaline water may cause the precipitation of both phosphate and iron and if it remains continuously above pH9, it may be harmful to fish.

Soil Type

Pond soil must retain water. Soils with a low infiltration rate are most suitable for fish pond. The filtration rate of different types of soils. The best soils for our purpose are thus the impermeable clay which can be easily compacted and made leak proof.

Table. Infiltration Rates of Different Types of Soil

Soil type	*Infiltration rate (mm/ha)*
Clay	1–5
Clay loam	5–10
Silty loam	10–20
Sandy loam	20–30
Sand	30–100

Loamy soils can also be used, but they need well compacting and may leak slightly in the early stages, although they tend to seal themselves with time. Sandy and gravelly soils should be avoided, but if they are the only ones available they must be made impermeable with a thick coating of clay or with polythene sheeting. Soil impermeability can also be achieved by soil compaction at the pond bottom and dyke with either a mixture of soil + 1–5 per cent cement or soil + 10–20 per cent cowdung. Treated areas should be kept moist for 2–3 days by gently sprinkling water to avoid cracking and finally the pond is filled with water. Peat soils have special problems, since they are usually very acidic in nature and need sufficient liming, while the organic matter decomposition may lead to dissolved oxygen deficiency. Soils rich in limestone also create special problems, since the excessive lime content tends to precipitate phosphate and iron. Such ponds would then have little plankton population and macrophytes and would be relatively sterile. This can be overcome by adding sufficient organic matter such as cowdung, poultry manure, etc.A general and convenient field test for the soil quality is to take a handful of moist soil from the test holes made at the proposed site and to compress it into a firm ball.

If the ball does not crumble after a little handling, it indicates that it contains sufficient clay for the purpose of pond construction. Accurate determination of the composition of the soil and its water-holding character is possible by hydrometer method.

Several test holes may be made across the site and soil samples may be collected vertically from every 0.5 m of depth reaching up to a level of 3–4 m in a test hole. Using the results of the soil tests, a soil profile chart for the proposed site may be drawn. An arbitrary soil profile chart is presented showing the presence of clayey soil up to a depth of 3.5 m.

DESIGNING

Based upon the survey on topography, soil type, water supply, etc., the detailed designing and layout of the ponds/farm are done. However, the following additional points are also to be considered.The production or stocking ponds are stocked with large size fingerlings of about 10–15 cm size in the case of composite fish culture.

To attain this size, the hatchlings are reared in much smaller and shallower ponds called nursery and rearing ponds for about 2–3 months. In the nursery ponds the hatchlings are reared up to fry stage and in the rearing ponds the fry are reared till fingerling stage. The ratio of water area among nursery, rearing and stocking ponds in a fish farm depend upon the basic objective of the farm. In case of a fish seed farm, only nursery and rearing ponds are to be constructed with a small area for few stocking ponds to be used for raising the brood fish, while in the case of fish production farm only stocking ponds are to be constructed for producing table size fish from fingerlings. There is no hard and fast rule regarding the size of a pond.

However, nursery ponds should be small and shallow. Ponds having 0.02–0.06 ha water area and 1–1.5 m depth are most suitable as nurseries. Rearing ponds are relatively larger, preferably between 0.06 to 0.10 ha in size and 1.5 to 2.0 m in depth. The sizes of stocking ponds vary tremendously. For newly constructed undrainable ponds, total water area of 0.25 to 1.0 ha is recommended.

In shallow ponds the water becomes heated easily. In deeper ponds light cannot reach the bottom. In very deep ponds thermal stratification may occur with colder deoxygenated bottom layer. Dead plankton and faecal matter from fishes may fall on the bottom layer where the nutrients may be locked up.

Table. Practical size and depth of nursery, rearing and stocking ponds

		Depth (m)*	
Pond type	*Size (ha)*	*Irrigated command/ water logged areas*	*Rainfed± /non -irrigatedareas*
Nursery pond	0.02 – 0.06	1.0 – 1.5	1.5 – 2.0
Rearing pond	0.06 – 0.10	1.5 – 2.0	2.0 – 2.5
Stocking pond	0.25 – 1.0	2.0 – 2.5	2.5 – 3.5

* Excluding the freeboard

\+ May vary depending on impermeable strata at pond bottom

However, in case of rain-fed areas where the water table goes down during the dry season, the depth should be kept around 3.0 – 3.5 m to store more water during the rainy season.Although a square pond is economical to construct for its minimum length of dyke, a rectangular shape of the pond (length:width in proportion of 3:1) is considered to be ideal.

In any case the pond width should not exceed 30 to 40 m as it is difficult to operate a fishing net in broader ponds. The nursery and rearing ponds may be square, since they are too small to pose any problem for netting. The corners must be curved to avoid fish escaping the net during harvesting.

CONSTRUCTION

Before initiating the construction work, proper estimates have to be prepared based upon the design details, which will include the cost of all the materials and the labour. Strict supervision is required at every step of construction to ensure the adherence to specifications laid down in the design.

Time of Construction

If the construction work is taken up at the most appropriate time or season of the year, the work becomes easier and economical. The best time of the year for constructing ponds in clayey soil is post-rainy period and winter when the soil is soft rather than at the end of the dry season when it is very hard.

For swampy and waterlogged areas the most desirable time is the late summer when the area becomes completely dry. However, if a pond is built during winter or early summer and is not filled immediately, weeds may grow and cover the bottom. In such cases deweeding is needed before filling the pond.

Preparation of Site

The site should be thoroughly cleared of all the trees, bushes, etc. Even the roots of trees should be removed. No woody material should be left because the same will eventually rot and cause leaks. Some tree trunks rot very slowly and may cause problems during netting.

Marking the Outlines

This operation involves laying out the features of ponds on the ground in order to mark out the areas from where the earth will have to be cut and removed and also where earth will have to be embanked. Initially, lines are drawn according to the layout, followed by pegging and fixing stakes or posts. Strings are stretched between the tops of pegs and posts to mark the complete profile of the dyke with its correct height, width and slopes

Pre-excavation Work

Prior to pond excavation and dyke construction, all loose surface soil should be removed from about 20 cm depth within the total outlined area of the dyke and the surface should be roughened by ploughing or digging. In order to unite the body of the dyke to subsoil, it is desirable to dig a small "V" shaped key trench.

When the dyke is to be made on a sandy, gravelly or marshy soil base, the construction of a key trench becomes essential and in such cases digging should be done until watertight foundations are reached.

The key trench is a small ditch or furrow dug along the line of the centre of the walls about 0.5 m – 1.0 m wide and 0.5 m deep. This trench is filled in with a good clayey soil and is well rammed. If good clayey soil is not available in the area, ordinary soil should be well compacted into the trench. The purpose of the trench is to stop seepage of water underneath the walls.

Pond Excavation and Construction of Dykes

The excavation work can be carried out within the area marked for the pond bottom either manually or mechanically. However, the final levelling of the pond bottom and sides should be done manually with proper ramming and finishing as per the original design. The construction of the pond becomes economical if earthen dykes are made around the pond using the excavated earth from the pond bed. All dykes should be raised, dumping the earth layer by layer stretching right across the whole section and in such cases each layer should not exceed 20 cm in thickness. All large clods should be broken and each layer should be thoroughly consolidated by watering and ramming. The sides and top of the dykes should be properly dressed and finished with wooden thappies (wooden block with handle for ramming). In case the soil quality is not suitable for making dykes, a clay core is provided in the dyke to make it watertight. A mixture of 1:2 of sand and clay is used to make the clay puddle. This should be consolidated, compacted and deposited in 10–15 cm thick layers. Each layer should be adequately moistened before the next layer is laid and precaution should be taken to prevent the puddle from becoming dry and cracking. Dykes must be well compacted to render them stable and the top should be rammed flat so that small vehicles can also run along when needed. Short creeping grass is recommended to be grown on the top and sides of the dyke. Trees are not desirable since their dense shade inhibits the productivity of the pond.

Water Inlet Structure

Since we are concerned here with static and undrainable ponds, a feeder stream running directly into the pond should be avoided. The feeder stream must therefore be diverted along the side of the pond and from a suitable point water is channeled to the pond when required. An inlet structure should be provided through which water can be let into the pond. A proper inlet enables the quantity of water flowing into the pond, to be regulated, preventing the entry of undesirable fish and other aquatic animals and the escape of stocked fish. For small ponds the best inlet structure is a galvanized iron pipe of about 10 cm diameter with a control tap and a screen basket The downstream end of the pipe should be 30–40 cm above the water level. A sluice is also suitable for this purpose, especially for larger ponds. A screen is also fixed to check the entry of undesirable fishes and other animals To avoid scouring when the pond is being filled, a concrete apron can be built at the sluice, or more cheaply, a layer of gravel laid down. Similarly, if water is let in with a pipe there should be a gravel bed laid down where the water stream falls into the pond. If gravity feed is not possible, water must be pumped from the supply source into the channel leading to the pond or even directly into the pond; but, in that case, the intake should be securely wrapped by a firm net to prevent undesirable fish and other animals from entering into the pond along with the water.

MAINTENANCE

Proper maintenance of the pond and pond structure is most essential. Most of the earthen structures, especially the dykes, are susceptible to weathering action and hence they need periodical checks. Attending to minor damages regularly avoids the chances of more costly repairs later.

The grass turfing needs special attention. Proper and timely mowing prevents the formation of weedy growth and tends to develop a root system more resistant to runoff. Erosion from the top during heavy rains causes grooving out of small channels and it is an indication that the top has not been properly consolidated.The area should be levelled with more soil and thoroughly rammed and then grass should be planted to bind it.

Side erosion at the dyke bottom may be due to a number of reasons. The worst damage is done by common carp.Erosion due to frequent wave action, particularly if the grass at the edge has been grazed by grass carp, can cause undercutting of banks and subsequent collapse of dykes. Some methods used to provide protection against such erosion are earth berms, stone or brick pitching, stakes/bamboo piling.

Surface washings and organic additions cause siltation which reduces the pond depth and pond fertility. The undrainable ponds should therefore be dewatered in the summer months at the interval of 5–7 years.

FISH SPECIES CULTURE IN PONDS

Although a large number of fish species grow successfully in ponds, only a restricted number of species are usually cultivated on commercial scale. Reasons for this restricted choice is obvious.Commercial pond culture basically aims at achieving maximum possible rate of fish production and profit through optimum utilization of the natural food and the supplementary feed which drastically limits the choice of fish species for pond cultivation.

SIGNIFICANCE OF FERTILISATION IN POND

China has a long history of pond fertilisation for fish culture ages ago. The farmers adopted the methods of manuring to rear fry. For instance, Dacao (green manure) has been applied in Guangdong and Guangxi Provinces and night soil has been used as manures in Jiangxi and Human Provinces so as to nurture fry into summer fingerlings. Rich experience has been accumulated. In fingerling rearing ponds and table-fish rearing ponds, fertilisation is aimed at developing natural food and saving artificial feeds. This method has been widely adopted in China. Phytoplankton, elementary producers in ponds take in all essential inorganic materials from the water and conduct phytosynthesis with sunlight, producing organic nourishment needed for their own growth and reproduction.

Fertilisation means to supply phytoplankton with nutrients for phytosynthesis and to promote the growth of phytoplankton, by which zooplankton and other aquatic animals are fed on for their growth and propagation. Fish feed on plankton and other hydrobios. Pond fertilisation lies in cultivation of various food organisms and their propagation in large quantities in fish ponds to provide fish with abundant natural feeds, by which they can grow faster. The yield of fish pond may be raised thereby.Among the fish food organisms in ponds, there are a series of interrelations between the predator and the prey. The biological term is "food chain". Fish is the last link of the food chain of hydrobios in ponds.

, *e.g.*, phytoplankton → Silver carp

phytoplankton → zooplankton?Bighead

aquatic plant → Grass carp

plankton → benthos → Black carp

Usually, animals could use only 5–20 per cent energy of both animal and plant feeds. Utilisation of energy is related to the length of food chain. The shorter the food chain is, the higher the rate of energy transfer will be, in other words, the higher the utilisation rate of energy transfer is, the higher the fish production will be. The hydrobios in ponds are in a constant progress of growth and death. The dead bodies of organsims will turn from complex organic materials into simple inorganic materials through decomposition by bacteria, then dissolve in water, which can again be utilised by phytoplankton to produce new individuals. Hence, the materials in ponds are in a constant state of circulation mainly through the food relationship between living organisms. This kind of circulation is called pond material circulation. The process of pond material circulation is just that of the production of fish and their food organisms in ponds. The circulation originates from the soluble minerals in water. Under certain conditions of light and heat, the propagation of phytoplankton depends upon the amount of nutrients in water. In turn, the production of all food organisms depends upon the proliferation of phytoplankton. Fertilisation is just to enhance the quantity of nutrients and to nurture food organisms in ponds so as to create favourable conditions for raising fish yield.

KINDS OF ORGANIC MANURES AND METHODS OF APPLICATION

Organic manures mainly refer to excrements of farm animals. It is a general term of manures containing organic matter. Nowadays, manures applied in fish ponds in China are mainly organic manures. In production, the following are often used, faeces and urine of livestock and poultry, night soil, green manure, compost and silkworm faeces, etc., Only through decomposition by microorganisms, organic manures may be converted to nutrients, which plants can absorb, and then, organic manures are full of nutritional elements with long effects.

Compost

Mixed compost is made of green manures and animal wastes. The mixture of several manures may make the ingredients of manure all the more suitable to the reproduction of plankton. The ratio of the constituents of manures depends upon the local sources of manures. From experiments, the following proportion of the ingredients of compost is suitable for plankton reproduction.

- Green grass 8:cattle faeces 8: human excreta 1:lime 0.17.
- Green grass 1:cattle faeces 1:lime 0.02

There are two methods to make compost, that is, heaping and soaking. Heaping method: The manure heap is to be made under aerobic conditions. Spread one layer of green grass, then sprinkle some lime on it. Then add another layer of faece manure and do it again, layer by layer upto 1.5---2 m. At last, cover the heap with 5---6 cm thick of mud. The ingredients of the compost will rot and decompose. After 3---4 weeks the compost can be used.

Soaking method: dig a pit near fish ponds, put in green grass, lime and faeces one layer after another and then add some water to soak all manures with no leakage at all. The use of lime is to neutralise the organic acids turned out in the process of rotting and decomposition of the compost so that they could not attenuate the activities of microorganisms, whereas they could promote the decompostion of organic materials. The compost can be taken out for use after 10---20 day's fermentation at a temperature of 20°---30°C.

Methods of the Application of Organic Manures

- Application of Dacao: In Guangdong and Guangxi Provinces, Dacao is habitually used to fertilise pond water. The so-called Dacao means several composite plants. In reality the grass is not limited to composite plants; some graminous plants and leguminous plants are sometimes used as well. In application, Dacao is just heaped at a corner of pond and is turned once every other day or every two days. The rotten part will go spreading in water and at last, the roots and stems which are hard to rot, will be dredged up out of ponds. The decomposition of green manure in water consumes a great amount of oxygen. In light of experiments, if 1000 kg of grasses are applied in a pond of 1 mu with a waterdepth of one meter, all fish may die of lacking oxygen as the pond is in a state of no oxygen from the second day to the sixth day. The peak of oxygen consumption is on the second and third day. And it will be reducing later on. It is appropriate to apply green manures frequently in a small amount, or to pour fresh water into ponds or adopt aerators, in order to guarantee plenty of oxygen in pond water.

Application of Nightsoil

In Jiangxi and Hunan Provinces, nightsoil is applied in fish ponds and the application amount varies with fish sizes. Before application, one portion of nightsoil is diluted with double portions of water. The dilution is then sprayed along the pond dikes. It is added once a day and the added amount is dependent on the fertility of water.

Application of Livestock Manures

The application of livestock manure as base manure is similar to that of green manures: heap the manures at a corner of pond or put them in small heaps in shallow water with a sunny exposure so as to make them decompose and spread gradually in water. If the manure is used as an additive, it is added in small heaps every 7---10 days.

Application of Mixed Compost

After fermentation, the compost is taken out, and given a flush. The liquid is collected and the residues are removed away. Spray the liquid evenly into ponds. In case the area of the pond is rather big, you may load the needed amount of manure on a boat, flush it with pond water in batches and then spray the liquid evenly in ponds. The manure dregs can be used to fertilise crops. For small ponds, there is no need to use a boat. The liquid is spread round the dikes. Another method is to put one side of the compost to the other in the pit and expose the liquid and then the manure liquid can be ladled out and spread into ponds according to the required amount.

The nutrients of the compost can be quickly absorbed by phytoplankton once the compost is applied in ponds. It consumes less dissolved oxygen in ponds since, the organic materials are already decomposed after full fermentation.

Effects of Manure Application on Food Organisms

First of all, the application of organic manure causes the propagation of bacteria in large quantities. On one hand, the bacteria decompose organic materials, mineralising them into nutritional inorganic materials, which can be utilised by phytoplankton; on the other hand, they themselves take part of the nutrients as their own structural materials, reproducing a large population of bacteria. After organic manure being applied, the organic detritus in water is increasing in quantity. The organic detritus carries dense bacteria on its surface; which are the important food for lower aquatic animals and filtering fish. After manuring, the predominant species of plankton at their initial appearance in water depend closely on the properties of the manure. If organic manure is applied, those that are fond of organic materials will show the first population bloom------phytoplankton: Ochromonas spp. Cryptomonas; zooplankton: Urotrichia spp., etc. For inorganic manure, the predominant species which will be seen at first are centric diatoms Centrales and Scenedesmus acuminatus.

There is a close relationship between the amount of manure applied and the plankton community. Huge amount of manure will lead to the presence of some species of green algae Chlorophyta and blue algae Cyanophyta while small amount of manure will lead to the presence of Navicula rostellata and Cyclotella stelligera , etc.

After each application, the nutrient content in the water increases, resulting in a planktonic peak. Phytoplankton which are easily digested by silver carp reach a peak after 4 days, whereas those that are not so easily digested will attain a climax in 5---10 days. Zooplankton reach a peak in 4---7 days.

Protozoans will be the first zooplankton to reach a peak, followed by rotifers, cladocerans, and finally copepods. Since, protozoans multiply by binary fission, by which the population increases very rapidly, it will, therefore, be the first to reach a peak. Rotifers undergo parthenogenesis under normal condition. The eggs produced is not very many, with an average of 10---20 eggs during their life span. Thus, the rotifer population reaches a peak slightly later than the protozoans. Cladocerans also reproduce parthenogenically, but the span between hatching and sexual maturity is longer. So the cladoceran peak appear later than that of the rotifer. The timing of manure application is most important. Ideally, the peak should appear at a time required to meet the demand of the fish fry.

Kinds of Inorganic Manures and Methods of Application

Kinds of Inorganic Manures

Inorganic manures mean chemical fertilisers. According to the composition, the chemical fertilisers can be divided into three groups, that is, nitrogenous, phosphoric and potash fertilisers. The advantages of inorganic fertilisers are exact constituents, fast effect, slight pollution, no consumption of dissolved oxygen in water, small application amount and convenient operation. However, when chemical fertilisers are applied in ponds, the first link of the food chain in ponds is principally phytoplankton which, if taken as food for zooplankton, is nutritionally not so nice as bacteria. Therefore, the zooplankton amount in ponds applied with inorganic manure often lags far behind that in ponds applied with organic manure. Moreover, in most ponds applied with chemical fertilisers, the predominant species of phytoplankton are Chlorophyta, which is nutritionally worse than the predominant species in ponds applied with organic manure, such as Chrysophyceae, Bacillatiophyceae and Cryptophyceae. In addition, the manure effect is rather short and it is difficult to control the water quality, therefore, the result of the application of chemical fertilisers there is not better than that of the application of organic fertilisers.

Nitrogenous Fertilisers

- Liquid ammonia : Molecular formula, NH_4OH or NH_3 x H_2O, with a nitrogen content of 12---16 per cent. It is a water solution of ammonia, which is an important product of small-scaled nitrogenous fertiliser factory with simple synthesising procedure and low cost. Ammonia is in an unsteady state when in water, easy to volatilise, so it could almost lose its effect through the volatilisation if it is exposed to the air for a long period of time.

- Ammonium sulphate: Molecular formula, $(NH_4)_2SO_4$, with a nitrogen content of 20-21 per cent. It is produced from the liquid ammonia directly neutralised with diluted sulphuric acid. It is white crystals when pure, apt to dissolve in water. 100 kg of water can dissolve 75 kg of ammonium sulphate. With a little absorption of moisture, it is convenient to preserve and apply.
- Urea: Molecular formula, $Co(NH_2)_2$, with a nitrogen content of 44-46 per cent. Ammonia and carbon dioxide are interacted and synthesised into urea under high heat and pressure. It is white crystals with a strong absorption of moisture. After dissolution in water, urea does not turn out ions and is unable to be absorbed directly by plants. It can be utilised by plants only after it is decomposed by urease excreted by urea-decomposing bacteria and transformed into ammonium carbonate. The conversion rate of urea is related with the temperature. It can be totally transformed into ammonium carbonate in 4---5 days at 20°C and in 2 days at 30°C.

Phosphoric Fertilisers

Calcium superphosphate: Main contents, $Ca(H_2PO_4)_2\ H_2O$ with 12-18 per cent of P_2O_5; subsidiary-contents, $CaSO_4.\ 2H_2O$, about 50 per cent. Usually, it is white powder. It is corrosive, and is apt to absorb moisture, smelling acidic since, there is some free acid in the products.

Methods of the Application of Inorganic Manure

Nitrogen is one of the essential nutritional elements in a plant. It is also the principal content in protein, and can accelerate the formation of plant chlorophyll and strengthen photosythesis. For this reason, the nitrogen content is one of the decisive factors of phytoplankton production. Nitrogen is always short in pond water, so the application of nitrogenous fertiliser may achieve better results. Nitrogenous fertiliser is generally better to be used as an additive because of its quick-effectiveness, and the nitrogenous fertiliser with an ammonium content must not be mixed with strong-alkaline materials lest it should convert into gas and lose effect through volatilisation. In the utilisation of nitrogenous fertiliser in ammonium form, due attention must be paid to toxicity of ammonia. Either ammonium or liquid ammonia will become ionic $(NH_4)+$ and non-ionic (NH_3) once they are placed in water. This phenomenon may be expressed in the following formula:

$$NH_3 + H_2O \rightleftharpoons NH^+{}_4 + OH^-$$

In an acidic state, the balance inclines to the right side with the concentration of ammonium ions enhancing; and in an alkaline state, NH_3

concentration will raise high. As water temperature comes up to 25°C with pH6, the concentration of NH_3 holds 0.05 per cent in total nitrogen; pH7, 0.49 per cent; pH8, 4.7 per cent; pH9, 32.9 per cent.

NH_3 is toxic to fish. It poisons juvenile rainbow trout at 0.3--0.4 mg/L. The Chinese carp have a higher toleration to ammonia anyhow, ammonia inhibits the growth of fish. The mortal concentration is beyond 13 mg/L. The highest concentration permitted in the water body for fish farming is 0.1 mg/L. For this reason, the amount must be strictly controlled in one application. What is more, caution must be given to pH value to avoid being applied in strong alkaline water (, *e.g.*, just after pond clearing with lime) since, liquid ammonia is alkaline itself. Besides, the amount of non-ionic ammonia is rising with the increasing of water temperature, so special care is needed when nitrogenous fertiliser in ammonia form is used in hot summer and antumn.

The application amount of nitrogenous fertiliser depends upon the nitrogen content of the fertiliser applied. In a pond with the area of 1 mu and the water depth of about 1.5 m, 1.5-2 kg may be given as base manure. From then on, the additive of the fertiliser is applied 3-4 times monthly with 0.5 kg/mu each time. Thus about 10 kg are needed for the whole year. For example, if nitrogen content in ammonium sulphate is about 20 per cent and 2 kg are given to 1 mu as base manure, the amount of ammonium sulphate needed is:

$$2kg \times \frac{100}{20} = 10kg$$

The annual amount needed is about 50 kg.

Applying method: make a solution of it and spread it near the dikes. In case of liquid ammonia, put the container with ammonia under water and then open the lid so as to let liquid ammonia diffuse out slowly. By this way, the volatilisation can be avoided during the operation on the bank.

Most waters are lacking phosphorus, so it is important to apply phosphoric fertiliser, which can accelerate the reproduction of azotobacteria, and enrich nitrogenous fertiliser in water. The application amount of phosphoric fertiliser is calculated in accordance with the content of phosphoric acid applied. If 0.5-1 kg/mu is used as base manure the annual amount might be about 5 kg. The method of calculation and application is the same as above.

Potassuum is also one of the principal nutritive elements for plants, but usually it is plentiful in water. There is no particular need to apply potash fertiliser in aquaculture.

10

Cultivation of Fish

Fish culture is the rational cultivation of fish in a confined water area where the practices of both agriculture and animal husbandry are applicable. The soil and water management aspect of fish culture practice involving application of organic manures and inorganic fertilizers for the production of microscopic plants, the phytoplankton, is basically similar to agriculture while husbandry of fish such as feeding, breeding and health care is more or less similar to a livestock farming system. This farming system is also unique in that the farmed animal is cold-blooded or poikilothermic and lives in a water medium. Although this fish farming is approximately 2000 years old, the importance of it has been realized only recently in the face of mounting pressure on land resources and scarcity of animal protein for the ever increasing human population.

FISH AS FOOD

Malnutrition and starvation are the two serious problems being faced by millions of rural poor in most of the developing countries. The problem of malnutrition is in fact more serious and of a bigger dimension than the starvation problem and is caused mainly due to animal protein-deficient diets. Animal protein is essential for proper growth, repair and maintenance of body organs and tissues.

Fish contain about 16–20% protein compared to about 12% in egg, 3.5% in milk and 6–8% in rice and wheat. Moreover, it is wholesome, tasty, highly nutritive and an excellent source of essential minerals, vitamins and essential amino acids. At present about 31% of the total animal protein supply in the Asian region is in the form of fish protein. For the poorest segments of the population, fish is not only the most important animal protein source, but often the only one.

HIGH MULTIPLICATION CAPACITY AND MINIMAL WATER REQUIREMENT

The reproductive potential of fish compared to any other farmed animal is also very high. A kilogram of female cultivable carp species yields on an average about 0.1 million eggs, each of which has the potential to become 1 kg fish in about a year. No livestock animal possesses this magnitude of fecundity. Although fish needs water as a medium to survive and grow, it consumes minimal quantity of water compared with any livestock or agricultural crop. Fish also enriches the water with its voided metabolites thus making the water more productive for agriculture.

Low Energy Requirement for Protein Production

Fish culture systems require a relatively less amount of energy for protein production than any other farming system. Carp culture, depending upon culture practices, requires energy at the rate of 22–468 KJ/g of protein production while a land animal farming system needs over 550 to 3 400 KJ/g.

Warm Water Favours Fish Growth

Fish are cold blooded or poikilothermic animals. In other words they cannot maintain a constant and high body temperature like other livestock animals. Instead, their body temperature fluctuates according to the surrounding temperature. In warmer climates, their metabolism accelerates and they grow faster, while in colder climates, the metabolic rate slows down, resulting in a reduced rate of growth. In this way they save energy by not spending it for maintaining a higher and constant body temperature.

AQUACULTURE PRODUCTION POTENTIAL

Although the world's total marine production now stands at more than 80 million tons per year, all trends indicate that a saturation point is quickly approaching. During the past decade the growth rate hovered around 2%, much lower than earlier decades. On the other hand, tremendous potential exists in aquaculture. Aquaculture presently produces over 8 million tonnes of fish and shellfish annually.

It is estimated that Asian aquaculture production could be raised to 20–30 million tonnes a year by the end of the century. Aquaculture production has increased at an annual growth rate of nearly 7% between 1975–84.

EMPLOYMENT POTENTIAL

Aquaculture is also considered to be a potential source of employment for poor farmers and displaced capture fishermen. Rapid development of aquaculture has already generated considerable employment through culture of marketable fish, fish seed production and marketing of fish and fish seed.

The National Agriculture Commission of India while estimating the employment potential of fish culture has indicated that every tonne of fish produced provides employment to 2.5 persons. The other important advantages of fish farming are that the production is carried out within easy reach of consumers and also the harvesting can be adjusted to demand, thus minimising distribution problems and spoilage.

Rural ponds in Asia, hitherto producing at subsistence level, have succeeded in increasing production per unit area through improved culture practices involving higher stocking densities, polyculture combinations, pond manuring and feeding. The switch over from monoculture practice to polyculture has significantly contributed toward higher production and the prospect of polyculture appears very bright as the fish seed of desired species is becoming easily available due to the establishment of a large number of hatcheries.

During recent years, advances have also been made in traditional aquaculture systems practised in rural India by the development of composite fish culture, a system of polyculture of a group of complementary and supplementary freshwater species of fast growing carps in undrainable ponds. With the successful demonstration in different agroclimatic zones of India, gradual improvements in technology have been made and it is now possible to obtain a production rate of over 10 t/ha/yr in experimental ponds and up to about 5 t/ha/yr in farmers' ponds against the traditional average rate of production of 600 kg/ha/yr. To meet the increasing demand of seed of culturable carps, hypophysation techniques have also been developed for both Indian and Chinese carps and as a result they are now being bred in captivity even by fish farmers in remote villages. The emergence of this culture technology suitable for undrainable ponds and the simultaneous development of hypophysation techniques for fish seed production has completely revolutionized fish farm productivity.

This manual intends to provide the basic concept and practical guidelines of fish culture in undrainable ponds. Since it is prepared especially for extension agents and field workers, certain important things have been repeated and at times experimental results have been simplified with a view to making it more practical, simple and illustrative. It outlines the practices of procurement and propagation of fish seed, rearing of spawn to fry and fingerling stage and production of table-size fish following simple sequential steps. Like other farming systems this culture system is also prone to certain unexpected hazards for which one has to be prepared and properly equipped.

Such hazards are disease outbreaks, oxygen depletion, pollution, flood, drought, poaching, etc. The content of this manual is a synthesis of the author's personal field experience, the information gathered from published literature and the observations of other workers in India.

Based upon Indian experience, water resources in other developing countries with similar agroclimatic conditions may be utilized for the development of fish culture. It is hoped that this manual will serve as a practical guide to extension workers in popularizing freshwater fish culture in undrainable ponds.

FATTY ACID COMPOSITION OF FISH

ENVIRONMENTAL INFLUENCES

Salinity

The difference between fatty acid compositions of marine and freshwater fish has been noted by several authors. Some examples of fatty acid patterns Although these fish lipids are higher in w 3 fatty acids, it is clear that freshwater fish have higher levels of w 6 fatty acids than marine species.

The average w 6/w 3 ratios are 0.37 and 0.16 for freshwater and marine fish, respectively. Fish in general contain more w 3 than w 6 polyunsaturated fatty acids (PUFA) and should have a higher dietary requirement for w 3 PUFA; thus the dietary EFA requirement of marine fish for w 3 PUFA may be higher than that of freshwater fish. The same type of difference in the w 6/w 3 ratio between freshwater and seawater is seen when some species of fish migrate from oceans to streams or vice versa. The PUFA ratio of sweet smelt (Plecoglosus altivelis) changes drastically in only one month as they migrate from the sea to a freshwater river.

A similar but reverse change occurs in the masu salmon (Oncorhynchus masu) as they migrate from freshwater to seawater. Even within the same species of fish, the salinity of the water seems to cause a dramatic change in the fatty acid pattern. The difference between marine and freshwater fish may be due simply to differences in the fatty acid content in the diet or it may be related to a specific requirement in fish related to physiological adaptations to the environments.

The phospholipids are generally considered to be structural or functional lipid, being incorporated to a large extent in the membrane structure of cell and subcellular particles. The triglycerides are more often storage lipids and reflect the fatty acid composition of the diet to a greater extent than do the phospholipids.

The fatty acid compositions of the triglyceride and phospholipid fractions of fish lipids are presented. It can be seen that the effect of changing environment on the fatty acid composition of the phospholipid is as great in the case of salmon and considerably greater in the case of the sweet smelt, than it is on the triglyceride composition.

Rainbow trout on diets containing either corn oil, which is high in w 6 but low in w 6 PUFA, showed a higher mortality and growth reduction in seawater than in freshwater over the twelve-week feeding period.

Temperature

There are several other factors besides the salinity of the water which affect the fatty acid composition and especially the PUFA of fish. The salmonids are mostly cold-water fish. The fatty acids from a number of marine animals from temperate and arctic waters show some significant differences in the general pattern; unfortunately analysis included fatty acids longer than 20:1. There are a number of other experiments demonstrating the effect of environmental temperature on fatty acid composition of aquatic animals. The general trend toward higher content of long chain PUFA at lower temperatures is quite clear. The ratio decreases with a decrease in temperature.

If the trends in fatty acid composition can be taken as clues to the EFA requirements of fish, the requirement would be greater for fish raised at lower temperatures. Fish raised in warmer waters, such as common carp, channel catfish and tilapia may do better with a mixture of fatty acids.

EFFECTS OF DIET

Some of the fatty acid compositions may be seriously affected by the dietary lipids. The mosquito fish and guppies were fed trout pellets which had an ratio of 2.75. The catfish were fed diets supplemented with either beef tallow or menhaden oil, with ratios of and respectively. These fish were able to alter the dietary ratio in favour of fatty acid incorporation into the flesh lipids even at the highest temperature.

Commercially available trout pellets are often low in w 3 PUFA and high in w 6 fatty acids. It is important not to ignore the effect of dietary lipid composition on fatty acid composition of fish fed artificial diets. When the dietary ratio is very high in fatty acids supplied by animal lard or vegetable oils, there is a tendency for fish to alter the ratio of PUFA incorporated in favour of fatty acids.

When the dietary oil is a fish oil high in to fatty acids, there is little change in the ratio of lipids incorporated into the fish. This is further suggestive evidence of an EFA requirement of fish for PUFA.

Seasonal Variation

Seasonal variations in the fatty acid composition of fish species have often been reported. Seasonal changes have been observed in total lipid and iodine values of herring oils. The iodine value or degree of unsaturation of the oil was minimal in April and maximal in June. The great increase in unsaturation corresponded to the onset of feeding in spring. The absence of a gas liquid chromatograph (GLC) at the time precluded identification of changes in individual fatty acids.

Flesh and viscera lipid content of the sardine Sardinops melanosticta vary from 3.9 to 10.77 percent and from 10.9 to 38.3 percent, respectively. The fatty acids of principal interest with respect to EFA metabolism are 20:4 20:5 and 22:6 3. There was considerable variation in all of these fatty acids in both neutral and polar lipid from both tissues. In the flesh, the 20:4 was consistently higher in the neutral lipid than in the polar lipid. The total consistently higher in polar lipid than in the neutral lipid. Thus, in spite of the major fluctuations in fatty acids caused by changes in diet and temperature throughout the seasons, there was a consistent preferential incorporation of PUFA of the w 3 series into the polar or phospholipid fraction of the lipids.

One of the best clues to the EFA requirements of a species can be gained from the fatty acid composition of the lipids incorporated into the offspring or egg. The act of reproduction or spawning also has a significant effect on the seasonal fluctuation of lipids in fish. Fatty acid composition of fish egg lipids is probably distinctive for each species and contains increased levels of compared to the liver lipids of the same female fish

Elevated levels of reduced in the ovary occurred compared to mesenteric fat of Pacific sardine fed a natural copepod diet. The blood fatty acids of the sardine fed the natural diet were similar to those of the ovary. When the sardines were fed trout food, both the blood and mesenteric fat responded to the diet with elevated 18:2w 6 and reduced 20:5 arid 22:6. The effect of the diet on ovary fatty acid content was considerably less, as relatively high levels of 20:5 and 22:6 were retained.

The ovary lipids of the sweet smelt show an increase in 16:0 and a reduction in the PUFA, especially in the phospholipids, compared to the lipids from the flesh of fish caught at the same time of year. The ratio of the ovary was lower than that of the flesh lipids, 0.21 and 0.17 for the ovary compared to 0.31 and 0.20 for the triglycerids and phospholipids of the flesh, respectively.

The hatchability of eggs from common carp fed several different formulated feeds is greatly reduced when the 22:6ω3 of the egg lipids is less than 10 percent. Further, the muscle, plasma and erythrocyte fatty acid compositions are more affected by dietary lipid than those of the eggs.

The EFA requirements of a number of species of fish have been investigated in nutritional studies. The fish themselves have given ample evidence for EFA preference by the types of fatty acids they incorporate into their lipids. Fish, in general, tend to utilize ω3 over ω6.

This is especially observed when the dietary lipids are high in ω6, as the fish tend to alter the w 6/ω3 ratio toward the ω3 fatty acids in the tissue lipids. The lipids of the egg must satisfy the EFA requirement of the embryo until it is able to feed. The fatty acid composition data suggest that the ω3 requirement is greater in seawater than in freshwater and higher in cold water than in warm water.

BODY LIPID COMPOSITION AND DIETARY LIPID REQUIREMENTS

Detailed information is still lacking on the dietary lipid requirements of many species of fish, but there is an abundance of information on the fatty acid composition of fish oils. Information on the lipid composition of fish can be used to make some guesses about dietary lipid requirements. Linolenic acid (18:3ω3) resulted in some sparing action and growth promotion in rats and fatty acids of the ω6 EFA prevented all of the EFA-deficiency symptoms.

Research with homeothermic land-dwelling animals showed that the ω6 series of fatty acids are the "essential fatty acids", while the ω3 series are considered to be non-essential or only have a partial sparing action on EFA-deficiency. The series of fatty acids have been shown to be essential to enough species that it began to become accepted that these are the essential fatty acids for all animals. It was assumed by many that fish also required ω6 fatty acids. Many researchers began by supplementing fish diets with vegetable oils, such as corn, peanut, or sunflower oil, which were rich in linoleic acid. The main sympton observed during the development of EFA deficiency in chinook salmon fed fat-free diets was a marked depigmentation that can be prevented by addition of 1 percent trilinolein, but not by 0.1 percent linolenic acid.

Although the ω6 fatty acids are considered to be essential, one of the general characteristics of fish oils is the low levels of ω6 series fatty acids and the higher levels of ω3 type fatty acids. There is evidence that polyunsaturated fatty acids (PUPA) of the ω3 series, which are present in relatively large concentrations in fish oil, play the role of essential fatty acid for fish.

POLYUNSATURATED FATTY ACID REQUIREMENTS OF FISH

When a test diet containing 13 percent corn oil and 2 percent cod-liver oil was fed to rainbow trout, subsequent deletion of the cod-liver oil from the diet depressed growth and produced some kidney degeneration which might be attributed to a lack of sufficient ω3 PUFA present in significant quantities in cod-liver oil.

Dietary fish oil is superior to corn oil in promoting growth of rainbow trout (Salmo gairdneri) and the yellow-tail (Seriola guingueradiata). Dietary linolenic acid or ethyl linolenate (18:3 ω3) gives a positive growth response for rainbow trout which may be attributed to a dietary requirement for ω3 fatty acids.

ESSENTIAL FATTY ACID REQUIREMENTS OF FISH

One of the most widely accepted theories explaining the presence of such high levels of 20:5 ω3 and 22:6 ω3 fatty acids in fish oils is related to the effect of unsaturation on the melting point of a lipid.

The greater degree of unsaturation of fatty acids in the fish phospholipids allows for flexibility of cell membrane at lower temperatures. The ω3 structure allows a greater degree of unsaturation than the ω6 or ω9. This theory is consistent with the fact that cold water fish have a greater nutritional requirement for ω3 fatty acids, while the EFA requirement of some warm water fish can be satisfied by a mixture of w 6 plus ω3.

Rainbow Trout

Rainbow trout, a cold water fish, requires ω3 fatty acids as EFA in the diet. The EFA requirement can be met by 1 percent 18:3ω3 in the diet.

Inclusion of 18:2ω6 in the diet may result in some improvement in growth and feed conversion compared to EFA deficient diets; however, the ω6 fatty acids will not prevent some EFA deficiency symptoms such as the "shock syndrome". Although it is clear that rainbow trout require ω3 fatty acids, it remains to be shown conclusively whether some dietary level of ω6 fatty acid is essential.

Either 20:5ω3 or 22:6ω3 is superior to 18:3ω3 in an EFA value for rainbow trout and the former two fatty acids in combination are superior to either alone. This is consistent with data for mammals, where 20:4ω6 has higher EFA value than 18:2ω6. The superior nutritional value of C-20 and C-22 carbon w 3-PUFA is further supported by the excellent growth promoting effects of dietary fish oils such as pollock liver oil and salmon oil for rainbow trout.

Channel Catfish

One of the most important warm water fish in North America is the channel catfish. The quantitative EFA requirement of the catfish has not yet been determined. However, the evidence is convincing that the w 3 requirement is not as great as that of rainbow trout. Analysis of fatty acids of lipids from catfish purchased at five processing plants showed very low levels of 20:4ω6, 20:5ω3 and 22:6ω3; 0.8-5.5, 0.2-1.3 and 0.6-6.1 percent of the total fatty acids, respectively. It was shown that corn oil added to a semipurified casein based diet initially resulted in a positive growth response and protein sparing, but later growth inhibition was observed.

The apparent repressive effects of corn oil may be due to its 18:2ω6 content since 20:5ω3 and 22:6ω3 present in menhaden oil had no apparent detrimental effects. The growth suppressing effects of 18:2ω6 were also noted when 3 percent corn oil was added to 3 percent beef tallow and 3 percent menhaden oil.

The growth suppression caused by unsaturated fatty acids does not appear to be limited to ω6 fatty acids. Linseed oil (high in 18:3ω3) in the diet of catfish resulted in growth suppression similar to that caused by corn oil compared with dietary beef tallow, olive oil and menhaden oil.

The Common Carp

The picture for another warm water fish, the common carp (Cyprinus carpio) is much clearer than that for the channel catfish. This fish has an EFA requirement for both ω3 and ω 6 fatty acids. The best weight gains and feed conversions are obtained in fish receiving a diet containing both 1 percent 18:2 ω6 and 1 percent 18:3ω3. With the carp, 20:5ω3 and 22:6ω3 at 0.5 percent of the diet are superior to 1 percent of 18:3ω3. Carp fed a fat-free, or EFA deficient, diet incorporated high levels of 20:3ω9 in their lipids, especially in the phospholipids.

The Eel

The eel, another warm water fish, has a requirement for both ω3 and w 6 fatty acids. Corn oil (high in ω6) and cod liver oil (high in ω3)in a 2:1 mixture are most favourable for the growth of eels. The eel requires ω6 and ω3 in the same proportion as the carp, but at a lower level in the diet; namely, 0.5 percent of each, rather than 1.0 percent of each PUFA.

The Plaice

The plaice becomes depleted of both ω3 and ω6 PUFA when fed a fat-free diet. The addition of 12:0 and 14:0 to the diet result in the synthesis of saturated and monoenoic fatty acids of chain lengths up to C18; however, increased levels of 20:3ω9 noted in trout and mammals have not been reported in plaice. Plaice fed dietary 18:2ω6 and 18:3ω3 will not produce significant amounts of 20:4ω6, 20:5ω3, or 22:6ω3.

The Turbot

The growth of turbot is much better with ω3 PUFA than with ω6 or saturated fat (hydrogenated coconut oil) in the diet. The turbot also appears to be unable to convert dietary 18:2ω6 to 20:4ω6 when fed corn oil, or to convert endogenous 18:1ω9 to 20:3ω9 when fed the EFA deficient diet. Although it appears to have an EFA requirement for ω3 fatty acids such as are present in cod liver oil, this requirement is not satisfied by 18:3ω3. The chain elongation and desaturation of 18:lω9, 18:2ω6, or 18:3ω3 has been found to be very limited (3-15 percent) in turbot compared to the rainbow trout where 70 percent of the 18:3ω3 was converted to 22:6w 3. The required level of long-chain ω 3 fatty acids for turbot is at least 0.8 percent of the diet.

The Red Sea Bream

The red sea bream (Chrysophyrys major) grows better when the dietary lipid is of marine origin (pollock residual oil) rather than a vegetable oil (such as corn oil). The EFA requirement of the red sea bream is not satisfied by either linoleic acid of corn oil or supplemented linolenate. A mixture of 20:5ω3 and 22:6ω3 supplemented to the corn oil diet has been shown to be effective in improving growth and condition of these fish. Thus, even in warm water, marine fish seem to require not just ω3 fatty acids but fatty acids of 20 to 22 carbon-chain length. A direct correlation between feed efficiency and the 18:1 level in the lipids of the red sea bream has been postulated.

Other Species

Among warm water marine fish, mullet and fundulus possess the ability to chain, elongate and desaturate 18:2ω6 or 18:3ω3 PUFAs. The process is, however, inhibited in fundulus by high levels (about 5 percent) of these PUFAs of 18:2ω6 or 18:3ω3 in the diet.

FATTY ACID METABOLISM IN FISH

It appears that high levels of 18-carbon ω6 or ω3 fatty acids inhibit the synthesis and metabolism of 18:lω9. It is interesting to note that the channel catfish, which also exhibits negative growth response to dietary 18:2ω6 or 18:3ω3, incorporates very high levels of 18:1 into its body lipids. The inclusion of either 18:2ω6 or 18:3ω3 in the diet reduces the levels of 18:1 fatty acids in body lipids. A similar reduction has also been observed in red sea bream liver phospholipid when either of the PUFAs is added to the diet.

The competitive inhibition of chain elongation and desaturation of members from one series of fatty acids for members of another series is well established, with ω3 > ω6 > ω 9 being the usual order of potency for inhibition.

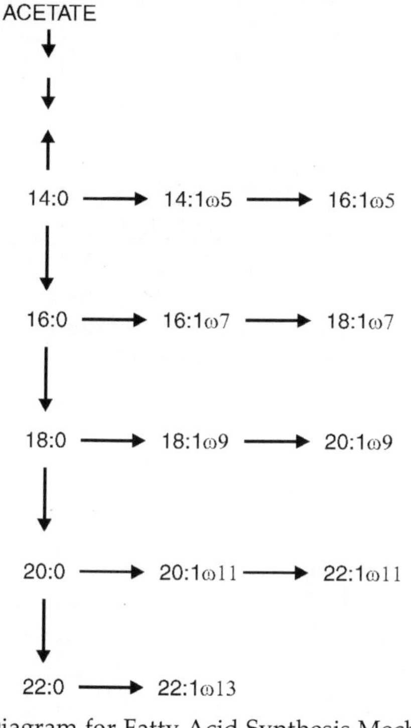

Fig. Flow Diagram for Fatty Acid Synthesis Mechanisms in Fish-Saturated and Monoenoic Fatty Acids

The pathways of fatty acid metabolism have been reviewed by Mead and Kayama (1967). Fish are able to synthesize, de novo from acetate, the even-chain, saturated fatty acids. Radio tracer studies have shown that fish can convert 16:0 to the ω7 monoene and 18:0 to the ω9 monoene. The ω5, ω11 and ω3 monoenes are proposed based on the identification of these isomers in the monoenes of herring oil.

Fish are unable to synthesize any fatty acids of the ω6 and u3 series unless a precursor with this ω structure is present in the diet. There is competitive inhibition of the elongation desaturation of fatty acids of one series by members of the other series. The ω3 fatty acids are the most potent inhibitors, the ω9 are the least.

The ability to elongate and desaturate fatty acids is not the same in all species of fish, as was noted earlier. The turbot was able to desaturate and elongate only 3-15 percent of 18:1ω9, 18:2ω6, or 18:3ω3, when given the C^{14} labelled fatty acid; in the rainbow trout, 70 percent of the label from 18:3ω3 (C^{14}) was found in 22:6ω3.

The essential fatty acids are not unique in their ability to supply energy. The b -oxidation of fatty acids in fish is basically the same as in mammals. The EFA and saturated and monoenoic fatty acids are all equally utilized by fish for energy production.

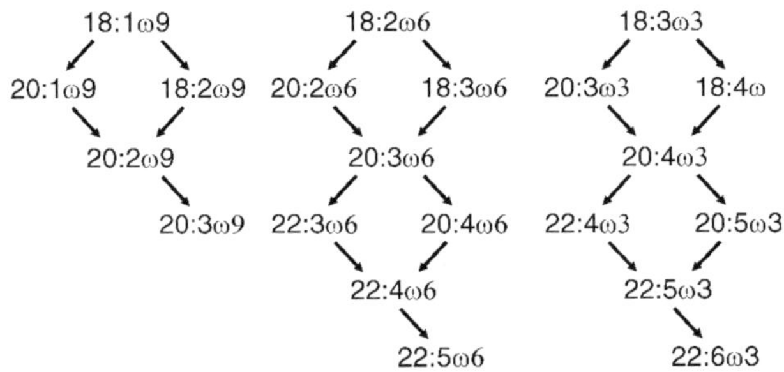

Fig. Flow Diagram for Fatty Acid Synthesis Mechanisms in Fish-Polyunsaturated Fatty Acids

Increased swelling rates of liver mitochondria occur in rainbow trout fed diets deficient in ω3 fatty acids. It is possible that EFA plays an important role in the permeability as well as the plasticity of membranes. The role of ω3 fatty acids in membrane permeability may be one of the factors accounting for differences in content of this family of fatty acids between freshwater and marine fish. Fish mitochondria with high levels of the ω3 PUFA and very low levels of ω6 fatty acids are very similar to mammalian mitochondria with respect to cytochrome content, b -oxidation of fatty acids, operation of the tricarboxylic acid cycle, electron transport and oxidative phosphorylation.

The ω3 PUFA may play the same role in fish that the ω 6 fatty acids play in rats. The EFA play another role in the mitochondria. In addition to their importance in membrane structure, the EFA are important in some enzyme systems.

Unsaturated fatty acids play an important role in the transportation of other lipids. It has been repeatedly shown that feeding PUFA will lower the cholesterol levels in animals with above-normal blood lipid and cholesterol levels. Fish oils are more effective in lowering cholesterol levels than are most dietary lipids.

The major portion of the fatty acids absorbed across the intestinal mucosa are transported as protein-lipid complexes stabilized by phospholipids. The low body temperature in fish probably results in a greater importance for unsaturation in transport of lipids than in homeothermic animals.

NEGATIVE ASPECTS OF LIPIDS IN FISH NUTRITION

The requirement by fish for PUFA of the ω 3 series creates problems with respect to feed storage. These types of fatty acids are very labile on oxidation. The products of lipid oxidation may react with other nutrients such as proteins, vitamins, etc. and reducing the available dietary levels or the oxidation products may be toxic. The effect of oxidized lipids on dietary proteins, enzymes and amino acids have been demonstrated. The use of oxidized menhaden oil in the diets of swine and rats caused decreased appetite, reduced growth, yellowish-brown pigmentation of depot fat and decreased haemoglobin and haematocrit levels. The negative effects of the oxidized fish oils were reversed by the addition of alpha-tocopherol acetate or ethoxyoquin to the diet.

Much of the use of vegetable oils in fish diets in the 1950s and 1960s might, in part, have been based on their greater stability in prepared diets. It has been demonstrated that rancid herring and hake meals in fish feeds caused dark colouration, anaemia, lethargy, brown-yellow pigmented liver, abnormal kidneys and small gill clubbing in chinook salmon.

The symptons can be alleviated by addition of alpha-tocopherol to the diets containing rancid fish meals. The addition of vitamin E would prevent the toxic or negative effects of adding 5 percent highly oxidized salmon oil to the diet of rainbow trout. This same sparing effect of alpha-tocopherol can also apply to rancid carp feed.

The positive nutritional value of ω 3 fatty acids in fish lipids for fish feeds can become a negative factor if adequate care is not taken in the preparation and storage of feeds. Only fresh oils with low peroxide values should be used in feeds. Fish feed ingredients such as fish meals should be protected against oxidation.

The level of vitamin E added to the diet should be increased as the PUFA level is increased. The finished feed, if possible, should be stored in air tight containers at reduced temperatures with minimum exposure to UV radiation and other factors accelerating the rate of lipid oxidation. The problems of rancidity or antioxidation of lipids in fish feeds should not be ignored.

Carbohydrates

Carbohydrates represent a broad group of substances which include the sugars, starches, gums and celluloses. The common attributes of carbohydrates are that they contain only the elements carbon, hydrogen and oxygen and that their combustion will yield carbon dioxide plus one or more molecules of Water.

The simplest carbohydrates are the three-carbon sugars which figure importantly in intermediary metabolism and the most complex are the naturally occurring polysaccharides, primarily of plant, origin. In the diet of animals and fish, two classes of polysaccharides are significant:

- Structural polysaccharides which are digestible by herbivorous species -cellulose, lignin, dextrans, mannans, inulin, pentosans, pectic acids, algic acids, agar and chitin; and
- Universally digestible polysaccharides-principally starch.

Carbohydrates make up three-fourths of the biomass of plants but are present only in small quantities in the animal body as glycogen, sugars and their derivatives. Glycogen is often referred to as animal starch because it is not present in plants. Derived mono-saccharides such as the sugar acids, amino sugars and the deoxysugars are constituents of all living organisms.

SPECIFIC TYPES OF FISH FARMS

Within intensive and extensive aquaculture methods, there are numerous specific types of fish farms; each has benefits and applications unique to its design.

FISH REARING IN CAGES AND PENS

As fish necessarily requires water, most people tend to think that for fish rearing they should have their own fish pond which may be situated near the streams, rivers, lakes, permanently flooded areas or reservoirs but do not have the ownership of any portion of any of them, or he may be having a water body which is linked with a lake, reservoir or river. Under such conditions one can go for rearing of fishes In cages and pens. The idea is to rear fish in a common and unbounded water body. This booklet explains the basic methods employed on cage and pen fish farming. Many poor farmers may get benefit from this method, at least partially meet their food and finance requirement.

If a farmer is living near a huge water body, such as canal, river, stream, lake, permanently flooded area or reserviors which does not belong to him or he has a water body which is linked with some other huge water bodies (rivers with lakes or sea, streams with big rivers or flooded areas) he can rear fishes in cages and pens. Of course under this system only a limited number of fishes or a. specific type of fish may be reared. However, the number will depend on the size of the cage or the pen used. Those who are able to establish bigger pens and cages will be able to rear proportionately large number of fish provided he can feed them.

Feeding of fish reared in cages and pens is very important among the management practices. Therefore the main thing one should consider is the sufficient availability of feed for the fish reared in the cages and pens. Of course the type of feed depends on the type of the fish being reared Some fish are carnivores, some herbivores and some others are omnivores. For more details about feeding of fish please consult Booklet No. 525 on "Fish Feeding".

Generally, a mixture of oil cakes and rice polish is a very good feed for most of the species of the fishes that are being reared by people.

The cages and pens can be constructed with any material conveniently available at the local level provided it is durable at least for one year. However, more the durability of the material greater will be economic benefit from this system of fish rearing though the initial cost may be more.

This type of fish rearing has been practiced since long in certain parts of the Asia. Perhaps this system of fish culture is originated with the view to control retrieval of the stocked fish. Its application is more relevant for those people for whom the suitable water supply and land for fish culture are becoming less and less available. Further to intensify this system of fish culture, scientific methods are being evolved for increasing the production of fish in cages by forced aeration or by using different types of filters and better feeding and management practices. In India this system can be practiced in the low lying and permanently flooded areas of Kerala, West Bengal, Orissa, Andhra Pradesh, Kamataka, Goa, Gujarat, North Bihar, Assam, Manipur, etc., though much work needs to be done still to perfect and standardise this system.

STRUCTURE OF CAGES AND PENS

The cages are structures which are covered and secured from every sides and the fishes are kept inside. They can be fixed stationary, suspended, partially or fully immersed, or floating, depending on the convenience. The cages are fitted with a door which can be opened in order to introduce the fishlings, to feed and care them and to harvest them at the end of the rearing season.

The best positioning of the cages is to fix them in such a way that the top portion of the cage is exposed to air and sunlight so that the fish can breath air whenever it is necessary. Also due to the constant contact with the atmospheric air the amount at oxygen dissolved into the water will be more to the benefit of the fishes.

On the other hand, the pens are fixed as fencing structures or enclosures covering an area of the water body. They are made by fixing the fencing material into the ground at the bottom and rising right through the water column and make them stand at about 1-2 ft above the water. The pens are open at the top and exposed to atmosphere. They cover much larger area than the cages. In a way pens are enclosed areas.

Pens also give the idea of several sections of an enclosed area each harbouring a certain number of fishes of different type, size, and age. For breeding purpose the breeding stock may be kept in small pens. Similarly for spawning and hatching also separate pens are required.

The pen looks like a fenced area. Wooden reapers, bamboo splits, reeds or any other suitable material can be used for the construction of the fence. The fencing material is fixed vertically on the ground, under the water and held very close to each other by ropes. Horizontally fixed fencing material is so closely arranged that the gap between them is reduced to the minimum required space, that would not allow the fish to come out and also they will remain safe from predators. Enclosed area denotes a larger area than the pen area. A portion of large water bodies such as lakes, reserviors, flooded areas, etc., is fenced around in order to rear large number of fish for commercial purpose as is practiced by some people living near the Chilka lake in Orissa.

The pens and enclosures are established only in shallow waters whereas the cages can be used even in deep water areas.

Material for Construction

There are a number of materials that can be used for making the cages and pens for fish rearing. Depending on the ingenuity of the farmers and availability of the things a farmer can use any type of material to construct the cages and the pens. Some of the common ones are mentioned here.

Bamboo and Choir Rope

Mature and seasoned bamboo splits tied closely using firmly twisted coconut fibre rope perhaps is the best material for making cages and pens in the rural areas where other commercially produced materials are not usually available. Both these materials are durable in submersed conditions. Thin bamboo splits of half inch thickness and width are held together closely by inter-twining them using thin coconut fibre ropes. The main frame of the cage is made of bigger bamboo splits or wooden reapers.

Nylon Threads

Strong nylon threads are better than any other material for making the cages and pens though this may be quite costly for an ordinary farmer. Nets are made out of this nylon thread. The mesh size of the net depends on the size of the fishlings which will be reared in the cage. For fry (up to 15 days old) the mesh size should be very small, for the fingerlings (15 to days) the size will be bigger, while for the growing or stocking fishes (reared for market size) the mesh size will be still bigger (half to one inch size).

Plastics

Tubes, sheets, ready made meshes made of strong plastic material can be used as cage and pen material. However the plastic should be very strong and non- degradable. Now-a-days very strong and non- degradable type of plastics are available in the market. One has to search and find out the most suitable material for his own purpose.

Wire Meshes

Wire mesh is another important material for the construction of cages and pens for fish farming. The mesh as already mentioned could be of any size depending on whether the farmer wants to keep the fry or fingerlings or fish grown for market purpose. It is highly advisable that the mesh is made of galvanised wire in order to reduce the possibility of rusting. Rust can be also controlled by coating with water proof paints on the wire mesh.

Wood

Wood can be used as a caging. or penning material. There are several types of wood which can withstand the submersible condition. There could be many other materials which are locally available and could be used for the construction of the cages and pens for fish rearing. The size and shape of the cage or pen is decided by the farmers. Those who have large space and want to rear fish in large number will need bigger cages and pens. But they could be made in any shape and size. Depending on the space available the cages and pens may be constructed big, medium or small, long or short. The shape may be rectangular, square, triangular, irregular or any other: The size of the cage and pen too vary a lot tram very big which harbours hundreds or thousands of fish, to very small in which only one or two fish may be kept.

Others

There may be several other materials which may be available locally or commercially for the construction of the cages and pens. It is for each one to look for the appropriate material available at cheaper rate and to use them for the construction.

There may be ready made modular type of material in the form of parts of the cage or pens. Using them we can construct, dismantle and reconstruct the cages and pens. Such materials may not be possible at the farmer's level in the case of cage construction but it is possible in the case of pen construction.

CONSTRUCTION AND FIXING OF CAGES AND PENS

Different methods are followed for the construction of the cages and pens for fish rearing. They are explained as follows under separate sub-heads. The construction of enclosures is very much similar to the construction of the pens. Only the enclosures are constructed in a bigger area while the pens cover comparatively a smaller area.

Construction and Fixing of the Cages

First of all the site where the cage is to be fixed is surveyed and measured to determine the size, shape and the way the cage is going to be fixed at the location. Arrangements for the materials are done accordingly.

The cage may be constructed in situ or constructed in parts at a different place.and brought them to be assembled at the place where It is going to be fixed. The most important point to be kept in mind is the mesh size. The mesh size is made bigger for keeping the bigger fishes whereas for the small fish smaller mesh work is provided. The cages are fixed in immovable or in floating manner. They can be fixed completely or partially immersed in the water. The cages are fitted with sufficient doors which will help in the feeding, caring and harvesting of the fishes grown. The cages are fixed in such a way that always the fresh water will flow through them and also enough sun light is available so that the fish inside will get the most optimum conditions to live.

Construction and Fixing of Pens

As in the case, first of all of cages the location is surveyed and the type and the amount of material needed is estimated before the construction of the pens and enclosures. The construction of the pens and enclosures are generally done in situ in the form of a fence. The fencing material consists of the main pole fixed firmly in the ground under the water at convenient distances. These are jointed by at least two horizontal bars or reapers one on the upper side and other on the lower side. If needed a third bar or reaper is fixed horizontally between the above mentioned horizontal bars. On to these horizontal reapers or bars are fixed the vertical bars closely to form the fencing for the pen or enclosure.

The vertical bars are fixed on the horizontal bars using nail or by tying with strong rope material. The distance between the two vertical bars will be adjusted to hold the minimum size of the fish that are released into the pen or enclosure. The pens can be constructed by fixing different segments of a fence. Each segment consists of a fixed number of vertical bars closely fitted by using two or three horizontal bars. The vertical bars at both ends and at the middle will be longer by one foot and pointed Because of these three vertical bars a segment can be fixed vertically by driving them into the soil till the other bars in the segment come into close contact with the soil surface.

After fixing the first segment fix the second segment close to the first one; then fix the third, fourth segments and go on till the whole fencing is completed to create an enclosed area. Using the same technique we can also separate the whole enclosure into several sections if needed. But even in fixing the segmented fences also we need the strongly fixed poles on to which the segments are fixed firmly. Pole can be fixed between each segment or at least between two segments each.

We can have segments without the above mentioned three longer and pointed vertical bars on each segment. In this case the segments are pressed down into the soil between two poles already fixed at an interval equal to the width of the segment and tied to tern. The pens are fitted with a door made of the same material as the pens. The door will be more like a segment of the fence which can be removed by lifting when one wants to open and fixed back when one wants to close it. The door is better secured on to the fence by tying rather than fixing it with hinges. Movement of the door fixed with the hinges does not work under the water and the door has to remain fixed into the soil. However one can adopt several ways to fix a door and it should be fixed in such a way that the door can be opened and closed like any other door fixed on hinges. The depth of the cages or the depth of the water in the pen should be sufficient for the fish to grow. The optimum depth for cage varies from 60 to 120 cm, and for the pens the depth of the water inside the pen should be at least 120 cm up to a maximum of 180 centimetres. The length of the cage should be minimum ten feet and the width should be at least three feet. However depending on the ability of the farmer the cages may be bigger or smaller than the above mentioned dimension. But bigger the cages better they are for the growth of the fishes. Fish require space for swimming around and the more we give them space greater will be the growth rate.

In the case of pens the only recommendation is to provide as much area (length and width) as possible. Obviously the depth, will be the same as that of the water body in which the pen is constructed. For the purpose of, hatching and fry raising cages and pens are unsuitable as the mesh size is bigger and the try can escape from the cages and pens. For this the cages and pens should be fitted with fine mesh net as that of a mosquito net. One can also use "Happas" (enclosures made of fine mesh net like the inverted mosquito net) for breeding, hatching and rearing of try up to the age of 90 days, by which time they grow up to the size of a fingerling. If the mesh size of the cage or the pen is sufficiently smaller, the fingerlings can be released into the cages or pens. Otherwise they have to be kept in suitable enclosures till they are grown big enough to be contained inside the cage or pen. Details about breeding and hatching of the fry and raising up to fingerlings stage are described in Booklet No. 543 on "Carp Breeding".

TYPES OF FISHES REARED IN THE CAGES AND PENS

There are a number of fishes that can be reared in cages and pens. In fact any type of fishes, prawns, frogs, crabs and shell fishes can be reared in the cages and pens. Some of the fishes that are suitable for cage rearing under mono-culture are: the air breathing fishes such as murrel (Channa marulius), striped murrel (Channa striatus), spotted murrel (Channa punctatus), magur of catfish family (Clarias batrachus), singhi (Heteropneustes tassilis), the fishes of tilapia family, the fishes of Indian carp family (catla, rohu and mrigal) and the Chinese carp family (silver carp, grass carp and common carp), trouts, etc.

Depending on the availability and the location any type of fish can be reared in the cages or pens. Since the cage arid pen rearing involves considerable amount of time, energy and finance it is advisable to rear the fast growing type of fishes rather than local types which are slow growing. Among the types of the fish above mentioned, grass carp needs special mention as they can be reared easily on vegetation. In places where there is plenty of marshy and water logged lands there air breathing fishes are most suitable for cage rearing In poly culture a combination of tilapia and catfish is more suitable for a poorly managed pen or enclosure conditions. They are also suitable to the cage conditions. Tilapia is a fast multiplying type of fish which is controlled by the carnivorous catfish.

In a well managed cage any single type of the carp fishes can be reared. The number of fish reared depends on the volume of the cage. A cage is too small for composite fish, farming (rearing a number of fishes together), but in the pens and enclosures one can go for composite fish farming as they replicate in the pond situations to a great extent. Rearing of frogs, prawns, shell fishes, etc., is not yet a common practice in India and a lot more research work needs to be done. The readers are aware that all types of ornamental fishes are reared in the cages for which a separate booklet is being prepared.

MANAGEMENT PRACTICES

After the cages and the pens are properly constructed the fishes are introduced into the cages or pens. The- fingerlings may be raised in one's own Lann or may be bought from some reliable nursery. Only the fingerlings bigger enough to be contained inside the cages or pens, are introduced. Others may be kept in special enclosures till they grow big enough to be introduced into the cage or pen. In the cage and pen system one can adopt monoculture or polyculture depending on one's liking, availability of the feed, availability of the fry, etc. The number of fish reared per cubic metre volume depends on the size of the fish. Roughly i we can say that one kilogram size fish per cubic metre volume of water may be held as a thumb rule. This is the minimum space needed. Accordingly two fish of half a kilogram each or four fish of 250 gm each or eight fish of 125 gm each or twenty fish of 50 gm each or forty fish of 25 gm each can be grown in one cubic meter volume of water. Though such mathematically arranged numbers may not always be feasible still such considerations gives one an idea of the number of fish that can reared in a given volume of cage and the pen.

In the cage rearing system the fishes do not get any natural feed though in the pen system, depending on the site there may be natural planktons of flora and fauna available; Hence the feeding becomes the most important aspect in fish rearing in the cages and pens. Regular feeding with sufficient quantity and quality feed should be carried out in cage rearing. Feeding should also be taken into consideration the nature of the fish reared whether carnivores or herbivores. Normally a mixture of oil cake and rice or wheat bran form the base in preparing feed for the fish. Along with this fish meal, meat meal, prawn meal, various types of fungi, worm meal, etc., should be included into this mixture may be raised in one's own farm or may be bought from some reliable nursery. Only the fingerlings bigger enough to be contained inside the cages or pens, are introduced. Others may be kept in special enclosures till they grow big enough to be introduced into the cage or pen. In the cage and pen system one can adopt monoculture or polyculture depending on one's liking, availability of the feed, availability of the fry, etc. The number of fish reared per cubic metre volume depends on the size of the fish. Roughly I we can say that one kilogram size fish per cubic metre volume of water may be held as a thumb rule. This is the minimum space needed. Accordingly two fish of half a kilogram each or four fish of 250 gm each or eight fish of 125 gm each or twenty fish of 50 gm each or forty fish of 25 gm each can be grown in one cubic meter volume of water. Though such mathematically arranged numbers may not always be feasible still such considerations gives one an idea of the number of fish that can reared in a given volume of cage and the pen.

In the cage rearing system the fishes do not get any natural feed though in the pen system, depending on the site there may be natural planktons of flora and fauna available; Hence the feeding becomes the most important aspect in fish rearing in the cages and pens. Regular feeding with sufficient quantity and quality feed should be carried out in cage rearing. Feeding should also be taken into consideration the nature of the fish reared whether carnivores or herbivores. Normally a mixture of oil cake and rice or wheat bran form the base in preparing feed for the fish. Along with this fish meal, meat meal, prawn meal, various types of fungi, worm meal, etc., should be included into this mixture.

The quantity of feed depends on the size of the fish bigger the fish greater will be the amount of feed. The more the intake greater will be the growth of the fish. For more details about feeding refer Booklet No.525 on "Fish Feeding".

Feed should be given on a flat vessel or pans placed in the cage or in the pens. This will avoid the loss of feed and it becomes easy to observe the condition of the fish while feeding. Fishes that feed actively indicate that they are in good health and those who do not eat or are sluggish in eating show that they are not in good physical conditions. Pan feeding also will give an idea whether the feed given is sufficient or not. The second aspect of the rearing of the fish in the cages and pens is to provide suitable environmental conditions to the fish in the cages and pens. Sometimes over crowding in the cages and pens may result in scarcity of feed and less availability of dissolved oxygen in the water. These negative effects should be over come by regular and sufficient feeding of the fish and allowing sufficient flow of water and enough sunlight through the cages or pens.

HARVESTING

Harvesting is done in the cages using a hand net. But in the pens and enclosures hand net will work only if the pen is small. If the pen is big we may have to use a small or medium type net depending on the size of the pen or enclosure. Compared to any other type of rearing methods cage and pen rearing of fish is the easiest system from the point of view of harvesting of the fishes. In the ponds and lakes when netting is done small and big and even the eggs may be caught and the fry and eggs may be destroyed. But in the case of cages one can easily select individual fish while harvesting and others are left free. However in the pens and enclosures harvesting is done by netting or by some other locally existing means.

The harvested fish are weighed and used for domestic purposes or sold out. It is advisable to maintain a record of the number of fish harvested and the yield by weight. Such records will help one to understand at what level of performance one is operating and how much net profit or loss is incurring.

11

Fish Pakaging Technology

PACKAGING EQUIPMENT

SEALING

A pack is only as good as its seal. Pouches are normally sealed on three edges during manufacture, leaving one edge to be sealed after filling. Heat sealing is the normal method of closing wet fish prepacks, usually by means of an electrically heated bar or strip; the metal may be heated continuously or heated and cooled alternately by an impulse technique. Some heat sealers employ high frequency heating, whereby rapid reversal of the current generates heat within the film. Some films, for example PVC and the polyamides, require the use of a special release coating on the jaws of the heat sealer to prevent sticking; polytetrafluoroethylene, PTFE, is often used for this purpose.

Automatic sealing machines usually employ a continuously heated bar which is clamped on the edge of the pouch for long enough to weld the films together. This type of machine is most suitable for PE and PP films. Some plastics, for example PVdC, cannot be sealed in this way, and here the impulse technique is used. The edge to be sealed is first heated and then allowed to cool while remaining clamped in the jaws of the sealer. Impulse sealing is slower, because of the cooling time required, and output is therefore reduced.

Small and comparatively cheap heat sealing machines can be bought or hired, so that a wet fish prepacking line can be started on a modest scale without investing large amounts of capital. Large machines handling thousands of packs an hour are expensive but they usually include a conveyor system.

Shrink Wrapping

Filled trays overwrapped with a shrink film are first held on a heated plate to seal the film under the tray, and are then passed though a hot air tunnel set at the correct temperature to shrink the film. The film becomes taut over the product, thus making a more attractive pack. The hot plate and air tunnel are relatively cheap and can have a high capacity.

PACK DESIGN AND PRINTING

Attractive presentation is one of the keys to success in marketing prepacked fish, and this means more than just arranging the fish neatly in the tray or pouch. Clarity of film, style, layout and colour of the printing on the film, and easily read information on price, weight and kind of product are all very important selling factors.

Artwork and lettering on the film can be either surface printed or sandwich printed. Surface printing is usually cheaper, but the print may become blurred or damaged by scuffing. With sandwich printing, one of the component films of a laminate is printed before the laminate is made up, so that the print is protected within the sandwich.

The print is usually applied to the surface of a flexible film either by the flexographic or the photogravure process. Flexographic printing, in which the ink is applied from a raised moulded surface of rubber, is generally cheaper for small orders, but the range of inks that can be used is limited to those that do not affect the rubber mould.

In the photogravure process the ink is applied directly from an engraved metallic roller, giving better reproduction than in flexographic printing. Photogravure can be as cheap as the flexographic method for large orders. Film manufacturers should be consulted about printing requirements, because some surfaces may require special inks and colours, or may need treatment before printing.

PACKAGING FOR DISTRIBUTION

Fresh fish packed in flexible film requires to be kept cool in transit and protected against physical damage. The product will spoil if it is too warm, and undue pressure may break the wrapping or squeeze fluid from the fish to make the appearance of the pack unattractive.

Packs should be transported in a protective outer container of a size that can be readily handled by shop staff. Cardboard outers with spacers between individual packs can be used provided insulated, refrigerated transport is available. Alternatively expanded polystyrene containers can be used to give some degree of insulation; non-returnable boxes of the kind used for carrying unwrapped fillets can be used, or specially designed boxes can be made in which built-in supports are provided for individual packs.

In the plain non-returnable polystyrene box, the packs stow best on edge and, if the packs are partially evacuated, the contents of the packs do not slide about within the pouch. The packs are kept cool in both types of polystyrene container by including ice in thick polyethylene bags; in the non-returnable box the ice bags are laid on top of the rows of packs, and in the special containers they are put in the slots at intervals between the packs. The type of box selected and the amount of ice used will depend on the length of journey and the nature of the distribution chain. In a well designed polystyrene container with a few ice packs it is possible to keep the temperature of the packs close to 0°C for up to 16 hours in an uninsulated vehicle, with an outside temperature of 16°C.

RETAIL EQUIPMENT

The retailer must have access to a chillroom capable of maintaining the packs within the temperature range 1-4°C, and must have a chilled display cabinet. Some existing chillrooms, designed to store bacon for example at about 7°C, may not be capable of running satisfactorily at a lower temperature.

The main stock of packs should be held in the chillroom, and the small number on display in the cabinet replaced as they are sold. The shop lighting and any lighting in the display cabinet should be chosen to show off the packs at their best; some smoked fish packs for example may appear to have un unattractive colour when seen in certain types of fluorescent light.

HYGIENE REQUIREMENTS FOR DIFFERENT PACKING MATERIALS

GENERAL

Boxes, containers and other equipment which come in contact with fish during catch, transport and production, have to be cleaned as often as possible and at least in between each use. Besides the natural micro flora found on fish and shellfish, external microorganisms from the surroundings are added. Bacteria and fungi come through the soil. the air and the water as well as from human contact with raw material.

Slime from fish provides a growth medium for most microorganisms. When the fish is dead, all defence mechanisms stop, and as a result the microorganisms living in the slime start to grow and attack the skin and the flesh. Offensive odours and flavours, discolouration and poor textures is the final result. The rate of spoilage can be reduced by quick cooling and careful handling of the raw material and use of clean equipment which minimize bacterial contamination.

WATER AND WATER QUALITY

Water is the cheapest and most convenient cleaning agent to use, but dried organic soil is difficult to remove by simply scrubbing with water, so detergents are added to dislodge, disperse and dissolve soil. The quality of water used in the fishing industry is an important consideration because both chemical and bacterial contamination of seafoods can occur if polluted water is used. The level of dissolved minerals can also affect the solubility of cleaning and sanitising agents in water. Hard water contains elevated levels of calcium and magnesium ions. This water usually comes from limestone areas or from deep wells. The ions in hard water complex out detergents and so either a water softener must be added to remove the ions or excessive amounts of detergents will be needed for effective cleaning. Rainwater and water from lakes and streams is usually softer, containing less mineral salts Contamination of lakes and rivers from various pollution sources can render it unsatisfactory for use. In fishing, this water is used not only for cleaning but also for ice manufacture and direct human consumption, so its quality must be known.

REQUIREMENTS TO WATER QUALITY

Bacteriological Requirements

- Total acceptable plate count, max. 50 bacteria per ml;
- If there are between 2-23 coliform bacteria per 100 ml, the water and water source must be given a detailed examination before being used;
- Water with more than 23 coliforms per 100 ml is unsuitable for use.

Physical Requirements

- Colour = without colour (max. 20 mg Pt/l)
- Smell and taste = none, or very little
- Appearance = clear and without sediments (max. 10 mg SiO2/l)

Chemical Requirements

- Permanganate-numbers (indicates content of organic substances) max. 20 mg kMnO4/l
- pH (acidity) 6-8.0
- Conductance, max. 300 microsiemen
- Sulphide, max. 0.1 mg/l
- Manganese, max. 0.05 mg/l
- Copper. max. 0.05 mg/l
- Hardness, preferably below 2° dH

These are standard values for a good water quality used for cleaning. Chlorination, filtration, and softening agents can improve water quality.

DETERGENTS

GENERAL

A great number of detergents are found in the market. These substances are often complex and contain many different chemical components. Most of the industrial detergents are corrosive and can, when used in a wrong way, cause damage to equipment as well as to persons.

The buyer of detergents has to make sure that the producer gives a thorough description of the chemical composition, areas of use and the correct dosing. Information about storing, safety requirements and marking as well as information on measures to be taken in case of accidents should be included.

REQUIREMENTS TO DETERGENTS

The cleaning solution must remove all dirt from the item being cleaned, without damaging it, even after it has been cleaned several times. The detergent must be stable for storage, *i.e.*, not change appearance form or quality when stored and be quickly and completely dissolvable. Liquid detergents are easy to use, but expensive. Powder detergents are cheaper, but require more time for accurate dosing and dissolving.

THE CLEANING PROCESS

The cleaning process can be divided into two:

1. Washing
2. Desinfection

It is not possible to use these two processes in combination and get a satisfactory result. This is clearly illustrated in the following example, from tests in the food technology industry:

Table. Effect of Cleaning Systems on Bacteria.

	Bacteria Numbers/cm^2
Before cleaning	2 900 000
After cleaning without detergent	40 000
After washing with detergent	5 500
After desinfection	10
Desinfection without cleaning	1 200 000

A well prepared washing programme has to include the following processes:

- Mechanical removal of dirt (brush or other equipment).
- Rinsing with tempered (or cold) water.
- Washing with hot water (40-70°C) with detergent.
- Rinsing with tempered/cold water.
- Desinfection of the clean surface leaning time for the sanitiser to work.

Even if a cleaning programme can be set up in general terms, the variations will be many because of types of contamination to be removed, type of equipment to be cleaned and available cleaning equipment.

CLEANING METHODS

Because dried soil is very difficult to remove, hasing and mechanical scrubbling should be performed as soon as possible before the soil dries. This will reduce the effort needed later on.

There are many methods which can be used, and a satisfactory result can be obtained in several ways.

According to the size of the equipment to be cleaned, the location and the frequency of cleaning needed, the following methods can be used:

- Manual cleaning.

- High pressure washing.
- Foam wash.

MANUAL CLEANING

Often, the working area is small or overloaded with other equipment so that manual cleaning is the only applicable method. Brushes of various types are simple to use and will give a satisfactory result in most cases. It is important that the equipment is kept clean and not used for other work.

Fig. Hand Brush for Washing Small Items.

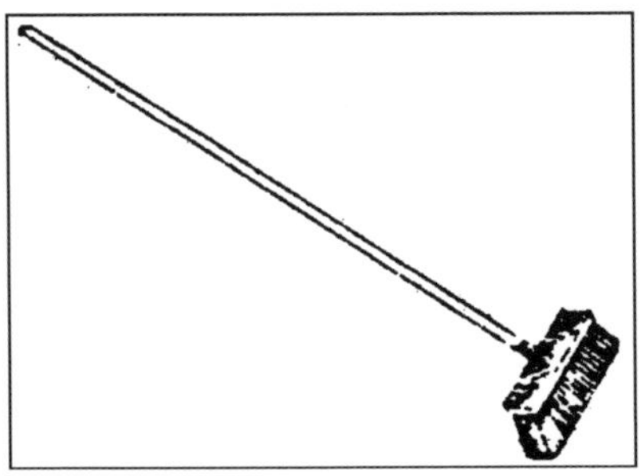

Fig. Brush with Handle for Washing Large Boxes, Containers, etc.

This simple equipment is low priced and well suited for small operations or when the labour cost is low. If dirt has been allowed to dry, the equipment to be cleaned should be put in water to soak for some time. Washing by hand will then be more simple and will in most cases give a good result.

FOAM WASHING

This process is often used in the food industry. Special foaming detergents are sprayed onto the surface to be cleaned and left in contact with that surface to allow the detergent to dissolve the soil. This is particularly useful on processing equipment where the complexity of design makes manual cleaning impossible. The foam and soil is removed by hasing or manual cleaning. Unfortunately the "airy" consistency of the foam means that it soon assumes the temperature of its surroundings so the extra cleaning power of hot detergent is not possible using this method. To make the foam, special equipment is needed. This is connected to a tank with special chemicals and the foam/water mixture is sprayed on the items to be cleaned.

HIGH PRESSURE WASHING

Equipment for this process is found in different shapes and sizes of both portable and stationary types

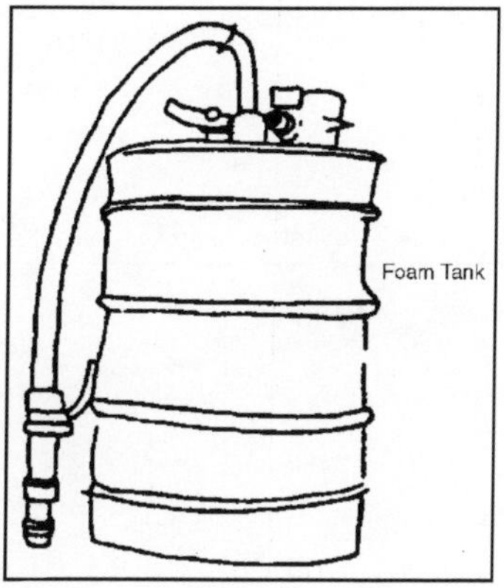

Fig. Foam Making Equipment.

The small portable units are often adapted to small production units and are self contained. The stationary ones will be connected to pipe lines leading to strategic places. These units usually supply hot detergent water at pressure. The small portable units give up to 40-110 litre liquid per minute and work under a pressure of 40 kg/m^2. Some of these units have mechanical equipment for dosing and mixture of detergents and water. The large stationary pumps can give up to 300-litre liquid per minute under a pressure of approx. 55 kg/cm^2. When using such equipment the detergent can be dosed/mixed manually or automatically.

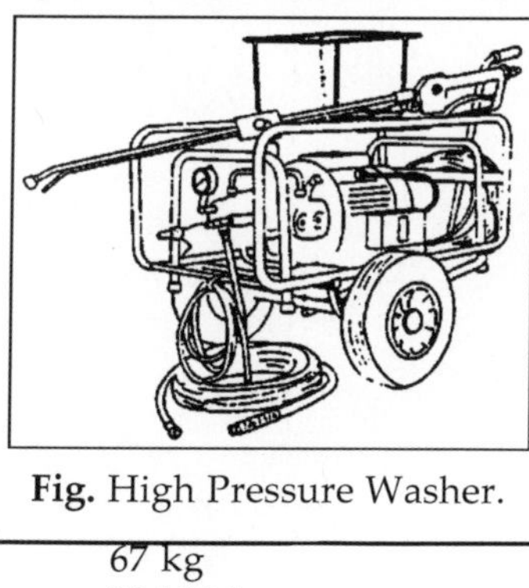

Fig. High Pressure Washer.

Total weight:	67 kg
Working pressure:	20.130 bar
Water capacity:	13 l/min
Water connection:	3/4 in hose
E1. Motor:	3 kW 220/380 V-3 phases
E1. cable:	8 m
H.P. hose:	8 m
Price:	approx. US$ 1 000

Specialised pressure cleaning equipment is also available

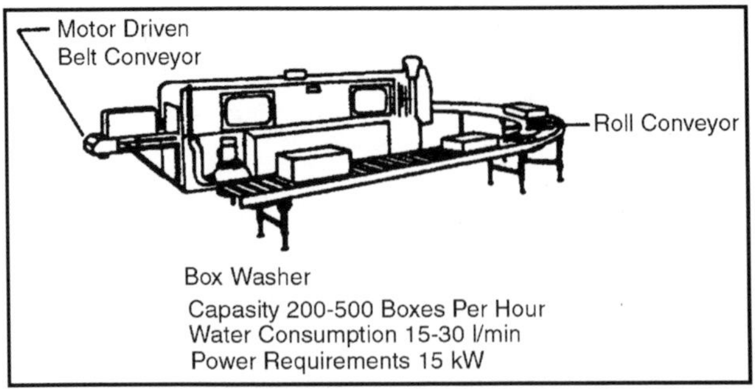

Fig. Box Washer.

DESINFECTION

Washing and desinfection must be performed separately for maximum effect. There are two reasons for this. Detergents and sanitisers are often chemically incompatible and most sanitisers are complexed not only by microorganisms but also by any organic soil present.

The washing process should remove all the soil and some of the microorganisms. This should be followed by thorough rinsing to remove all detergent then application of a sanitiser or disinfectant which will kill most microorganisms and inhibit the growth of others. This should be left some time to allow it to work before rinsing off.

Desinfection is not only done to boxes and containers, but to all other equipment as well as floors and walls.

Below follows a rough list of existing types of desinfection products:

- Chloride products
  - Hypochlorite (10-20 ppm/litre of water)
  - Chlorine amines
  - Clorisocyanur acids
- Jodine products
- Phenoles
- Alcohols
- Quaternary ammonium compounds

A variety of methods are available for application of sanitisers; these include manual spreading, spraying and fogging for enclosed tanks containers, etc..

Cleaned boxes made of modern materials like plastic, ought to have no more than 10.20 nos of bacteria per cm^2 on the surface when being examined.

PACKAGING USED IN FRESHWATER FISH PROCESSING

THE ROLE OF PACKAGING

The previous section discussed the processing methods most often used by small freshwater fish processing plants. Quality assurance is essential in each technological process, and suitable packaging materials and methods are of great importance. If these requirements are not met all efforts made during processing could be of little avail, which could lead to serious economic losses.

Packaging should protect the product from contamination and prevent it from spoilage, and at the same time it should:

- Extend shelf life of a product.
- Facilitate distribution and display.
- Give the product greater consumer appeal.
- Facilitate the display of information on the product.

The quality of freshwater fish which is delivered to the consumer or the processing plant as live fish greatly depends on correct handling during transport and, when processed, on suitable packaging. For short distances, the live fish can be transported in insulated containers with lids, capacity varying from 300 to 1 000 kg of fish. Fish can also be transported in normal lorries, but for long distances the water in the containers must be aerated and cooled by portable devices.

In order to maintain good quality of fresh fish during transportation, fish boxes made of suitable materials should be used

When purchasing fish boxes the six following requirements should be remembered; they should:

1. Be of a suitable size for the range of fish to be handled or the product to be put into them.
2. Be of a convenient size for manual handling or lifting by mechanical equipment.
3. Be stackable such that the weight of the containers on top rests on the containers underneath and not on the fish.
4. Be constructed of impervious non-staining materials.
5. Be easy to clean.
6. Provide drainage for melted ice.

Fish boxes are usually made of high-density polyethylene. Although this offers many advantages, such as duration, lightness, ease of cleaning, there are also disadvantages, *e.g.*, high price and the fact that they are not returnable. That is why disposable fish boxes of about 25 kg capacity (fish and ice) are more often used: these include fibreboard cartons, waxed and waterproof boxes. In the case of transport by lorries with no cooling system, insulated cartons, *e.g.*, boards made of moulded polyestyrene should be preferred. The latter is commonly used for delivery of chilled and frozen fish and fish products to wholesale and retail outlets. In the case of fillets, each layer of fillets should be packed thin and separated from the ice with a plastic foil.

Styropor boxes are normally sold with lids, which fit very closely and can be with or without drainage holes. In a typical range, wall thickness varies with box size; *e.g.*, a 6 kg capacity box has a 15 mm thick wall, a 10 kg box a 19 mm wall, a 25 kg box a 25 mm wall. The main disadvantage of moulded polyestyrene fish boxes is their lack of strength. They are easily damaged or broken by rough handling. This limits their size and use.

Polyestyrene is difficult to clean. Polyestyrene boxes are difficult to re-use, and are usually non-returnable. They may cause disposal problems due to their bulk.

The packaging industry improves its products by using new materials with better insulating properties or by introducing new leakproof designs. The new containers are often lighter and less bulky. For example, the Therma Gard packing system consists of a metallised plastic bag (which reflects practically all radiant heat). This is then wrapped in a waterproof and leakproof carton. The metallised bag, together with a bubble-pack wrapper, provides a double-pack insulation. The Therma Gard bag can be sealed airtight and thus be used for carrying live fish. The Stratech aluminised boxes have a wall thickness of only 5mm and it is claimed that these boxes have similar insulating characteristics as polystyrene boxes with 30 mm wall thickness.

The future use of expendable packages is becoming questionable as there is a growing discussion, for example in some states of the USA, on imposing a ban on these packages.

The main drawbacks in using returnable containers are freight costs for returning empty containers. Use of "knock-down" returnable containers will reduce freight costs.

RETAIL PACKAGING FOR FRESHWATER FISH PRODUCTS

The main role of packaging is described above but in respect of retail presentation it should also reduce the smell and the drip, and enable the product to be tucked into shopping baskets with other purchases. Moreover, the packaging of fish products should ensure attractive presentation among other food products without contaminating them.

Basic packaging materials include paper, cartons, sheets of metal, metal foils and many kinds of plastics. Despite the rapid growth in use of plastics, the role of paper and carton as packaging materials does not decrease.

Kraft paper or carton are often laminated with polyethylene or aluminium foil which render them waterproof. Such material is used for production of trays for packaging of fresh or frozen products.

More often, trays are made of plastic materials such as polyestyrene or expanded polyestyrene. Expanded polyestyrene is frequently used but it is partly oxygen-permeable and so those products which are sensitive to rancidity have to be additionally overwrapped or skin-packed with suitable film.

The materials mentioned above are not stable at high temperatures and hence are not suitable for trays to be used in an oven. Polyester can be used as a packing material for heating of the product in the traditional and microwave ovens, but this material cannot be used for microwave cooking. Trays used for packing are generally overwrapped with a protective film, often with PE wrapping which shrinks. The film shrinking is achieved by use of hot air or hot water. Stretch wrapping is often used for products which are heat-sensitive. The film is stretched over the product manually (very often in the supermarket) or by machine. Foils used as wrapping or bags for packing of trays with product must be puncture-proof, extensible and impervious to gases like oxygen.

Hundreds of different films are used in the packaging industry.

These can be broadly categorised into two groups:

1. Basic films consisting of a single layer of film.
2. Laminate consisting of two or more basic films glued together or bonded together by heat or by adhesives.

Plastics such as polyethylene film or copolymer of ethylene and vinyl acetate are very often used for packing of frozen products. Polyethylene packs can be produced manually using pre-made bags. An impulse or bar sealer is used to seal the bags which are hand-filled.

In order to improve the barrier properties of packages laminates are used, for example polyester/polythene. Products which are particularly sensitive to oxygen are vacuum-packed. During the sealing operation, air is removed from the package. A laminate nylon/polythene is commonly used as packaging material. This type of packaging is used, for instance, for smoked trout which are arranged on a board with, for example, a coated texture. Numerous machines exist for vacuum-packing with single, double or continuous chambers. Vacuum-sealing machines can additionally be equipped with a modified atmosphere packing system (MAP). Immediately on removing the air from the package a mixture of gases is pumped in. Usually this mixture consists of 30 per cent nitrogen, 40 per cent carbon dioxide, and 30 per cent oxygen. In the case of fat fish the oxygen is replaced by nitrogen. This method is increasingly used for packing fresh fish. The MAP products have to be stored at the temperatures lower than 3° C because of *C. botulinum* hazard. MAP packages consist of two kinds of foil. The bottom film is foil-rigid or semi-rigid. This foil is formed by, for example, extrusion and the resultant tray is moved to the packing section. Because of product drip it is placed on an absorbing board. The top web is drawn over the filled trays and sealed round the edges. The pack may be evacuated or gas-flushed before sealing. Vacuum-skin packaging is becoming more common for packing smoked fish. In this process the wrapper is heated and wrapped over the product, the film moulding completely to the product shape and sealing the product completely, forming an extra skin.

LABELLING REQUIREMENTS FOR FRESHWATER FISH PRODUCTS

Lack of detailed standards and existence of only limited regulations concerning wholesomeness and sanitary conditions for production and trade of food products characterise the market economy. Here, the problem of labelling is of a particular importance. Regulations in this regard are very detailed and are aimed at protecting the health of the consumer and providing the best information. These requirements enable the consumer to decide which products to buy. A label placed on the product should inform the consumer about the raw material used, method of preparation and form of consumption, shelf life, etc.

Product labelling is of prime concern in the European Union. Directive 79/112/EEC of 18 December 1978 was revised several times, and in 1990 there came into force a new Directive 90/496/EEC which concerned labelling and providing information on nutritive and energetic values (kcal or kJ/100 g or 100 ml), the amount of basic ingredients and nutritive compounds such as: proteins, carbohydrates, fat, fibre, sodium and vitamin content (EEC, 1979). These requirements were supplemented in Directive 89/396/EEC recommending the labelling of batches of product which would make it easier to withdraw the batch from commodity turnover in the case of health hazard.

Taking into account the necessity to ensure complete information on the product to facilitate the selection of a healthy and economic diet, the US Food and Drug Administration (FDA) recently proposed a voluntary Nutritive Labelling Programme which covers, *inter alia*, a proposal for placing on the product for example information concerning the percentage of recommended daily intake of protein, vitamin A and C, iron, calcium, etc.

FRESHWATER FISH PROCESSING -EQUIPMENT AND TECHNOLOGICAL LINES

In small freshwater fish processing plants only limited preservation methods are used as compared with marine fish processing establishments.

CHILLING AND STORAGE OF CHILLED PRODUCTS

Decreasing the temperature of the fish to about 0° C slows down the microbiological, chemical and biochemical decomposition processes and extends fish stability. Thus when the raw material is cooled quickly, just after capture, and kept at low temperature during transport, processing and distribution, it meets the basic processing requirements. Its usefulness is extended and at the same time fish quality is maintained.

In freshwater fish processing the raw material, and semi-products and final products are almost exclusively ice-cooled. The heat exchange process between fish and ice is complex as it takes place between the fish surface and the ice, between the surface of fish and the melting ice water, and also between the fish and the cool air in spaces between the pieces of ice.

Overall, it is a dynamic process, changing minute by minute. Water from the melting ice plays the most important role as it causes a typical convective exchange of heat. But the direct exchange of heat between ice and fish is also important, and thus the ice granulation is very important for the whole process.

In modern fish processing plants, especially the small ones, flake ice generators dominate as flake ice ensures major contact surface with fish and its production cost is low. Flake ice production consists in freesing a thin layer of water on the cooled surface of a cylindric evaporator and then scraping off the ice with a knife.

Modern ice generators generally comprise a vertical cylindric evaporator. Ice is formed on the outer, inner, or on both the surfaces of evaporators (Figure).

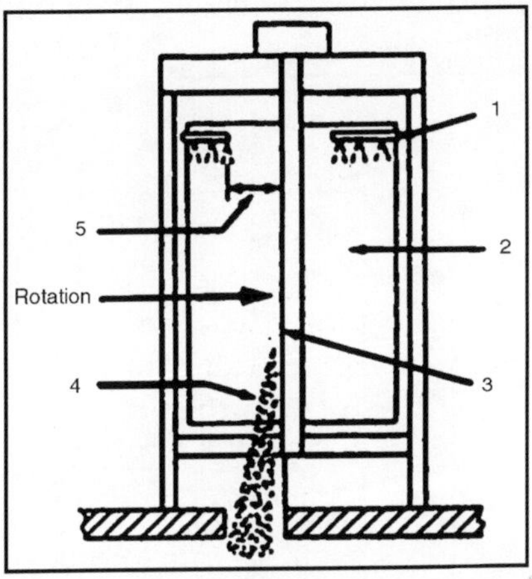

Fig. Flake Ice Maker. 1. Water Spray, 2. Ice Freesing Drum, 3. Ice Breaking Device, 4. Ice to Store, 5. Subcooling Zone.

Ice production is a continuous process and ice is collected in an insulated container. When the container is full the mechanism stops functioning. Capacities of flake ice generators vary from 100 kg/24 h to 60 t/24 hours. However, due to the high cost of equipment, fish producer should rather consider purchasing flake ice from the nearest cold store plant.

When the producer decides for organisational reasons (*e.g.*, production unevenly distributed in time) to buy an ice generator it is advisable to buy two small capacity generators instead of one of a greater capacity. The effectiveness of temperature exchange depends on the thickness of the layers of fish and the distribution of ice. For example, an 80 mm layer of fish requires two hours to decrease the temperature from 10° C to 17° C when exposed to double-sided cooling, and about 24 hours when exposed to one-sided cooling. To evaluate optimal conditions for fast cooling of fish, many parameters (degree of ice granulation, temperature of the fish and the environment), which influence the activity of the process, should be known.

Greater amounts of ice do not shorten the process. It was ascertained that use of 25 per cent ice in relation to the amount of fish causes temperature to drop to 5° C after 3.3 hours, for 50 per cent ice - cooling down to 1° C takes 6 hours, but for 75 per cent ice - 2.25 hours. Standards for use of ice should be set individually for different types of fish and fish products, different conditions, seasons, etc. The ambient temperature does not affect the cooling rate of the fish, but considerably affects the amount of ice necessary to maintain a low temperature. It is difficult to determine the exact amount of ice needed to keep the fish temperature at about 0° C. In short-distance transportation (up to 24 hours) during the cold season (up to 10° C) 1 kg of flake ice is sufficient to cool 8 kg of fish. When ambient temperature exceeds 10° C, 1 kg of flake ice suffices for 4 kg of fish.

The following diagrams show the flow of technological processes for chilled products:

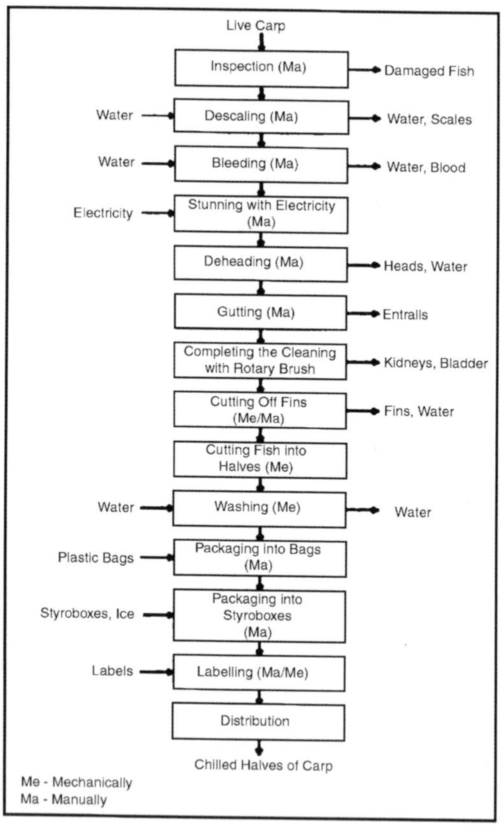

Fig. Production of Chilled Halves of Carp (Technology Used in Czech Republic).

Proper handling of freshwater fish as raw material and its products ensures continuous cooling with ice and maintenance of temperature. All processing phases should be as short as possible and if for any reason a surplus of raw material occurs this should be sent to the cold stores. Raw materials and products should be transported so as to ensure the maintenance of temperature close to 0° C; this involves both the most simple isothermal vehicles and mechanically-cooled containers.

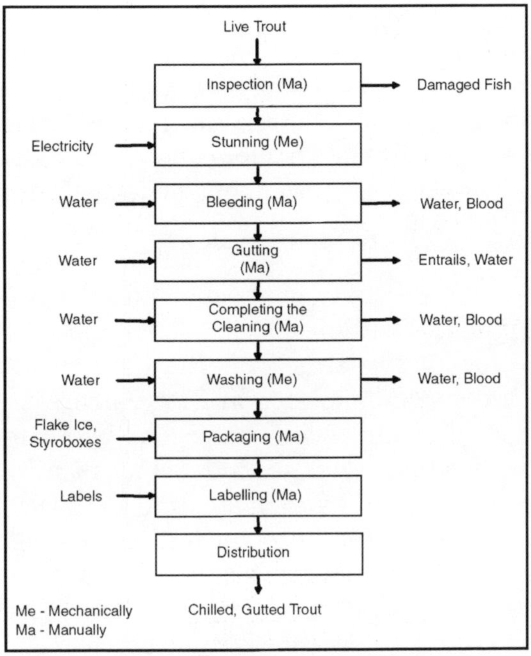

Fig. Production of Chilled, Gutted Trout.

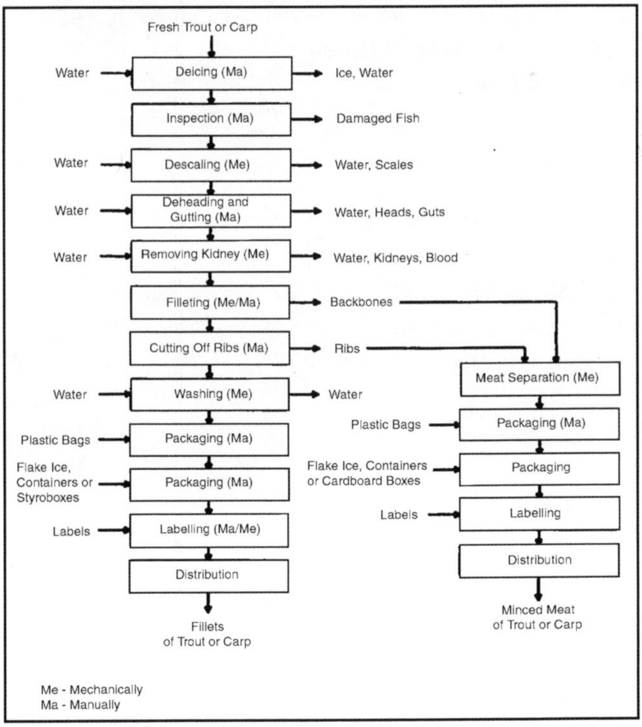

Fig. Production of Chilled Fillets of Trout and Carp (Technology Used in Poland).

Fish and fish products should reach the buyer without delay. In practice, in freshwater fish processing the wholesale storage phase is omitted due to the small scale of this kind of production. Products are delivered direct to shops where they should be placed in cold stores and if necessary ice should be added. Good trade practice indicates that retailers should only keep a one-day stock of cooled fish or fish products such as fillets, deheaded and gutted fish.

FREESING AND REFRIGERATED STORAGE

Even when the most effective chilling methods and further chilled storage of raw fish and fish products are applied, shelf life is limited. Freesing is needed to extend shelf life for long periods. This can be achieved by changing two parameters: first, a considerable decrease in temperature, and second, by freesing the water in the fish tissue. The second is of particular importance because water in the fish tissue acts as a solvent for many organic and mineral compounds which are a suitable environment for the growth of micro-organisms and also because they influence the biochemical processes. At the same time, the frozen water in the tissue causes changes in muscle tissue as a result of damage of cell structure during the formation of ice crystals. Further, the denaturation of proteins takes place during this process. An increased drain of tissue fluids, fat oxidation and dehydration are the effects of denaturation which are visible after the defrosting process. During the freesing process the majority of micro-organisms is inactivated and only psychrotrophic bacteria can develop in such conditions and to a limited degree. A temperature of about -10° C is a limit for growth of such micro-organisms. Some moulds and yeasts multiply very slowly at -15 to -18° C.

Fish should be frozen rapidly in order to produce the highest quality frozen products. Quick freesing implies a fast change from cryoscopic temperature to -5° C. During this period (about 2 hours) the main changes take place in fish tissue. A faster freesing process is linked to the formation of smaller ice crystals which damage the cellular membranes to a lesser degree, especially if freesing takes place before *rigour mortis* sets in.

The size of the ice crystals depends on the duration and temperature at which the fish was chilled/stored prior to freesing. The longer the time and the higher the temperature the bigger the crystals. Changes of proteins and oxidisation of lipids in muscle tissue are the results of slow freesing process and unsuitable conditions (time, temperature) of fish storage before freesing. These affect the quality of the final product.

In small fish processing plants there are usually two kinds of freesing equipment: chamber freezers and contact-plate freezers. The simplest is the chamber freezer-batch air blast freezer which consists of a battery of evaporators, a ventilator for air circulation and a rack for trays with fish products or for unpacked raw material. Versatility is the main advantage of such freezers as they make it possible to freeze different kinds of products, for example, regular shape blocks of fish/fillets and individual fish/fillets on the wire nets.

For that reason such freezers can be used in small plants; but high energy consumption and their large size are the main disadvantage. Contact freezers are far less common in fish processing plants with low daily production. Their operation consists in placing the fish for freesing between two plates which are cooled mechanically. This device is installed exclusively for freesing fish which is in regular blocks. In these freezers, good contact between the plates and the fish is essential to ensure rapid removal of heat from the product. Many kinds of such freezers are available including those with limited capacity, *e.g.*, 1 500 kg/24 h, and requiring little space, about 1.2 m^2.

Even properly frozen fish has limited storage life. Low temperatures inhibit processes of microbiological decomposition but do not protect against fat oxidation and loss of water. The stability of frozen fish depends on the initial quality of the raw material, the rancidity, the drying process and the storage temperature.

Glasing is the simplest and cheapest method which effectively prevents water loss of from fish tissue and prevents rancidity. Glasing consists of forming a very thin adherent layer of ice on the fish's surface. This method is used especially for freesing of whole fish or in fish/fillet blocks. Individual portions of fish or individual fillets are packed in plastic material characterised by low permeability of water vapour and oxygen. This prevents rancidity and loss of water.

The storage temperature of frozen products is the next factor which influences the quality and stability of frozen products. Unfortunately, industrial practice shows that the basic principles of freesing process are often not complied with, especially in small and poorly equipped establishments. Fish is frequently frozen in store chambers, home freezers, etc. The capacity of such chambers is limited, temperature is not stable and generally lower than required. Further, no temperature recording is made. Low quality of products results from such practice, particularly texture and flavour; fish becomes dry and very often discoloured.

Table. Practical Storage Life (PSL) of Fish Products in Relation to Storage Temperature.

Fish product	Storage Life in Months		
	–18 °C	–24 °C	–30 °C
Fat fish glazed	5	9	> 12
Lean fish fillets	9	12	24

SMOKING OF FRESHWATER FISH

Smoked freshwater fish such as eel or trout, and less often carp, are the most popular fish products. Saturation of raw material with wood smoking is the main principle of the smoking process. During this process, some water is removed from the tissue and changes of proteins occur.

The smoked fish is then ready for consumption without further culinary treatment. There are two methods of fish smoking: hot and cold, which give very different products. The difference lies in stability and sensory properties which in turn depend on a degree of fish dryness and saturation with smoke components.

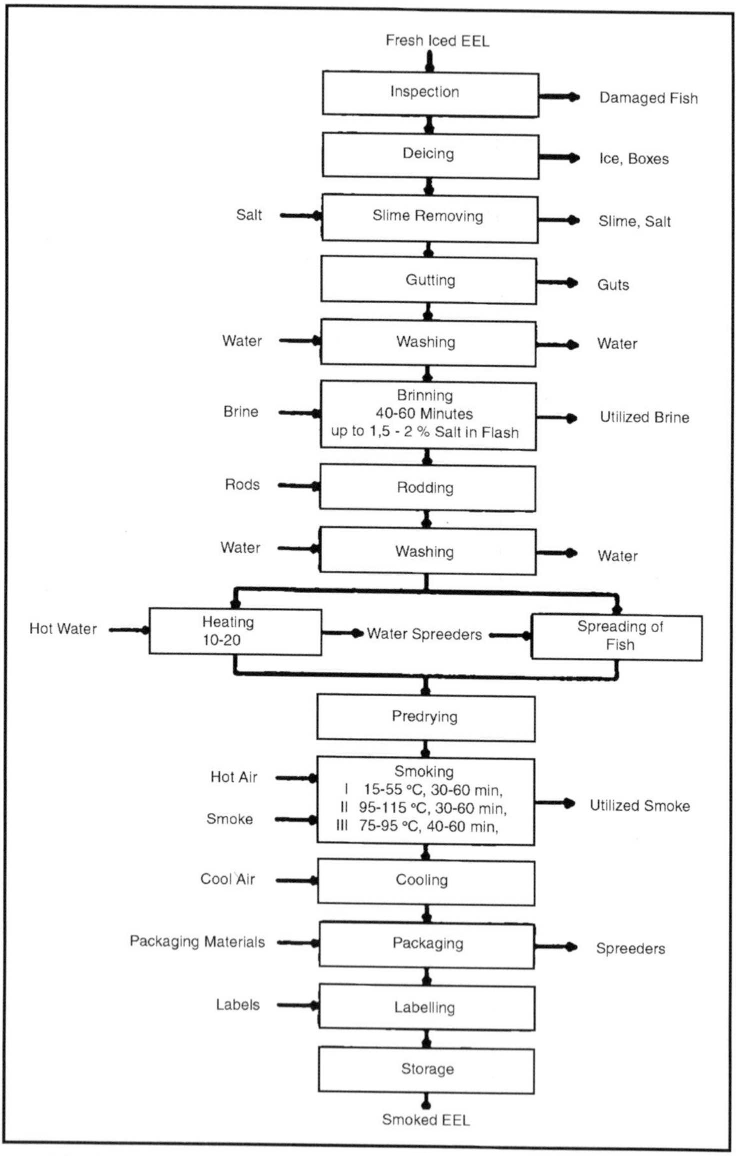

Fig. Flow Diagram of Hot Smoking of EEL: I–Preliminary Drying, II–Thermal Treatment, III–Smoking.

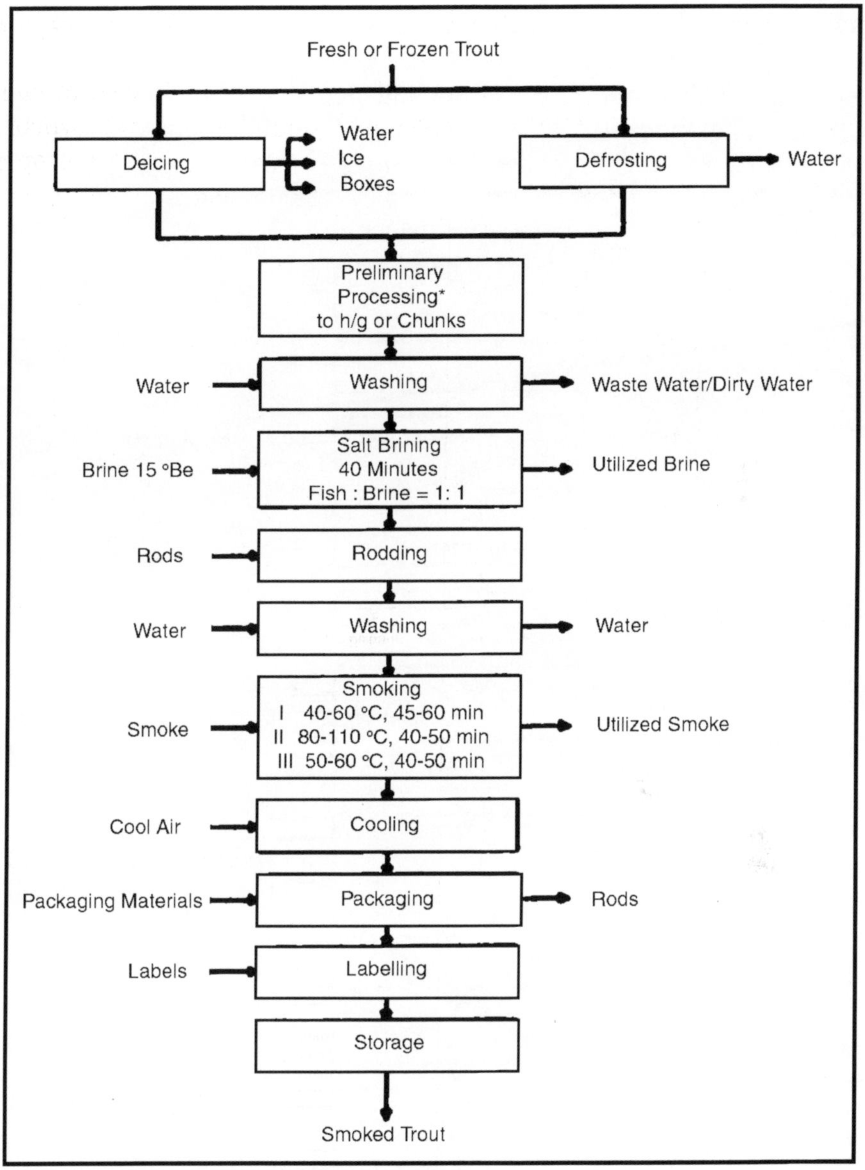

Fig. Flow Diagram of Hot Smoking of Trout: I–Preliminary Drying, II–Thermal Treatment, III–Smoking.

Smoke is produced by a not complete burning of some type of wood and is a mixture of more than a hundred chemical components. The chemical composition of smoke depends on the type of wood and traditionally deciduous tree wood is used.

During the smoking process sensory features such as colour and flavour undergo changes. The colour of properly smoked fish depends on the quantity and composition of the smoke components absorbed through the fish surface;

the higher the smoke density the darker the colour of the fish. Smoke density and humidity inside the smokehouse influence smoked fish characteristics. Flavour is the most typical feature of smoked products. It is generally considered that phenol compounds and other components soluble in water are the most important criteria in creating flavour in smoked products.

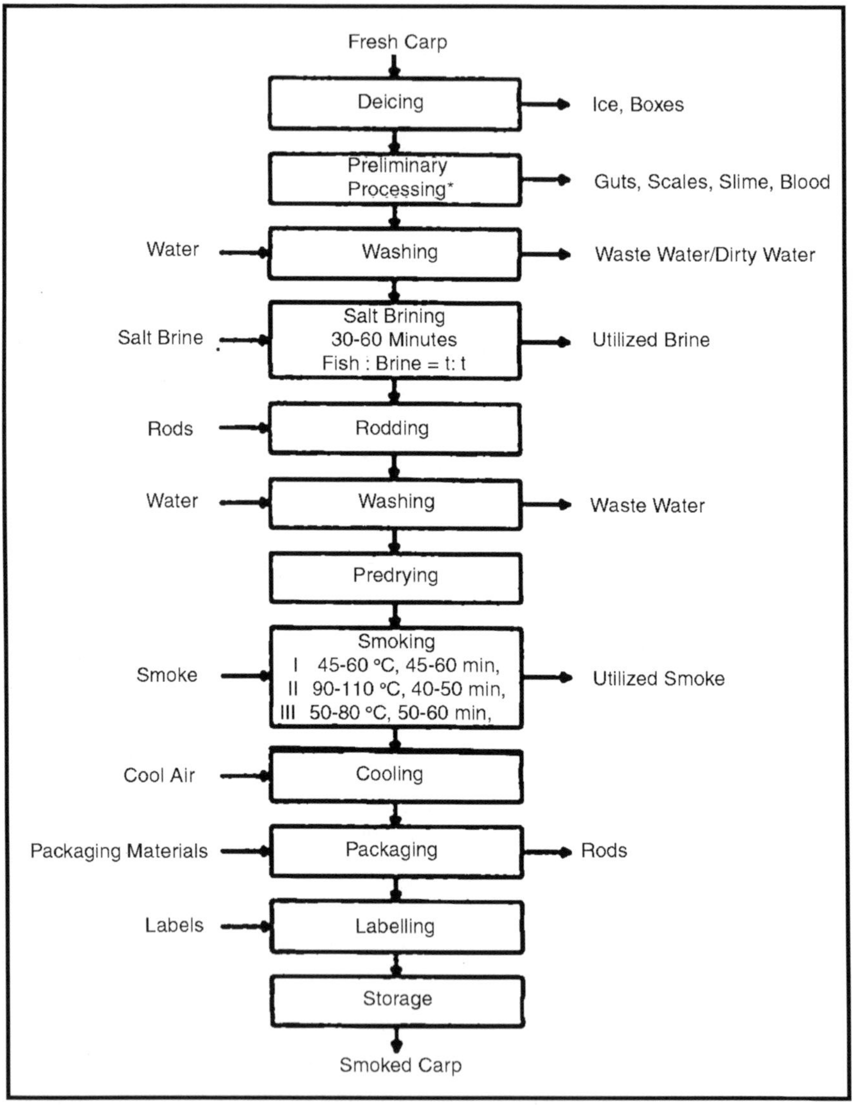

Fig. Flow Diagram of Hot Smoking of Carp: I–Preliminary Drying, II–Thermal Treatment, III–Smoking.

The presence of antioxidants in smoke renders smoked products resistant to rancidity. Hot-smoking reduces microbiological growth thanks to high temperature (close to 80° C in tissue) and the antiseptic components of smoke. Generally, after hot-smoking fish products contain only meso- and thermophilous micro-organisms, resulting from heating the product and not the antiseptic action of smoke components and salt content. Cold-smoking enables preservation of the product by smoke components. Their concentration in the product is higher than in hot-smoked fish and the product is drier. The vegetative forms of micro-organisms are the most sensitive to smoke treatment but spores of moulds are relatively resistant. For that reason, smoked products often grow with mould - the main disadvantage. The hot-smoking process includes the preliminary processing of raw material, brining, drying to a certain loss of water content, the actual smoking process and thermal treatment at temperatures above 30° C, usually 70-80° C.

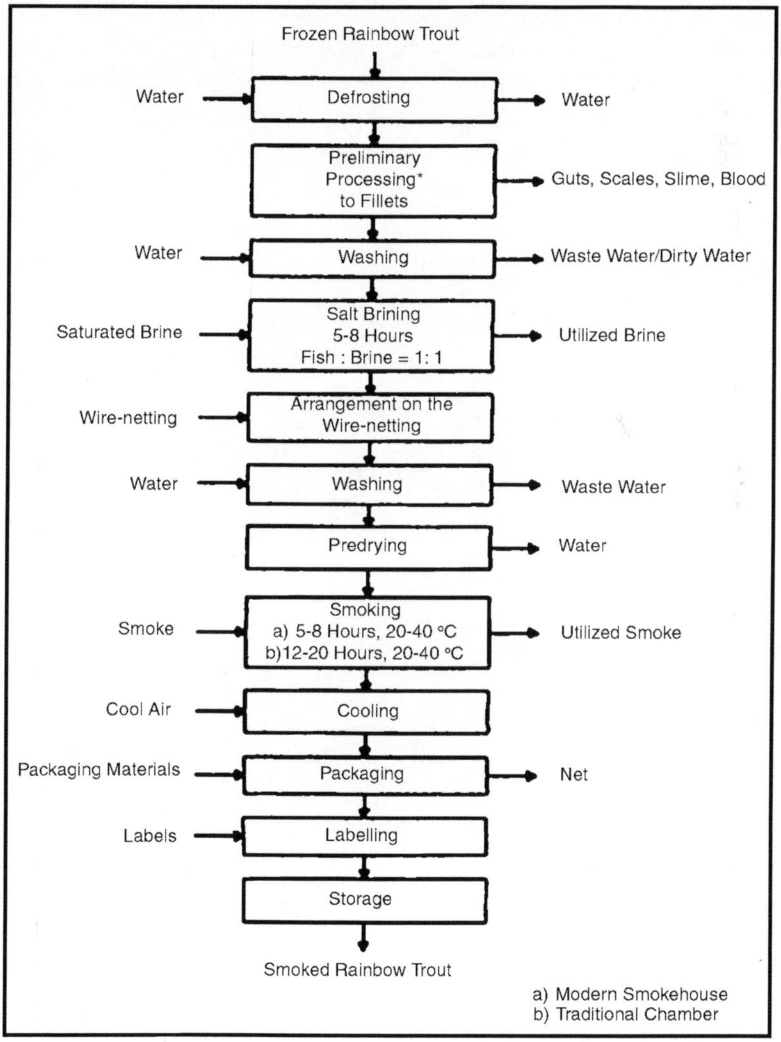

Fig. Flow Diagram of Cold Smoking of Rainbow Trout Fillets: I–Preliminary Drying, II–Thermal Treatment, III–Smoking.

The cold-smoking process involves no thermal treatment and the entire process is carried out at temperatures below 30° C.

During hot-smoking, brining is carried out to ensure penetration of about 2 per cent of salt into the fish tissue; the salt gives the desired taste to the product. During cold-smoking, salt is required for the conditioning process which favours the action of the enzymes. However, the brining process can be a source of microbiological reinfection. It was shown that multiple use of brine, 20 per cent salt content, may produce a source of many micro-organisms including spores of *Clostridium botulinum*. The brine thus needs to be changed frequently. Drying is carried out in order to reduce the water content in fish tissue to a level which ensures product stability and texture. Usually 25-30 per cent weight loss takes place during hot-smoking and 40-45 per cent during cold-smoking.

During hot-smoking thermal treatment should be continued until the temperature inside the thickest part of the fish reaches about 70° C. This ensures the denaturation of proteins and destruction of micro-organisms to a high degree. In some countries, *e.g.*, the USA, fish originating from the Great Lakes could be infested with *C. botulinum*. Thus fish with minimum 3.5 per cent salt content should be heated up to 82.2° C and thermal treatment continued for about half an hour.

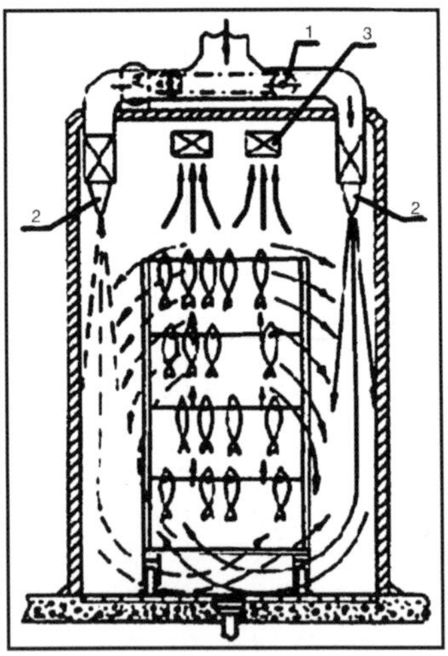

Fig. Modern Smoking Chamber: 1–Valve, 2–Smoke Inlet, 3–Smoke Outlet.

That process should be followed by very rapid cooling and storage at temperature below 4° C or preferably freesing. Thermal treatment should be conducted at humidity lower than 70 per cent because of bacteriological effect. Thermal treatment in the modern smoking house, very often equipped with an automatic control stem and adjustment of processing parameters, like air and smoke, can be programmed to maintain optimum temperature. Traditional methods of smoking do not ensure the same results but the traditional process, carried out in smoking chambers, is much cheaper. Wood is a source of smoke and energy necessary for this process. The effectiveness of traditional method depends on the experience of the operator.

PRODUCTION OF FISH SILAGE FROM OFFAL

During fish processing, a large quantity of offal is produced and its proper utilisation poses a problem, particularly for smaller processing plants. Fishmeal production is not profitable because of a low supply of the raw material, and thus production of a liquid form of this fish product is the only simple solution. Production of fish hydrolysate (silage) to be used as feed is the cheapest way of utilising offal. Considering the capital needed and the operating costs for fishmeal and hydrolysate production, production of the liquid form of this by-product is very profitable and it can be done by small plants. It is a simple technological process, but several rules must be observed to obtain a satisfactory final product. The raw material, the, offal, must be fresh; decomposing offal should not be processed. The main phases of offal processing are: grinding of offal or whole fish, acidifying of the pulp and liquefying it which results from a self-digestion (autolysis) process. Adequate grinding is a basic operation of the process.

The following preservatives are used to produce pyrosilage:

- Sodium pyrosulphite ($Na_2S_2O_5$), 1 per cent for fatty and medium fatty offal, and 1.3 per cent for lean product,
- Sulphuric or hydrochloric acid, both at 1 per cent concentration in the mix.

The measured pH should always be the final indicator of a proper level of acidification and should range from 3.5 to 4.5. The pH should never exceed 4.5.

The basic requirement of the process is to obtain homogeneity of the mix consisting of the fish, inorganic acid and sodium pyrosulphite. Homogeneity can be achieved by using slowly revolving mixers or other methods (turbulent mixing causes aeration of the mix and consequently oxidation of fatty acids). When mixing is too gentle, pockets of mix occur which do not contain preservatives, and decomposition of the product by the bacteria may begin. Each day the end-product is pumped into the retaining tank(s). These tanks should be equipped with mixers or recirculating systems powered by pumps. The tanks should be located under a roof to avoid solar radiation.

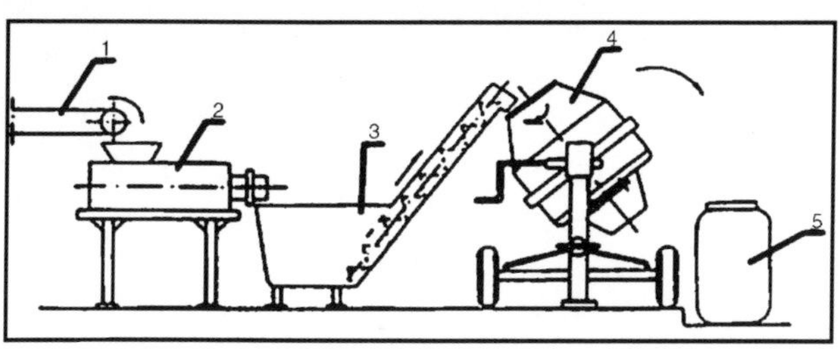

Fig. Simplified Layout of Silage Production: 1. Feeder, 2. Grinder, 3. Tank with Elevator, 4. Mixer, 5. Plastic Barrel.

The silage can be stored for up to 6 months if it is stirred periodically and kept at about 15-20 °C. In small freshwater fish processing plants where the volume of offal and fish not used for consumption is low (*i.e.*, 1-2 t/shift), the production of fish hydrolysate is simplified. The processing equipment consists of a grinder (sieve openings 6-10 mm in diameter, processing capacity *circa* 400 kg/hour), dispenser with a worm-wheel unloading conveyor, rotating mixer made of suitable materials with a 150-l volume drum, and 120-l plastic barrels. This equipment (Figure) is manned by an operator who can produce 2 t of liquid feed per shift.

A production cycle consists of the following stages:

- Grinding of the raw product in a grinder.
- Loading of *circa* 100 l of ground product from the dispenser into the mixer drum, and adding 1.6-2.0 l of sulphuric acid at density 1.28-1.3.
- Mixing for about 10 minutes and adding a solution of sodium pyrosulphite (1 kg of pyrosulphite dissolved in 3-4 l of water).
- Additional mixing for 5-7 minutes and pouring of the product from the mixer drum straight into the 120 l barrels.

An approximate chemical composition of fish silage is:

- Protein - about 15 per cent.
- Fat - 6-14 per cent (depending on raw material).
- Ash - 2.4 per cent.
- Micro-elements and vitamins.

Different forms of fish hydrolysate are used for feeding pigs, poultry, fur animals and fish. Hydrolysates contain very valuable, easily assimilated proteins and fatty acids, unaltered vitamins, micro-elements and digestive enzymes. For pig and poultry feed, fish hydrolysates can be substituted for fish meal, meat and bone meal, and powdered blood. Experiments showed a 10-20 per cent increase in weight and a lower feed use per weight gained by an animal. It was determined that 1 kg of hydrolysate equals 0.3 kg of fish meal, and its use reduces the need for feed by 0.66 kg per 1 kg of weight gained. Polish scientists reported 0.7 kg/day of weight gained when bacon-type pigs were fed fish hydrolysates.

According to Danish researchers, no more than 15 per cent of the total feed given to pigs should consist of fish hydrolysates, and these should be detracted from the diet several weeks before slaughter. The Polish and Danish experiments confirmed the positive results of feeding poultry with fish hydrolysates instead of fish meal (chickens were fed hydrolysates in amounts equal to 50 per cent of the daily protein requirement).

Substitution of dry animal and fish meal with hydrolysates gave very good production results:

- Use of feed per 1 kg of weight gained equal to 2.54 kg.
- Mean body weight of an 8-week old chicken was 1.20 kg.

- Slaughter efficiency higher by 23 per cent.
- Costs of the components used in the feed lowered by 20 per cent.
- Content of additional animal feed lowered by two-thirds, that is, by 110 kg/1 t of combined feed, fish and meat meal, and powdered milk.

PRECAUTIONARY APPROACH TO FISHERY TECHNOLOGY

OBJECTIVE

Recognising that many aquatic resources are overfished and that the fishing capacity presently available jeopardise their conservation and rational use, technological changes aimed solely at further increasing fishing capacity would not generally be seen as desirable.

Instead a precautionary approach to technological changes would aim at:

- Improving the conservation and long-term sustainability of living aquatic resources;
- Preventing irreversible or unacceptable damage to the environment;
- Improving the social and economic benefits derived from fishing, and
- Improving the safety and working conditions of fishery workers.

INTRODUCTION

Fishery technology consists of the equipment and practices used for finding, harvesting, handling, processing and distributing of aquatic resources and their products.

Different fishery technologies will have different effects on the ecosystem, the social structure of fishing communities, the safety of fishery workers and the ease, effectiveness and efficiency of management of the fishery. It is the amount and context in which fishery technology is used (*e.g.*, when, where and by whom) that influence whether the objectives of fisheries management are reached, and not the technology. For instance, the current overfishing of many aquatic resources is the product of both the efficiency of the finding and catching technologies and of the amount used. Similarly, building a fishmeal plant might involuntarily result in severe changes in the way the fishery is conducted, and in the community's social structure.

Fishery technology is constantly evolving and its efficiency in catching fish will increase over time. For example, a 4 per cent increase in efficiency per year would cause a doubling of the fishing mortality rate in 18 years if the fishing effort remained constant. A precautionary approach to management should take such increases into account.

A precautionary approach should be adopted for the development of new technologies or the transfer of existing technologies to other fisheries to avoid unplanned abrupt changes in fishing pressure or social structures. Certain technologies will be considered undesirable, if they create unacceptable effects (*e.g.*, poison or explosives) or if their adoption leads to wasteful use (*e.g.*, at sea, sorting machines have been banned where they might increase discarding). Fishery technologies produce side effects on the environment and on non-target species. These effects have often been ignored but, in the context of a precautionary approach, some technologies may warrant a review. Similarly, a precautionary approach would encourage careful consideration of the side effects of new fishery technologies before they are introduced.

Each fishery technology has advantages and disadvantages that should be balanced in a precautionary approach, and it may be better to have a mixture of technologies. When new fishery technology is introduced, it should be carefully evaluated to assess its potential direct and indirect effects. If a mix of fishery technology representing "best current practice" in an area can be identified, precautionary management would encourage its adoption while it would discourage damaging ones. Responsible fishery technology achieves the specific fishery management objectives with the minimumdamaging side effects. These concepts (of responsible fishing and best current practices) were addressed by the UN General Assembly and in the Cancun Declaration.

A precautionary approach would provide for a process of initial and on-going review of the effects of fishery technology as it is introduced or evolves in local practice. However, the extent to which a precautionary approach can be applied to the management of technological changes depends on the existing level of management. In some cases, education of fishermen and consumers towards responsible practices may be the only possible approach. Where elaborate research, management and enforcement systems are in place, a wider variety of options are available for application of the precautionary approach. However, although some gears and practices are prohibited they may continue to be used. The adoption of a precautionary approach to the management of new fishery technology depends on the ability to achieve compliance through education and/or enforcement. The following sections assume that institutional arrangements exist to achieve compliance.

EVALUATING THE IMPACTS OF TECHNOLOGIES

A precautionary approach to developing and selecting responsible technologies for fishing requires an appropriate understanding of the consequences of their adoption and use. These consequences, particularly the impacts on non-target species and ecosystems, may be highly uncertain. Nevertheless, some information exists and more can be obtained. The problem of evaluating impacts is relevant both to the use of existing technologies and to the development of new ones, as well as to the introduction of existing technologies to new areas. The description of a given technology would state its relative impacts and advantages for a given species in a specific environment. Target fishery, environmental and ecosystem, socio-economic and legal factors should be considered when evaluating the impacts of fishery technologies.

The factors to consider when evaluating the impacts of fishery technology include:

- Target-fishery factors such as selectivity by size and species (*e.g.*, target, non-target, and protected species; discards; survival of escapees; "Ghost fishing"; and catching capacity);
- Environmental and ecosystem factors such as bio-diversity; habitat degradation; contamination and pollution; generation of debris and rubbish disposal; direct mortality; predator-prey relationships;
- Socio-economic factors such as safety and occupational hazards; training requirements; user conflicts; economic performance; employment; monitoring and enforcement requirements and costs; and techno-economic factors (*i.e.*, infrastructure and service requirements; cost and technological accessibility; product quality; and energy efficiency), and
- Legal factors such as existing legislation; need for new legislation; international agreements; and civil liberties.

These factors could be used to identify beneficial new technologies or damaging ones, to assess the ability of a fishery to accommodate increased use of an established technology and to help direct monitoring and special reporting procedures towards important questions. Technologies for aids to navigation, fish-locating devices, processing and distribution could also be described and evaluated using the above criteria. This will require a suitable description of technologies, cross-referenced against a range of possible impacts. Other elements relevant to the specific technology/area evaluated would also be included.

The approaches used to evaluate impacts will vary according to the human and financial resources available to collect information. If resources are limited, it may be possible to make decisions based on existing information on the impacts of similar technologies in similar environments. Monitoring of existing fishing practices (for example recording of bycatch) will provide additional information.

Where financial and human resources are limited, existing information on impacts could be used to do desk studies following the approach to evaluation suggested above. Although some general guidelines can be given, based on known characteristics of types of resources and technology, the most appropriate mix of technologies to be used in a particular fishery should be established on a case-by-case basis, following evaluations made at appropriate regional and national levels. Such evaluations could be refined with practical experience and weighed in accordance with local social and economic values.

In the case of new technologies, or technologies new to an area, pilot studies may be cost-effective in evaluating the impacts and can be useful in demonstrating the benefits of new technology. For example, the introduction of escape ports in lobster traps for undersised individuals demonstrated to fishermen that catch rates of large lobsters increased. On the other hand, pilot studies cannot demonstrate long-term gains such as increased yield per recruit, but they will show the short-term losses.

Considerable resources are required for major experiments to measure effects of fishery technology on the marine environment, but well-designed experiments of this type (either as research projects or via experimental management) will provide the most useful information on which to judge the impacts of technologies in particular areas or habitats. This information may be relevant in other areas than the study sites or fisheries from which the data were derived.

Procedures developed in other contexts for protecting the environment could also be suitable when evaluating new technologies in fisheries or major alterations to existing ones. This would be particularly necessary when there are vulnerable resources or fragile ecosystems, that must be protected. In a precautionary approach, proponents of new fishery technology would be required by the State to provide for a proper evaluation of the potential impacts of new techniques before authorisation is given.

The maximum cost that could be justified for evaluating new fishery technology or practices should be in proportion to the expected benefits and impacts.

IMPLEMENTATION

In a precautionary approach to managing fishery technology, a designated lead authority should have the mandate to evaluate and decide on the acceptability of a proposed new technology, or changes to existing technology, and oversee the impact evaluation procedure. Proponents and other stakeholders should be able to appeal if the proper procedure has not been followed or if the decision by authorities does not appear to agree with the conclusions of the review. As authorisation procedures in the majority of cases would be for minor technical improvements, the procedures could be kept simple and administration costs held at a relatively low level.

However, minimal progressive improvements will accumulate over time and periodic reviews of the impacts of existing technology will be necessary. Increases in catching efficiency result from the rapid growth in the use of modern information technologies in most fisheries around the world (acoustic fish detection and identification, gear and vessel monitoring, satellite-based environmental sensing and navigation, and easy inter-vessel communication). However, information, formally treated as "a measure of the reduction of uncertainty", can also potentially improve selectivity, safety and profitability of fishing operations and thus create beneficial effects. Restricting the use of improved information technologies will rarely be justified or successful and there should be a positive attitude towards technical progress in fisheries in general especially with regards to safety at sea and fishermen's health.

The benefits of technological improvements need adequate extension work and education to encourage their adoption. The promotion of the best technology would benefit from improvement in international cooperation regarding technology transfer, as underscored in UNCED's Agenda 21. The successful international efforts in the Eastern Central Pacific in training crews in effectively avoiding bycatches of dolphins through the use of specifically designed technology is a good example of what can be achieved in this respect.

TECHNOLOGY RESEARCH AND DEVELOPMENT

Fishery technology research in support of a precautionary approach would encourage the improvement of existing technologies and promote the development of appropriate new technologies. Such research would not just concentrate on gears used for capture; for example, research into the cost-effective purification of water supplies to ice plants might considerably reduce post-harvest losses and improve product quality and safety.

Technological developments such as satellite tracking may also help precautionary management by improving monitoring of commercial operations and by enabling research to reduce uncertainty about relevant aspects of fisheries science.

IMPLEMENTATION GUIDELINES

The following measures could be applied in order to implement a precautionary approach to fishery technology development and transfer.

Authority

- Effective mechanisms to ensure that the introduction of technology is subject to review and regulation should be established.

Evaluation Procedures

- A first step in the evaluation procedure is the documentation of the characteristics and amount of the fishery technology currently used.
- Procedures for the evaluation of new technologies with a view to identify their characteristics in order to promote the use of beneficial technologies and prevent usage of those leading to difficult-to-reverse changes should be established.
- These procedures should evaluate with appropriate accuracy the possible impacts of the proposed technology in order to avoid wasteful capital and social investments.
- Authorities should ensure that proponents and other stakeholders understand their obligations and their rights regarding such procedures.

- The extent of the evaluation procedures should match the potential effects of the proposed technology, *e.g.*, from desk study through full scale impact studies, possibly including or leading to pilot projects.

Implementation

- Authorities should implement technology gradually to minimize the risk of irreversible damage or overinvestment.
- Existing technologies and their effect on the environment should be reviewed periodically.
- Technological developments may modify the practices of fishery workers. To achieve the full benefits of the technology and to ensure the safety of fishery workers, training in the proper use of the new technology should be provided.
- In fisheries that are being rehabilitated, the opportunity should be taken to review the mix of technologies used.
- Research into responsible fishery technology should be encouraged.
- Technology research for the reduction of uncertainty in stock assessment and monitoring should be encouraged.

SOFTWARE USED IN FISHERIES FOR STATISTICAL ANALYSIS

INTRODUCTION

Fish is an important source of protein and its harvest, handling, processing and distribution provide livelihood for millions of people as well as providing valuable foreign exchange earnings to the country. It is a highly perishable food, requires proper handling, processing and distribution, if it is to be utilised in a cost effective and efficient way. Today, fisheries are estimated to provide 16 per cent of the world population's protein, and that figure is considerably elevated in some developing nations and in regions that depend heavily on the sea. The world capture fisheries production in 2008 was about 90 million tonnes, with an estimated first-sale value of US$ 93.9 billion, of which about 80 million tones from marine waters and a record 10 million tonnes from inland waters. It has been relatively stable in the past decade, with the exception of marked fluctuations driven by catches of anchoveta – a species extremely susceptible to oceanographic conditions determined by the El Niño Southern Oscillation – in the Southeast Pacific. In 2008, China, Peru and Indonesia were the top capture fish producing countries. China remained by far the global leader with production of about 15 million tones (FAO-2010).

With its 8129 km long coastline, India has immense potential for developing and producing a variety of sea foods, highly sought after throughout the world, directly leading to a huge amount of export in this field. It forms more than 70 per cent of sea food all over the world. This requirement has created a big fishing industry. The increased requirement of sea food all over the world and relatively easy breeding of certain varieties of fish, has led to creation of sea breeding farms in the shallow coastal areas of our country's eastern coast. Fish breeding is a new industry in our country.

The exploited fish populations and the fishery dependent on them are highly dynamic that require regular monitoring and periodical stock assessment. The resource evaluation is essential for evolving appropriate management strategies for rational exploitation and long-term sustenance of the exploited stocks. The validity and efficiency of the management formulations depend on the quality of database of which information on the catch and fishing effort is an important component.

Thus, the data on catch and the expended effort besides other fishery-related statistics (including socio-economic status, infrastructure, etc.) are essential for effective management of the fisheries. The compilation of accurate, relevant and timely data in a standard form that makes it comparable, and the considered analysis of this data, is essential to underpin the development and utilisation of the fisheries sector.

As per the FAO Technical Guidelines for Responsible Fisheries, "States should ensure that timely, complete and reliable statistics on catch and fishing effort are collected and maintained in accordance with applicable international standards and practices and in sufficient detail to allow sound statistical analysis. Such data should be updated regularly and verified through an appropriate system".

Fishery statistics are the primary means to measure the performance of a fishery within the social, economic, biological and environmental framework in which it is conducted. The collection of fishery data is based on a relatively small group of concepts and approaches, including most importantly the quantities harvested (catch), the related type and duration of fishing operations (fishing effort), the economic costs and returns of fishing and the distribution of these in time and space.

SPSS

Analytics plays an increasingly important role in helping your organisation achieve its objectives. The IBM SPSS Statistics family delivers the core capabilities needed for end-to-end analytics. To ensure that the most advanced techniques are available to a broader group of analysts and business users, we have made enhancements to the features and capabilities of IBM SPSS Statistics Base and its many specialised modules.

IBM SPSS Statistics 20 continues to increase accessibility to advanced analytics through improved tools, output, and ease-of-use features.

This release focuses on increasing the analytic capabilities of the software through:

- Mapping, which adds a geographic dimension to analysis and reporting

- Improvements to existing procedures
- Enhancements that increase analysts' productivity

Highlights:

- New mapping capabilities - add a geographic dimension to analysis
- Faster tabular outputs - allow viewing of results much faster
- Improved Generalised Linear Mixed Models (GLMM) - can now be run with ordinal values
- Updates to IBM® SPSS® AMOS™ scripting, enabling greater customisation of complex models
- Save large files within the sort procedure to save time and boost performance
- Compression of temporary files within the sort procedure to save space and improve performance

STATISTICAL ANALYSIS WITH SAS/STAT® SOFTWARE

Overview

From traditional statistical analysis of variance and predictive modeling to exact methods and statistical visualisation techniques, SAS/STAT software is designed for both specialised and enterprisewide analytical needs. SAS/STAT software provides a complete, comprehensive set of tools that can meet the data analysis needs of the entire organisation.

Benefits

- *Take advantage of all data in order to uncover new business opportunities and increase revenue:* SAS/STAT software is designed to handle large data sets from disparate sources, enabling you to take advantage of all data that is available for your analyses. Analysts are freed to focus on analysis rather than data issues.
- *Move the scientific discovery process forward by applying the latest statistical techniques:* Statistical procedures in SAS are constantly being updated to reflect the latest advances in statistical methodology, thus enabling you to go beyond the basics for more advanced statistical analyses. In addition, technical support is provided by experienced master's- and doctorate-level statisticians who provide a level of service not often found with other software vendors.
- *Achieve corporate and governmental compliance:* SAS has more than 35 years of experience developing advanced statistical analysis software and a proven reputation for delivering superior, reliable results. With SAS/STAT software, you can produce repeatable code that is easily documented and verified for corporate and governmental compliance issues.

Features

Analysis of Variance

- Balanced and unbalanced designs.
- Multivariate analysis of variance and repeated measurements.
- Linear and non-linear mixed models.

Mixed Models

- Linear mixed models.
- Non-linear mixed models.
- Generalised linear mixed models.

Regression

- Least squares regression with nine model selection techniques, including stepwise regression.
- Diagnostic measures.
- Robust regression; Loess regression.
- Non-linear regression and quadratic response surface models.
- Partial least squares.
- Quantile regression.

Categorical Data Analysis

- Contingency tables and measures of association.
- Logistic regression and log linear models; generalised linear models.
- Bioassay analysis.
- Generalised estimating equations.
- Weighted least squares regression.
- Exact methods.
- Zero-inflated Poisson regression.
- Zero-inflated negative binomial regression.

Bayesian Analysis

- Bayesian modeling and inference for generalised linear models, accelerated life failure models, Cox regression models and piecewise exponential models.
- General procedure fits Bayesian models with arbitrary priors and likelihood functions.

Multivariate Analysis

- Factor analysis.
- Principal components.
- Canonical correlation and discriminate analysis.
- Path analysis.

- Structural equations.

Survival Analysis

- Comparison of survival distributions.
- Accelerated failure time models.
- Proportional hazards models.

Psychometric Analysis

- Multidimensional scaling.
- Conjoint analysis with variable transformations.
- Correspondence analysis.

Cluster Analysis

- Hierarchical clustering of multivariate data or distance data.
- Disjoint clustering of large data sets.
- Non-parametric clustering with hypothesis tests for the number of clusters.

Non-parametric Analysis

- Non-parametric analysis of variance. Exact probabilities computed for many non-parametric statistics.
- Kruskal-Wallis, Wilcoxon-Mann-Whitney and Friedman tests.
- Other rank tests for balanced or unbalanced one-way or two-way designs.

Survey Data Analysis

- Sample selection.
- Descriptive statistics and t-tests.
- Linear and logistic regression.
- Frequency table analysis.
- Cox proportional hazards model.

Multiple Imputation for Missing Values

- Regression and propensity scoring for monotone missing patterns.
- MCMC method for arbitrary missing patterns.
- Combine results for statistically valid inferences.

Study Planning

- Power and Sample Size application provides interface for computation of sample sizes and characterisation of power for t-tests, confidence intervals, linear models, tests of proportions and rank tests for survival analysis.

SYSTAT

Systat is a statistics and statistical graphics software package, developed by Leland Wilkinson in the late 1970s, who was at the time an assistant professor of psychology at the University of Illinois at Chicago. Systat was incorporated in 1983 and grew to over 50 employees.

In 1995 Systat was sold to SPSS Inc., who marketed the product to a scientific audience under the SPSS Science division. By 2002, SPSS had changed its focus to business analytics and decided to sell SYSTAT to Cranes Software in Bangalore, India. Cranes formed Systat Software, Inc. to market and distribute SYSTAT in the US, and a number of other divisions for global distribution. The headquarters are in Chicago, Illinois.

By 2005, SYSTAT was in its eleventh version having a revamped codebase completely changed from Fortran into C++. Version 13 came out in 2009, with improvements in the user interface and several new features.

STATISTICA

STATISTICA is a statistics and analytics software package developed by StatSoft. STATISTICA provides data analysis, data management, data mining, and data visualisation procedures. STATISTICA product categories include Enterprise (for use across a site or organisation), Web-Based (for use with a server and web browser), Concurrent Network Desktop, and Single-User Desktop.

STATISTICA originally derives from a set of software packages and add-ons that were initially developed during the Mid 1980's by StatSoft. Following the 1986 release of CSS (Complete Statistical System) and the 1988 release of MacSS (Macintosh Statistical System), the first DOS version of STATISTICA (trademarked in capitals as STATISTICA) was released in 1991. In 1992, the Macintosh version of STATISTICA was released.

STATISTICA 5.0, was released in 1995 which automatically configured itself for new 32-bit Windows 95/NT or the older version of Windows (3.1) and featured a large number of new statistics and graphics procedures, a word-processor-style output editor of unlimited capacity (combining tables and graphs), and a built-in professional development environment that enabled the user to easily design new procedures (*e.g.*, via the included, comprehensive STATISTICA Basic language) and integrate them with the STATISTICA system. STATISTICA 5.1 was released in 1996 followed by STATISTICA '97 and STATISTICA '98 editions.

In 2001, STATISTICA 6 was based on the COM architecture and high-end technologies (such as multithreading and support for distributed computing). STATISTICA 9 was released in 2009, supporting 32 bit and 64-bit computing. The most recent release of STATISTICA is STATISTICA 10. This release features further performance optimisations for the 64-bit CPU architecture, as well as advanced multithreading technologies, integration with Microsoft Sharepoint, Microsoft Office 2010 and other applications, the ability to generate Java and C# code, and other GUI and kernel improvements. Localised versions of STATISTICA (including the entire STATISTICA family of products) are available in Chinese (both Traditional and Simplified), Czech, English, French, German, Italian, Japanese, Polish, Russian, and Spanish. STATISTICA documentation is available in Arabic, Chinese, Czech, English, French, German, Hungarian, Italian, Japanese, Korean, Polish, Portuguese, Russian, Spanish, and other languages.

ARTFISH APPROACHES, RULES AND TECHNIQUES FOR FISHERIES STATISTICAL MONITORING

ARTFISH stands for Approaches, Rules and Techniques for Fisheries statistical monitoring. It has been developed as a standardised tool adaptable to most fisheries in the developing countries. Its design was driven by the need to provide users with robust, user-friendly and error-free approaches with computer software, and achieve the implementation of cost-effective fishery statistical systems with minimal external assistance.

Software Components

The first ARTFISH system appeared in 1993 as an MS-DOS application. ARTFISH was re-written for Windows in 2000. Its present upgraded version was released in November 2007.

The ARTFISH outputs consist of documents, guidelines, manuals, case studies, training kits and computer software. Its components are ARTBASIC for handling sample data and producing monthly estimates and ARTSER for the integration of monthly estimates into annual databases.

The basic variables involve catch, effort, CPUE, prices, values and average fish size. ARTFISH utilises technology that increases user cognition and decreases training time. It also provides detailed statistical diagnostics on data quality and allows insights into the fisheries for which the data is being collected.

ARTBASIC

It is used during the production phase of a data collection programme. Its purpose is to organise the system standards and classifications as well as primary sample data into standardised database structures and produce estimates of total effort, catch and values.

The ARTBASIC generic approach operates within the context of a calendar month, a logical stratum and a specific boat/gear type. Its architecture provides for decentralised entry of data and production of estimates with multi-lingual capability.

ARTSER

This component is used to integrate the monthly estimates produced by ARTBASIC into standardised tables showing monthly estimates of catch, CPUEs, effort, prices and values during a reference year. Its functions allow for flexible data extraction and grouping and provide users with reporting, exporting and graphical representation services. It also provides a multi-lingual capability for performing data mining and analysis.

FREESING METHODS AND EQUIPMENT

DIRECT AND INDIRECT SYSTEMS

In the direct expansion system, the refrigerant absorbs heat directly from the material to be cooled. In the indirect or brine system, the refrigerant absorbs the heat that the brine absorbs from the material to be cooled. Both these systems are commonly employed in commercial freesing operations.

Freesing methods are broadly classified into:

- Freesing in air
- Indirect contact freesing
- Immersion freesing
- Cryogenic freesing

Freesing in Air

Air is the most common freesing medium. Air freesing systems are of two types- still air freesing and forced air freesing.

Freesing in Still Air

The freezer consist of an insulated room or a cabinet maintained at –28 to –45°C. Fish, packaged or otherwise,placed in aluminum trays, is kept on shelves made of pipes or coils through which the refrigerant is circulated. Time taken for freesing may be 12 hours or more.

Freesing in still air is the least expensive method of freesing; however, is the slowest method as well.

Air Blast Freezer

Air blast freezer consists of an insulated room or a tunnel. Air is cooled by blowing a fan through the finned cooling coil of the refrigeration system. Cool air passes over the fish to be frozen and picks up the heat from the product, walls of the freezer, etc. Temperature is maintained at –35 to –40°C.

Advantages:

- Versatile method - applicable to any size and shape.
- Easy for producing IQF fish, fish fillets or coated products.
- Suitable for operation in batch or continuous process.

tests, confidence intervals, linear models, tests of proportions and rank tests for survival analysis.

SYSTAT

Systat is a statistics and statistical graphics software package, developed by Leland Wilkinson in the late 1970s, who was at the time an assistant professor of psychology at the University of Illinois at Chicago. Systat was incorporated in 1983 and grew to over 50 employees.

In 1995 Systat was sold to SPSS Inc., who marketed the product to a scientific audience under the SPSS Science division. By 2002, SPSS had changed its focus to business analytics and decided to sell SYSTAT to Cranes Software in Bangalore, India. Cranes formed Systat Software, Inc. to market and distribute SYSTAT in the US, and a number of other divisions for global distribution. The headquarters are in Chicago, Illinois.

By 2005, SYSTAT was in its eleventh version having a revamped codebase completely changed from Fortran into C++. Version 13 came out in 2009, with improvements in the user interface and several new features.

STATISTICA

STATISTICA is a statistics and analytics software package developed by StatSoft. STATISTICA provides data analysis, data management, data mining, and data visualisation procedures. STATISTICA product categories include Enterprise (for use across a site or organisation), Web-Based (for use with a server and web browser), Concurrent Network Desktop, and Single-User Desktop.

STATISTICA originally derives from a set of software packages and add-ons that were initially developed during the Mid 1980's by StatSoft. Following the 1986 release of CSS (Complete Statistical System) and the 1988 release of MacSS (Macintosh Statistical System), the first DOS version of STATISTICA (trademarked in capitals as STATISTICA) was released in 1991. In 1992, the Macintosh version of STATISTICA was released.

STATISTICA 5.0, was released in 1995 which automatically configured itself for new 32-bit Windows 95/NT or the older version of Windows (3.1) and featured a large number of new statistics and graphics procedures, a word-processor-style output editor of unlimited capacity (combining tables and graphs), and a built-in professional development environment that enabled the user to easily design new procedures (*e.g.*, via the included, comprehensive STATISTICA Basic language) and integrate them with the STATISTICA system. STATISTICA 5.1 was released in 1996 followed by STATISTICA '97 and STATISTICA '98 editions.

In 2001, STATISTICA 6 was based on the COM architecture and high-end technologies (such as multithreading and support for distributed computing).

STATISTICA 9 was released in 2009, supporting 32 bit and 64-bit computing. The most recent release of STATISTICA is STATISTICA 10. This release features further performance optimisations for the 64-bit CPU architecture, as well as advanced multithreading technologies, integration with Microsoft Sharepoint, Microsoft Office 2010 and other applications, the ability to generate Java and C# code, and other GUI and kernel improvements. Localised versions of STATISTICA (including the entire STATISTICA family of products) are available in Chinese (both Traditional and Simplified), Czech, English, French, German, Italian, Japanese, Polish, Russian, and Spanish. STATISTICA documentation is available in Arabic, Chinese, Czech, English, French, German, Hungarian, Italian, Japanese, Korean, Polish, Portuguese, Russian, Spanish, and other languages.

ARTFISH APPROACHES, RULES AND TECHNIQUES FOR FISHERIES STATISTICAL MONITORING

ARTFISH stands for Approaches, Rules and Techniques for Fisheries statistical monitoring. It has been developed as a standardised tool adaptable to most fisheries in the developing countries. Its design was driven by the need to provide users with robust, user-friendly and error-free approaches with computer software, and achieve the implementation of cost-effective fishery statistical systems with minimal external assistance.

Software Components

The first ARTFISH system appeared in 1993 as an MS-DOS application. ARTFISH was re-written for Windows in 2000. Its present upgraded version was released in November 2007.

The ARTFISH outputs consist of documents, guidelines, manuals, case studies, training kits and computer software. Its components are ARTBASIC for handling sample data and producing monthly estimates and ARTSER for the integration of monthly estimates into annual databases.

The basic variables involve catch, effort, CPUE, prices, values and average fish size. ARTFISH utilises technology that increases user cognition and decreases training time. It also provides detailed statistical diagnostics on data quality and allows insights into the fisheries for which the data is being collected.

ARTBASIC

It is used during the production phase of a data collection programme. Its purpose is to organise the system standards and classifications as well as primary sample data into standardised database structures and produce estimates of total effort, catch and values.

The ARTBASIC generic approach operates within the context of a calendar month, a logical stratum and a specific boat/gear type. Its architecture provides for decentralised entry of data and production of estimates with multi-lingual capability.

ARTSER

This component is used to integrate the monthly estimates produced by ARTBASIC into standardised tables showing monthly estimates of catch, CPUEs, effort, prices and values during a reference year. Its functions allow for flexible data extraction and grouping and provide users with reporting, exporting and graphical representation services. It also provides a multi-lingual capability for performing data mining and analysis.

FREESING METHODS AND EQUIPMENT

DIRECT AND INDIRECT SYSTEMS

In the direct expansion system, the refrigerant absorbs heat directly from the material to be cooled. In the indirect or brine system, the refrigerant absorbs the heat that the brine absorbs from the material to be cooled. Both these systems are commonly employed in commercial freesing operations.

Freesing methods are broadly classified into:

- Freesing in air
- Indirect contact freesing
- Immersion freesing
- Cryogenic freesing

Freesing in Air

Air is the most common freesing medium. Air freesing systems are of two types- still air freesing and forced air freesing.

Freesing in Still Air

The freezer consist of an insulated room or a cabinet maintained at –28 to –45°C. Fish, packaged or otherwise,placed in aluminum trays, is kept on shelves made of pipes or coils through which the refrigerant is circulated. Time taken for freesing may be 12 hours or more.

Freesing in still air is the least expensive method of freesing; however, is the slowest method as well.

Air Blast Freezer

Air blast freezer consists of an insulated room or a tunnel. Air is cooled by blowing a fan through the finned cooling coil of the refrigeration system. Cool air passes over the fish to be frozen and picks up the heat from the product, walls of the freezer, etc. Temperature is maintained at –35 to –40°C.

Advantages:

- Versatile method - applicable to any size and shape.
- Easy for producing IQF fish, fish fillets or coated products.
- Suitable for operation in batch or continuous process.

Disadvantages:

- Occupy large space
- Consume more energy
- Excessive dehydration if the fish is unpacked.
- Fish will swell giving a bulged appearance if not confined between rigid surfaces during freesing.

Continuous Air Blast Freezer

This is an improvement in the air blast freezer where a conveyor belt is used to move the fish continuously through the room or the tunnel. The speed of the conveyor can be adjusted to suit the type of the fish to be frozen. Air flow can be either countercurrent to the movement of the material or across the belt. Air velocity maintained at 150- 300 m/ sec. Makes intimate contact with the fish and accomplishes freesing at a fast rate. Freesing is quicker and any type of fish can be frozen in large quantities in shorter periods. Spiral belt freezers are an example.

Fluidised Bed Freesing

Fluidised bed freezer is an improvement over the continuous belt freezer. Fluidisation is a method of keeping the solid particles partially supported in a rising column of cold air. The particles kept on a mesh are individually suspended in an upward stream of cold air at a velocity sufficient to float the particles in the cold air. Each particle is surrounded by air and is kept suspended separated from each other. Freesing is rapid as the best possible heat transfer is ensured between the cold air and the product. Air velocity of 120m/min. And operating temperature of –35 to –40°C is common in fluidised bed freesing. This is most suitable for small sized and uniform products like shrimp, small fish, etc.

Advantages:

- Rapid rate of freesing
- Less product dehydration
- Short freesing time
- No distortion of shape.

The major disadvantage is that large and non-uniform products cannot be fluidised at reasonable air velocity.

INDIRECT CONTACT FREESING

Indirect contact freesing can be defined as freesing a product by keeping it in contact with a metal surface which is cooled by some refrigerant. They are of two types—The horizontal plate freezers and the vertical plate freezers.

Horizontal Plate Freezer

These freezers in general have 15-20 plates. The product to be frozen contained in metal freesing trays are loaded between freesing plates and

Bibliography

A. Solaimalai, N. Ravisankar and B. Chandrasekar: *Farming Systems: Theory and Practice*, International Book Distributors, 2005.

A.S. Rawat: *Alternative Farming Systems in Dry Temperate Zone of Himachal Pradesh*, Indus, Delhi, 2002.

Amarjeet Singh: *Animal Husbandry*, Rajat Publication, Delhi, 2008.

Aruna T. Kumar: *Handbook of Animal Husbandry*, Indian Council of Agricultural Research, 2008.

Ashok Kumar: *Animal Husbandry*, Discovery, Publication, Delhi, 2006.

B.L. Jana: *Farming Systems*, Agrotech Publication, Academy, 2014.

B.R. Selvamani and R.K. Mahadevan: *Fish Farming Systems*, Campus Books International, Delhi, 2008.

D. Thyagarajan, A. Ashok, G. Raj Manohar and M. Murugan: *Emerging Opportunities in Alternative Poultry Farming Systems*, Satish Serial Publishing House, 2014.

David Norman and Mancolm Douglas: *Farming Systems Development and Soil Conservation*, Daya, Publication, Delhi, 2007.

Devendra Thakur and D.N. Thakur:*Tribal Life in India, Vol. 3. Tribal Agriculture and Animal Husbandry*, Deep and Deep, Publication, Delhi, 2010.

Dinesh Arora: *Biotech's Dictionary of Animal Husbandry*, Biotech, Publication, Delhi, 2005.

Gyan Deep Singh: *Animal Husbandry*, Anmol, Publication, Delhi, 2008.

L. Lal: *Animal Husbandry (Objective Fundamentals)*, Agrotech, Publication, Delhi, 2010.

Manoj Kumar Rai: *Handbook of Animal Husbandry*, Oxford Book Company, 2012.

Manoj Kumar Rai: *Textbook of Animal Husbandry*, Oxford Book Company, Delhi, 2012.

Nishamani Kar: *Animal Husbandry and Rural Development*, Deep and Deep, Publication, Delhi, 2002.

P K Saraswat: *International Encyclopaedia of Dairy Technology and Animal Husbandry*, Anmol, Publication, Delhi, 2010.

P Mathialagan: *GTextbook of Animal Husbandry and Livestock Extension*, International Book Dist, 2007.

P.P. Joglekar and Pankaj Goyal: *Animal Husbandry and Allied Technologies in Ancient India*, Pentagon Press, Delhi, 2015.

Pratap Narain; M P Singh; Amal Kar; S Kathju and Praveen Kumar: *Diversification of Arid Farming Systems*, Scientific Publication, Delhi, 2008.

R Renaville and A Burny: *Biotechnology in Animal Husbandry*, Springer Publication, Delhi, 2008.

Rajendra Reddy and J.P. Abhay Shankar: *Animal Husbandry*, Commonwealth, Delhi, 2008.

Renaville: *Biotechnology in Animal Husbandry*, Springer, Publication, Delhi, 2004.

S.C. Panda: *Cropping and Farming Systems*, Agrobios, Publication, Delhi, 2004.

Shagufta Jamal and H P S Arya: *Participatory Rural Appraisal in Agriculture and Animal Husbandry*, Concept, Publication, Delhi, 2004.

Shagufta: *Cropping and Farming Systems*, A.P.H., Publication, Delhi, 2011.

U.K. Behera: *A Textbook of Farming Systems*, Agrotech Publishing Academy, 2014.

U.K. Behera: *Advances in Farming Systems*, Agrotech Publication, Delhi, 2012.

William J.A. Payne and R. Trevor Wilson: *An Introduction to Animal Husbandry in the Tropics*, Wiley, Publisher, Delhi, 2013.

Yasin Khan: *Animal Husbandry and Farming Management*, Agrotech Press, Delhi, 2013.

Index